C0-AVX-571

ECOLOGY and BIOMECHANICS

A Mechanical Approach to the Ecology of Animals and Plants

ECOLOGY and BIOMECHANICS

A Mechanical Approach to the Ecology of Animals and Plants

Edited by
Anthony Herrel
Thomas Speck
Nicholas P. Rowe

A CRC title, part of the Taylor & Francis imprint, a member of the Taylor & Francis Group, the academic division of T&F Informa plc.

Published in 2006 by
CRC Press
Taylor & Francis Group
6000 Broken Sound Parkway NW, Suite 300
Boca Raton, FL 33487-2742

Printed in the United States of America on acid-free paper
10 9 8 7 6 5 4 3 2 1

International Standard Book Number-10: 0-8493-3209-5 (Hardcover)
International Standard Book Number-13: 978-0-8493-3209-8 (Hardcover)
Library of Congress Card Number 2005024415

Library of Congress Cataloging-in-Publication Data

Ecology and biomechanics : a mechanical approach to the ecology of animals and plants / editors, Anthony Herrel, Thomas Speck, Nicholas Rowe.
p. cm.
Includes bibliographical references.
ISBN-13: 978-0-8493-3209-8 (hardcover : alk. paper)
ISBN-10: 0-8493-3209-5 (hardcover : alk. paper)
1. Ecology. 2. Biomechanics. I. Herrel, Anthony. II. Speck, Thomas. III. Rowe, Nicholas P.

QH541.14.E243 2006
577--dc22 2005024415

Taylor & Francis Group
is the Academic Division of Informa plc.

Visit the Taylor & Francis Web site at
http://www.taylorandfrancis.com

and the CRC Press Web site at
http://www.crcpress.com

Preface

Ecology and Biomechanics: A Mechanical Approach to the Ecology of Animals and Plants. When we decided upon this title there were two elements we wanted to emphasize: first, that biomechanical approaches have a lot to offer to ecological questions and second, that the approaches included in this book are independent of the organism being studied. Indeed, the present collection of state-of-the-art papers beautifully highlights how biomechanics can provide novel insights into long standing ecological and evolutionary questions (e.g., Chapters 4 and 14). As illustrated, for example, in the chapter by Wendy Griffiths (Chapter 5) on grazing in ruminants, there is tremendous scope for applying engineering principles to the understanding of foraging strategies used by animals. Although most of the examples included in the book emphasize distinct organism–environment relationships, it is foreseeable that in the longer term these kinds of approaches will span larger temporal and spatial scales to achieve wider application across ecosystems. The chapter by Karen Christensen-Dalsgaard (Chapter 13) nicely illustrates this, and highlights how microbial ecosystems can be understood from the mechanics, morphology, and motile responses of the individual organisms. The range of topics covered clearly demonstrates that increasing numbers of workers have begun to combine biomechanics and ecology to generate novel insights into questions of an ecological nature.

We hope that this book will highlight the important cross fertilization that can occur by combining approaches from two — at first sight — very disparate subdisciplines within the general field of biology and will stimulate other researchers to follow these kinds of approaches. The subjects covered include research based on both plants (Chapters 1 to 4) and animals (Chapters 10 and 12) as well as the interaction between plants and animals (Chapters 5 to 9). By highlighting both theoretical concepts and practical approaches, we hope that this book will be an important reference for students and researchers alike. Most of the chapters included in this book were originally presented at a symposium entitled "Ecology and Biomechanics" held at the annual meeting of the Society for Experimental Biology in Edinburgh, U.K., in April 2004. Generous support by the Biomechanics Group of the Society for Experimental Biology enabled us to invite many outstanding speakers, most of whom decided to contribute to the present volume.

Anthony Herrel is a postdoctoral researcher of the Fund for Scientific Research, Flanders, Belgium (FWO-Vl). Thomas Speck is Professor for Functional Morphology and Director of the Botanic Garden of the Albert-Ludwigs-Universität Freiburg (Germany). Nick Rowe is Chargé de Recherche (CNRS) at the Botanique and Bioinformatique Research Institute, Montpellier, France.

Editors

Anthony Herrel earned his degree at the University of Antwerp under the supervision of Professor Dr. F. De Vree with a thesis on lizard herbivory. He then became a postdoctoral researcher at the Fund for Scientific Research — Flanders, Belgium (FWO-Vl) with a home base at the University of Antwerp in Belgium. During his first postdoctoral appointment he was awarded a mobility grant, which enabled him to conduct research at the laboratory of Dr. K. Nishikawa at Northern Arizona University on motor control during feeding in lizards using nerve transection experiments. Later, during his second appointment, he spent a year at Tulane University in New Orleans in the laboratory of Dr. D. Irschick working on the evolution of cranial morphology and diet in *Anolis* lizards. He is currently on his final appointment as postdoctoral researcher at the FWO-Vl and is based at the Functional Morphology Laboratory at the University of Antwerp, working under the tutelage of Dr. P. Aerts.

Thomas Speck graduated in 1986 from the University of Freiburg with a diploma thesis on the biomechanics of early land plants. In 1990 he completed his doctorate thesis on biomechanics and functional anatomy of various plant growth forms and received his habilitation in botany and biophysics in 1996. Since 2002 he has been professor of functional morphology and director of the Botanic Garden of the University of Freiburg (Germany). He has written over 130 papers covering many aspects of plant biomechanics, functional morphology, biomimetics, and phylogeny of plants. He is a member of the board of the Competence Networks Plants as Concept Generators for Biomimetical Materials and Technologies and BIO*KON* e.V. He is president of the Society of Botanic Gardens, Germany.

Nicholas P. Rowe graduated from the Department of Botany, University of Bristol, where he went on to complete his doctorate thesis in fossil plant biology in 1986. He has since carried out research in plant biology in London, France, and Germany. Over the last 10 years he has been a research scientist for the Centre National de la Recherche Scientifique in Montpellier, France. A relative latecomer to the world of biomechanics, his research has combined plant biomechanics with evolutionary and ecological studies, and integrating such approaches with both laboratory experimental work and field work, particularly in the tropics. His main interests are functional and evolutionary studies in plants, particularly the evolution of complex structures, major innovations, and in particular, the evolution and biomechanics of various plant growth forms such as trees, shrubs, and climbers.

Editors

Acknowledgments

We would like to acknowledge the much appreciated help of Bieke Vanhooydonck and Katleen Huyghe with the final editing and formatting of the chapters presented in this book. Additionally, we would like to thank the following referees for critical comments on the chapters and the many helpful suggestions for improvement: P. Aerts (University of Antwerp, Belgium), B. Borrell (University of California, Berkeley), T. Buckley (Australian National University, Canberra), A. Davy (University of East Anglia, Norwich, United Kingdom), R. Dudley (University of California, Berkeley), M. Edmunds (University of Central Lancashire, Preston, United Kingdom), S. Eigenbrode (University of Idaho, Moscow), T. Fenchel (University of Copenhagen, Denmark), F. Gallenmüller (University of Freiburg, Germany), A. Goodman (University of Lincoln, United Kingdom), D. Irschick (Tulane University, New Orleans, Louisiana), G. Jeronimidis (University of Reading, United Kingdom), M. Jervis (Cardiff University, United Kingdom), R. Jetter (University of British Columbia, Vancouver, Canada), J. Losos (Washington University, St. Louis, Missouri), K. Lunau (University of Duesseldorf, Germany), B. Moon (University of Louisiana, Lafayette, Louisiana), C. Neinhuis (Technical University of Dresden, Germany), J. Pilarski (Northern Arizona University, Arizona), B. Vanhooydonck (University of Antwerp, Belgium), J. Vincent (University of Bath, United Kingdom). Finally we would like to thank Amy Rodriguez, David Fausel, and John Sulzycki at Taylor & Francis for all their help with the practical aspects of putting together the book.

Acknowledgments

List of Contributors

Peter Aerts
Department of Biology
University of Antwerp
Antwerpen, Belgium

Brendan J. Borrell
Department of Integrative Biology
University of California
Berkeley, California

Tanja Bruening
Department of Zoology II
Biozentrum, Am Hubland
Würzburg, Germany

Karen K. Christensen-Dalsgaard
Department of Biology
Aarhus University
Aarhus, Denmark
and
School of Biological Sciences
The University of Manchester
Manchester, United Kingdom

Regine Claßen-Bockhoff
Johannes Gutenberg Universität
Institut für spezielle Botanik
Mainz, Germany

Catherine Coutand
U.M.R. Physiologie Intégrée de l'Arbre Fruitier et Forestier
INRA, Site de Crouël
Clermont-Ferrand, France

Roland Ennos
School of Biological Sciences
University of Manchester
Manchester, United Kingdom

Walter Federle
Department of Zoology II
Biozentrum, Am Hubland
Würzburg, Germany

Thierry Fourcaud
U.M.R. Laboratoire de Rheologie du Bois de Bordeaux
Domaine de l'Hermitage
Cestas, France
and
U.M.R. Botanique et Bionformatique de l'Architecture des Plantes
Montpellier, France

Meriem Fournier
U.M.R. Ecologie des Forets de Guyane (Mixed Unit: CIRAD/CNRS/ENGREF/INRA)
Campus Agronomique de Kourou
Kourou, France

Friederike Gallenmüller
Botanischer Garten
Universität Freiburg
Freiburg, Germany

Yvonne Golding
School of Biological Sciences
University of Manchester
Manchester, United Kingdom

Elena Gorb
Evolutionary Biomaterials Group
Max-Planck-Institut für Metallforschung
Stuttgart, Germany

Stanislav Gorb
Evolutionary Biomaterials Group
Max-Planck-Institut für Metallforschung
Stuttgart, Germany

Wendy M. Griffiths
AgResearch Limited
Invermay Agricultural Centre
Mosgiel, New Zealand

Deane L. Harder
Botanischer Garten
Universität Freiburg
Freiburg, Germany

Andrew P. Hendry
Redpath Museum and Department of Biology
McGill University
Montreal, Quebec, Canada

Anthony Herrel
Department of Biology
University of Antwerp
Antwerpen, Belgium

Catriona L. Hurd
Botany Department
University of Otago
Dunedin, New Zealand

Duncan J. Irschick
Department of Ecology and Evolutionary Biology
Tulane University
New Orleans, Louisiana

Sandrine Isnard
U.M.R. Botanique et Bioinformatique de l'Architecture des Plantes
Montpellier, France

Rob S. James
Coventry University
School of Science and the Environment
Coventry, United Kingdom

Harald W. Krenn
Department of Evolutionary Biology
Institute of Zoology
University of Vienna
Vienna, Austria

Katherine A. McCulloh
Department of Biology
University of Utah
Salt Lake City, Utah

Bruno Moulia
U.M.R. Physiologie Intégrée de l'Arbre Fruitier et Forestier
INRA, Site de Crouël
Clermont-Ferrand, France

Carlos A. Navas
Departamento de Fisiologia
Instituto de Biociências
Universidade de São Paulo
São Paulo, SP, Brasil

Jeffrey Podos
Department of Biology
University of Massachusetts
Amherst, Massachusetts

Martin Reith
Botanischer Garten
Universität Freiburg
Freiburg, Germany

Nicholas P. Rowe
U.M.R. Botanique et Bioinformatique de l'Architecture des Plantes
Montpellier, France

Thomas Speck
Botanischer Garten
Universität Freiburg
Freiburg, Germany

John S. Sperry
Department of Biology
University of Utah
Salt Lake City, Utah

Craig L. Stevens
National Institute for Water and Atmospheric Research
Kilbirnie, New Zealand

Alexia Stokes
U.M.R. Laboratoire de Rheologie du Bois de Bordeaux
Domaine de l'Hermitage
Cestas, France

Bieke Vanhooydonck
Department of Biology
University of Antwerp
Antwerpen, Belgium

Robbie S. Wilson
Physiological Ecology Laboratory
Department of Zoology
The University of Queensland
St. Lucia, Australia

Table of Contents

1 Tree Biomechanics and Growth Strategies in the Context of Forest Functional Ecology

Meriem Fournier, Alexia Stokes, Catherine Coutand, Thierry Fourcaud, and Bruno Moulia

CONTENTS

1.1 INTRODUCTION

Whereas the mechanical performance of plant organs has often been discussed in evolutionary biology [1,2], tree biomechanics has rarely been considered in the context of functional ecology. Functional ecology aims at understanding the functions of organisms that result in fluxes of biomass or energy within an ecosystem, e.g., a forest. This discipline studies the processes controlling these fluxes, at either the scale of an individual, community, or ecosystem, with their response to natural or anthropic environmental variations.

Ecological differences among vascular land plant species arise from different ways of acquiring the same major resources of light, water, CO_2, and nutrients. An ecological strategy is the manner in which species secure carbon profit, i.e., both light and CO_2 absorption, during vegetative growth, and this also ensures gene transmission in the future [3]. At the present time, the relationship between biodiversity and ecosystem functioning is one of the most debated questions in ecology, and it is of great importance to identify variations in ecological strategies between species [3–6].

In this context, the field of tree biomechanics is concerned with the manner in which trees develop support structures to explore space and acquire resources, and, by feedback, to allocate biomass to the support function. The purpose of this chapter is to discuss how an understanding of the solid mechanics of materials and structures has contributed to functional ecology with examples taken from current studies in tree biomechanics.

Mechanics gives physical limits to size, form, and structure because living organisms must follow physical laws [7]. This discipline also allows several relationships between function and size, form, or structure to be explored. Solid mechanics provides the relationships between supported loads (inputs) to outputs such as displacements, strains, stresses (local distribution of loads), and safety factors against buckling or failure through given parameters [8]. These parameters may be structural geometry (shape) and material properties, e.g., critical stresses or strains leading to failure, or the relationship between stresses and strains given, e.g., in the simplest case by the modulus of elasticity (or Young's modulus) [8]. Biomechanics is much more ambitious than solid mechanics. Biomechanics aims at analyzing the behavior of an organism that performs many not explicitly specified functions using geometry and material properties fabricated by processes shaped by the complexity of both evolution and physiology. Thus, to use the framework of solid mechanics to solve biological problems concerning form and function, biomechanics involves different steps. Initially, a representation of the plant and of the supported loads using a mechanical model is necessary. This step means that initial choices must be made because models are nearly always simplifications. For example, can wind be considered as a static or a dynamic force for the problem considered? Are stems paraboloids or cylinders? Is root anchorage perfectly rigid or not? These initial choices can have huge consequences on the subsequently discussed outputs, in particular concerning the functional significance or the adaptive value of mechanical outputs, e.g., safety factors [9] or gravitropic movements. The subsequent discussions tend to be biological in nature and therefore out of the scope of engineering science.

Before dealing with several ecological questions, we first present some biomechanical characteristics of trees and develop questions concerning height growth strategies. We then discuss successively the underlying mechanical problems and associated models, i.e., the representation of supported loads, plant shape, and material along with the biological problems, data, and hypotheses, especially those tackling the biological control of size, shape, and material properties. The practical application of biomechanics in eco-engineering [10] is then discussed.

1.2 SOME BIOMECHANICAL CHARACTERISTICS OF TREES

Trees are among the largest living organisms and are the tallest self-supporting plants. Growth in height incurs high costs because of the investment in safe and stable support structures [11]. For the engineer, the understanding of tree biomechanics represents a challenge to current knowledge because trees can be very tall and very slender, and yet display a long life span. As we see later on, secondary growth, i.e., growth in thickness or radial growth, occurs in the cambial meristem located just underneath the bark [12] and is the main process contributing to the survival of such structures during their long life span.

1.2.1 Wood as a Lightweight, Cellular- and Fiber-Reinforced Material

Secondary growth produces an efficient support tissue: wood. Wood has been used by human beings for many years — dried wood has been used to construct buildings or make furniture. Such dried wood has a moisture content that depends on air temperature and humidity, and is made up of wood cells possessing empty lumina. Living trees, however, possess green wood. In green wood, cell walls are saturated, and additional water also fills up the lumina [13]. Because the mechanical properties of wood depend on the moisture content of the cell wall, the drier the wood, the stiffer and stronger it is [13]. Caution should thus be taken when using engineering literature in wood sciences because databases are not always suitable for biomechanical analyses dealing with moist, green wood. However, mechanical properties do not vary significantly beyond a moisture content of approximately 30% (on a dry weight basis) when cell walls are saturated and lumina empty [13,14]. In living trees, water transport affects lumen water content with cell walls in the sapwood being completely or partially saturated. As a consequence, although wood moisture content varies in living trees, e.g., according to seasons, species, and ontogeny, the variations of mechanical properties of green wood, i.e., stiffness and strength, during the growing season can be neglected.

Rheological data concerning green wood are scarce (but see, for example [15,16]), and there is a need for more systematic studies in this area. Meanwhile, whenever comparisons are made, there is usually a good correlation between the properties of green and dry wood used to estimate green wood properties [14]. Because of its complex structure at different scales, wood can be considered to be a very "high tech" material. An analysis of specific properties, i.e., ratios of mechanical properties to density, reveals that at the cellular level, wood is a "honeycomb-like" lightweight material of high performance. This cellular structure is also the origin of the close relationship between dried wood specific gravity, which represents the amount of supporting material characterized by its porosity, and mechanical properties [17,18]. For instance, using the regressions established by Guitard [17] at an interspecific level on a wide sample of species with a large range of densities, and transforming mass, volume, and modulus of elasticity of air-dried wood to green wood and oven-dried properties, we can approximate the parallel to the grain modulus of elasticity of green wood for angiosperms by:

$$E = 10400\left(\frac{D}{0.53}\right)^{1.03} \tag{1.1}$$

where E is the modulus of elasticity of green wood (MPa) (pooling together estimations by several methods: tension, compression, bending) and D is the basic density, i.e., the ratio of the oven-dried biomass to the volume of green wood.

These relationships show an approximately constant ratio between E and basic density. Thus, as pointed out by several authors [19–21], wood's mechanical efficiency relative to stiffness and dry biomass available is almost constant, no matter how porous the wood. However, dried biomass does not represent the true weight supported by a living tree, and the ratio of E to humid density changes as the more porous wood can absorb more water (Figure 1.1). Furthermore, an exhaustive dis-

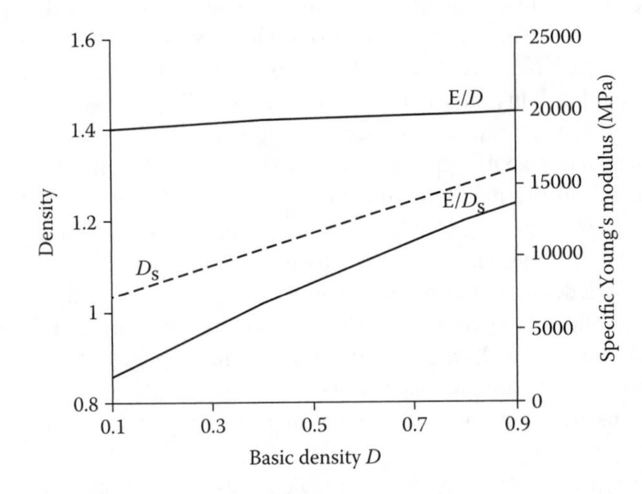

FIGURE 1.1 Evolution of the specific modulus of elasticity for angiosperm green wood (ratio of the modulus of elasticity E to wood density) with basic density D. D is the amount of dried biomass per unit of green volume, i.e., the cost of support. D_S (dotted line) is the density at full saturation, i.e., cell lumens are entirely filled with water, for wood density obtained in functioning sapwood, i.e., maximal self-weight of support organs. E/D is almost constant while E/D_S increases significantly with wood basic density.

cussion about design should also include additional branch and leaf weights. Thus, wood mechanical performance relative to design against, for example buckling, can change from light to dense woods. Such a distinction between mechanical efficiency, i.e., the cost of support per unit of dried mass, and performance (design safety relative to supported, humid mass) has never been considered.

At the level of the cell wall, wood is a multilayered material and can be considered as a reinforced composite made up of microfibrils composed of crystalline cellulose embedded in a matrix of lignins and hemicelluloses [22,23]. This composite structure is the major reason for the high anisotropy of wood: mechanical stiffness and strength are much greater along the grain, in the direction more or less parallel to the stem axis. This longitudinal direction is usually the most loaded direction and is held in bending in beamlike structures, such as trunks and branches. Because the cellulose microfibrils are very stiff, one important structural feature at the cell wall level is the angle between cellulose microfibrils and the cell axis in the S_2 layer [22]. Significant changes in this microfibril angle (MFA) can be observed, such as in juvenile and compression wood, which have a much greater MFA [22,24]. Therefore, these types of wood are much less stiff than can be expected from their density, e.g., by using standard formulas to estimate the modulus of elasticity from wood dried density [17,18].

1.2.2 Wood Variability

Wood structure and properties vary between and within species [25]. The adaptive mechanical performances of wood structure among different species in relation to tree phylogeny and other functional traits have rarely been discussed [26]. Among the huge diversity of tropical species, wood density (of dried biomass) has often been used as a measure of maximal growth rate and of relative shade tolerance. Fast-growing, shade-intolerant species have lower wood densities [27,28]. Within a species, faster growth is usually associated with lower density, especially in softwoods, although many exceptions can be found, e.g., in oak, faster growth is associated with higher density [29]

Another complicating factor when considering wood structure is that wood is not homogeneous within the radial cross section [25]. Variability due to the presence of several different types of wood can be observed. These different types of wood include: reaction wood (see below), early and late wood (specializing, respectively, in transport and support), juvenile wood (the wood formed from a juvenile cambium [25]), and heartwood (the central wood that does not conduct sap and is impregnated with chemicals as a result of secondary metabolism occurring in the sapwood) [30]. Although such variability within the cross section is very common, the specific geometrical pattern of these different types of wood depends on species and genetic backgrounds, as well as environmental conditions and the stage of ontogeny. For example, juvenile wood is often less dense and stiff than normal wood [22], but the contrary can also be found [31]. The adaptive interest for tree mechanical safety of such radial variations in wood density has been discussed by Schniewind [32], Wiemann and Williamson [33], and Woodcock and Shier [34].

1.2.3 Mechanics of Secondary Growth

Secondary growth, or the peripheral deposition of load-bearing tissue over time, is not a well-known feature in mechanical engineering. This phenomenon thus requires a careful analysis because inert structures are considered by engineers to exist before being subjected to loading. However, in the case of plants and trees, the structure is already loaded before the new material is laid down, and even during the formation of this new material, mechanical loading continues to occur. For example, when dealing with the local distribution of stresses induced by self-weight in both compression and bending, the solid mechanics theory of homogenous materials would predict a linear distribution from the upper to the lower side. This theory can be modified to take into consideration material heterogeneity within the cross section [35]. In both cases, using formulas from standard mechanical engineering textbooks [8,35] allows us to calculate stresses from the total self-weight and the whole cross-sectional geometry without any data about growth history [7]. However, this analysis implicitly supposes that the total weight has been fixed after the formation of the cross section, whereas in trees, both the weight and cross section grow simultaneously. Taking into account the relative kinetics of cross section and weight growth, Fournier and coworkers [36] emphasized the huge discrepancies when classical engineering theories are used. For example, peripheral wood that is very young supports only a small amount of self-weight, i.e., the weight increment in the above stem and crown since peripheral wood, even when the tree is leaning and self-weight acts as a bending load [37]. This consideration is also of great importance when analyzing successive shapes of growing stems that are continuously bent by gravitational forces (see Section 1.3.1.2).

1.3 Biomechanical and Ecological Significance of Height

Height is recognized universally as a major plant trait, giving most benefit to the plant in terms of access to light, and therefore makes up part of a plant's ecological strategy [3]. Nevertheless, as pointed out by Westoby et al. [3], different elements should be separated from an ecological point of view: the rate of height growth associated to light foraging, the asymptotic height, and the capacity to persist at a given height. Moreover, investment in height includes several trade-offs and adaptive elements. The question of the coexistence of species at a wide range of heights has been studied in a mathematical framework using game theory [38]. Whether maximal asymptotic height is constrained by physical limitations, e.g., mechanical support or hydraulics, or only by the biological competition for light, i.e., height growth stops when it ceases to offer a competitive advantage, is still an open question [39]. Hydraulic limitations of tree height have been discussed [39–41]. Although some kind of trade-off may be involved between these different functions, we discuss only the biomechanical aspects of the question.

1.3.1 Biomechanical Environmental Constraints on Tree Height and Their Ecological Significance

Although growth in length permits the stem to grow higher, the stem also needs to be self-supporting. Mechanical instability can occur under the effects of self-weight, wind forces, or the combination of both. When such instability occurs, it can produce failure or not, with obviously distinct ecological consequences. To assess whether these risks are or are not ecological constraints and which mechanical load (if any) is limiting for height growth, researchers find that a mechanical representation, i.e., a model of the geometry, shape, loads, and boundary conditions, is an extremely useful tool.

Furthermore, these mechanical models can provide a basis for the understanding of several biomechanical aspects of the dynamics of forest communities. Not only is forest dynamics concerned with tree mechanical stability in communities because storm damage to trees can induce gaps that are the motor processes of forest growth dynamics, but mechanical stability is also influenced by forest dynamics. Competition for space in communities can induce huge variations in tree form and architecture with, in particular, a modification of allometric relations [42] as well as changes in wood quality linked to tree growth rate [25]. Ancelin et al. [43–45] developed an individual tree-based mechanical model of this feedback between tree biomechanics and forest dynamics.

1.3.1.1 Safety Factor

Safety factors are the nondimensional ratios between a characteristic of the present situation and the critical non–self-supporting one [9,46]. A safety factor of 1 (or lower than 1) means that the critical situation is reached. The higher the safety factor, the lower the risk. An important point to be assessed is whether the mechanical risk can be linked to material failure due to increasing bending or buckling because either could be limiting, but each requires distinct analyses that can lead to different conclusions. Bending occurs when a force component is acting perpendicular to the trunk, such as wind drag in a straight tree, or self-weight in a leaning tree. When bending stresses exceed the material strength, failure occurs. In a standing tree, the safety factor is then defined as the ratio of the material strength to the actual bending stress. Buckling is caused by a loss of stability of an equilibrium. For example, if a straight column is loaded under compression and at some critical point, the compressed equilibrium state becomes unstable, then any mechanical perturbation would induce a high degree of bending (see [7] for a more complete introduction).

In other words, the column is no longer self-supporting. Safety factors can be defined as the ratio of the critical weight to the actual weight. In plant biomechanics, interest is rather on what can be achieved for a given amount of aerial biomass. Safety factors for buckling are then usually defined as the ratio of the critical height to the buckling height, assuming relations, usually allometric, between weight and height. Mechanical models have been developed to calculate critical situations for both bending failure and buckling (e.g., [7,9,19,47–53]).

Such criteria are useful to compare the mechanical constraints between species or environmental situations. Many authors have also discussed the optimality of phenotypes at an individual (optimal stem taper) or population level (optimal stem slenderness), assuming that the optimal shape maximizes the height for a given diameter [19,48,50,54] or results in a constant breakage risk along the stem [47,53,55]. Slenderness rules, i.e., relationships that are usually allometric, between height and diameter within a population of trees are usually derived from the assumption of constant safety factors among the population (see [51] for a critical review and [56,57] for a general discussion about adaptative interpretations of allometries from mechanical and alternative hypotheses). However, several authors have discussed the values of safety factors when they are close or not to the critical limit, and their variability with tree ontogeny [21,49,58–60]. All of them found that safety factors against buckling decrease with growth in saplings as the competition for light became more intense and material resources that could be used for trunk growth become less available. A few authors have also studied safety factors in relation to species' shade tolerance and light conditions [58,61]. However, these approaches have always considered that trees have to avoid any critical situation and have never discussed the postcritical behavior of a tree nor the cost of height loss and its possible recovery. Nevertheless, buckling can lead to breakage or permanent, plastic stem lean, which is recoverable through the tree's gravitropic response (see Section 1.4.2). Breakage itself does not necessarily result in tree death and recovery can occur through healing of wounds or resprouting. Determining the conditions for buckling to occur is thus not sufficient, and the assumption that buckling is a catastrophic biological event remains to be tested in each particular case.

1.3.1.2 Analysis of Successive Shapes Occurring during Growth Due to the Continuous Increase of Supported Loads

Growth is by itself a mechanical constraint. Indeed, from a mechanical point of view, a small initial bending should be amplified by growth because in any cross section of the trunk, growth increases bending loads due to self-weight. Thus, bending curvature is increased and stiffened by continuing radial growth in an amount depending on the relative rates of bending moment and cross-sectional stiffness increases [37,62–64]. This dynamic and continuous growth constraint has rarely been analyzed carefully and has never been considered in ecological studies. In some cases such constraints may be considerable, such as sudden increase of loads (e.g. leaf flushes or heavy fruit production) on slender flexible stems, which is followed by cambial growth that adjusts the curved shape [62]. However, it is clear that without any biological control of verticality, e.g., a selection of the most vertical trees, or the action of gravitropism to restore verticality (see Section 1.4.2), any given degree of stem lean at a given height should increase significantly with growth. Studying two populations of saplings of *Goupia glabra* Aubl. (shade-intolerant species of the rainforest in French Guiana) in understory and full light conditions, we found that the lean never increases and even decreases in the most competitive (understory) environment (Figure 1.2). Therefore, these data provide

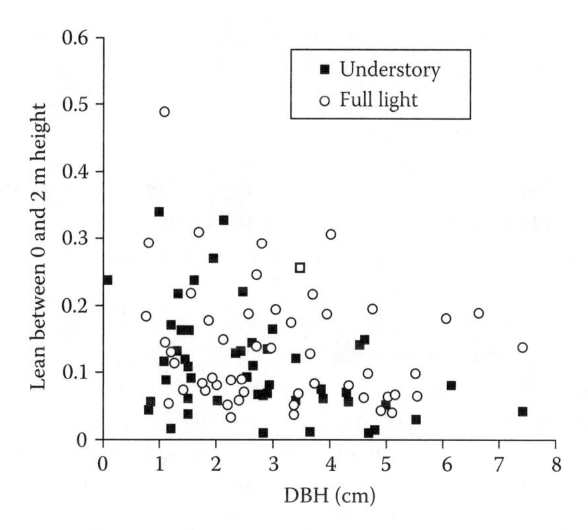

FIGURE 1.2 Variation in stem lean (%) between 0- and 2-m height in two populations of *Goupia glabra* Aubl. saplings from the French Guiana tropical rainforest (Fournier and Jaouen, unpublished data). Lean in seedlings grown in full light (white circles) does not increase with diameter breast height (DBH) (Spearman R is not significant); in understory seedlings (black squares), the lean was found to decrease (Spearman $R = 0.40$, $P = 0.006$).

evidence of the existence of biological reactions to the gravitational mechanical constraint at the population level.

1.3.2 Biomechanical Functional Traits Defined from Risk Assessment

Biomechanical functional traits are the combination of morphological, anatomical, and physiological characteristics that define the height growth strategy. When focusing on the mechanical constraints on this strategy, the functional traits are combinations of the size, shape, and material properties that influence the risk of tilting, bending, or breakage, and are analyzed as inputs of the mechanical model designed to describe the mechanical constraint.

1.3.2.1 Buckling or Breakage of Stems

Tree mechanical design against buckling [48] or breakage [47] has been studied for over a century. Most existing models (see [51] for a synthesis) have considered the tree as a vertical, tapered pole of a homogeneous material, loaded either by static, lateral wind forces, or by its own self-weight, with a perfectly stiff anchorage. Therefore, the functional traits involved and analyzed with regards to their contribution to the risk of mechanical instability are typically: the characteristics of pole size (volume, diameter, or height), pole shape (slenderness, taper, cross-sectional shape), material properties (modulus of elasticity, occasionally torsional modulus, failure criteria usually given by a single critical stress), self-weight (density of the pole material and additional weight of the crown), or structural parameters that define wind forces (drag coefficients and crown area [44]).

1.3.2.2 Root Anchorage

In many cases, failure due to mechanical loading often occurs in the root system. Thus, an understanding of root biomechanics is of crucial interest, not only because the anchoring capacity of a plant is an important factor for survival with regards to external abiotic stresses, such as wind loading or animal grazing, but also because roots are a major component in the reinforcement of soil. Whereas many studies have been carried out on the morphological development of roots with regards to their absorption capacity [65–67], very few investigations have focused on the mechanical role of roots [68,69]. Nevertheless, these pioneer studies have provided a sound base for a better understanding of root anchorage efficiency in both plants and trees.

Root anchorage has largely been investigated at the single root level [70,71] or at the scale of whole root systems [72–76], whereas soil reinforcement by roots has generally been considered at the population scale [77–79]. To better understand the biomechanical role of specific root elements and in particular plant adaptation to mechanical stresses, a distinction must be made between small roots, i.e., roots that resist tension but which have a low bending stiffness, and large roots, i.e., roots that can resist both tension and bending. The first category can be compared to "cable" structural elements, whereas the second type can be considered as "beam" elements. This latter category is mainly encountered in adult trees or shrubs and the former in herbaceous species. Such a distinction between these two categories of roots is necessary to avoid confusion when considering the consequences of root mechanical properties on uprooting efficiency, as discussed in the next paragraph.

Over the last 30 years, an increasing awareness of the role of fine roots (defined as less than 25 mm in diameter) in soil reinforcement has led to several studies being carried out on the mechanical properties of roots [80–83]. Soil shear strength is enhanced by the presence of roots due to the increase in additional apparent cohesion [71,84,85]. When roots are held in tension, such as pull-out or soil slippage on a slope, root tensile strength is fully mobilized and roots act as reinforcing fibers in the surrounding soil matrix [86,87]. In studies where the tensile strength of small roots has been measured, it is usually shown that the strength, as well as the modulus of elasticity, decreases with increasing diameter d, following an exponential law of the type $\beta \exp(-\alpha d)$ (Figure 1.3) (values of root resistance in tension, bending, and compression are given for different woody species in [72]). This decrease in tensile strength is due to a lower quantity of cellulose in small roots ([83]; see Figure 1.3). Although this type of information is invaluable when studying the mechanism or root reinforcement, especially on slopes subject to instability problems [77,86,88], it is also of extreme interest to researchers trying to understand the specific role of small roots on tree anchorage. It could be suggested that for a fixed amount of invested biomass, a network of several small roots is more resistant in tension than a few large structural roots [89,90]. However, a large number of small roots may be also detrimental to anchorage because a group effect could result in more failure occurring in the soil [89,91].

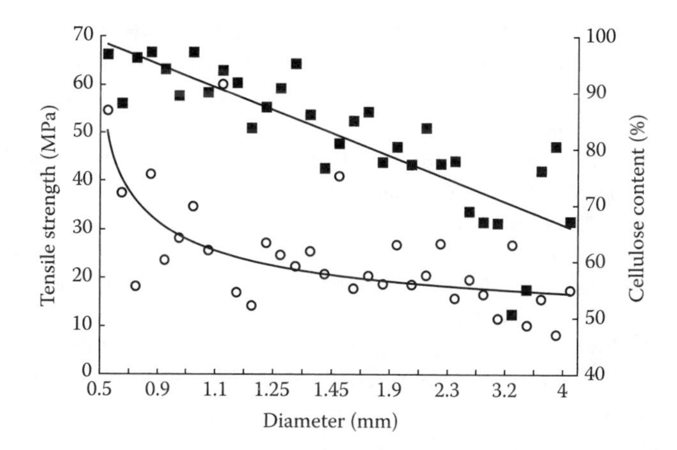

FIGURE 1.3 Tensile strength increased significantly with decreasing root diameter ($y = 28.96x^{(0.57)}$, $R^2 = -0.45$, $P < 0.05$) and cellulose content ($y = 0.47x^{(-14.42)}$, $R^2 = 0.23$, $P < 0.005$) in roots of sweet chestnut (*Castanea sativa* Mill.) (after [83]).

Fine roots have often been ignored when investigating the root anchorage of forest trees. This neglect is mainly due to the difficulty in extracting them from the soil, in particular those roots far away from the trunk. Nevertheless, these distal roots determine the boundary conditions of the whole structure and can be very important from the biomechanical point of view. However, this observation cannot be applied to all plants, e.g., it has been shown in leek (*Allium porrum* L.) seedlings that the distal part of a long, single, fine root is not stressed before failure of its proximal part [70]. Therefore, the failure mechanism in trees is probably significantly different than that observed in herbaceous species.

The difficulty of investigating root anchorage is not only due to the complexity of the mechanisms occurring in both roots and soil, but also to their multifactorial aspect [88,89]. A good alternative to difficult and time-consuming field experiments can be found in numerical modeling. Dupuy et al. [89] carried out such numerical analyses using the finite element method. These authors determined the mechanical response of small ramified roots to pull-out in tension. Parametric studies showed that the number of lateral ramifications and their diameter were both major components affecting the resistance to pull-out for a given soil pressure.

Plant anchorage efficiency must be investigated taking into consideration not only the mechanical behavior of single roots, but more importantly, the whole root architecture. A number of studies were carried out in the 1990s on annual or herbaceous plants [70,76]. Ennos and Fitter [92] proposed an alternative hypothesis concerning root system shape and function. These authors suggested that creeping and climbing plants develop fibrous root systems because the only mechanical stress transferred to the roots is tension. However, root systems of single-stemmed, free-standing plants tend to develop a more plate-like or tap-like morphology [93]. Based on mechanical assumptions, Ennos and Fitter [92] also showed that these different anchorage strategies can have an impact on the biomass allocation ratio between above- and below-ground parts during plant growth. Other studies on dicots have since been carried out to quantify the root biomass investment according to the external loading on the plant [94].

Investigating the anchorage of adult trees is rather more complicated because of the morphological complexity encountered in tree root systems. Several structural elements of importance for mechanical stability can coexist. Root topology, i.e., the way branches are linked together, is of major interest when trying to understand how external forces are transmitted throughout the whole system and into the soil [90,91,95,96]. Root system depth and the number of root branches have also been identified as highly significant components of tree anchorage [88,96]. Coutts [97,98] identified the main components that play a role in root anchorage of Sitka spruce (*Picea sitchensis* Bong) by order of importance, i.e., the weight of the root–soil plate, the windward roots in tension, the soil cohesion, and the bending strength of leeward roots. A further component to consider in shallowly rooted species is the presence of buttressing around the stem. Although several hypotheses exist concerning the development and function of buttresses in tropical trees [99–102], in temperate species at least, buttresses tend to develop in trees with shallow, plate-like systems [103]. The presence of such buttresses will help external loading forces be transmitted more smoothly along the lateral roots and into the soil, thereby improving anchorage [104]. Particular attention has also been paid to tree species that develop tap root systems [88,101,105,106]. Specific experiments carried out by Mickovski and Ennos [107] on Scots pine (*Pinus sylvestris* L.) showed that in tap root systems, lateral roots are not a major component of root anchorage. However, in separate studies, Niklas et al. [108] illustrated the lack of efficiency of a massive tap root if not associated with thick lateral roots, and Tamasi et al. [109] showed that in oak (*Quercus robur* L.) seedlings subjected to artificial wind loading, lateral root growth was increased at the expense of tap root length.

Contradictory results are often encountered in the literature concerning the relationship between root architecture and anchorage efficiency, and one explanation may lie in the underestimated role of soil characteristics on uprooting [110]. Numerical analyses may help fill this gap in knowledge [95,111]. Fourcaud et al. [95] and Dupuy et al. [96] developed methods allowing morphological data from real or simulated root systems to be subjected to virtual uprooting tests. Soil mechanical properties could be changed easily, therefore, allowing a rapid assessment of root architecture efficiency in different soils [112]. In a clay soil, the root and soil system of a heart root system rotates around an axis that is situated directly beneath the stem, whereas in sandy soil, the same system rotates around an axis that is shifted leeward. Heart and tap root systems [93] also behave similarly in clay soil but are over twice as resistant to overturning than plate or herringbone [66] systems in the same soil. However, between the four root types that were studied, anchorage in sandy soil was less variable between the four root types; the most efficient anchorage in sand was found in the tap-rooted system and the least, the plate root system [112].

In conclusion, although the study of root biomechanics has been neglected until recent years, a large number of studies exist that elucidate the mechanisms by which roots are anchored in the soil. Modifications in root system architecture due to external loading will have consequences not only for anchorage efficiency but for the ability of root systems to absorb nutrients [66,67]. Future studies need to incorporate both root and soil mechanical properties into numerical models, which are in turn validated by field experiments.

1.3.3 Biomechanical Functional Traits and Processes Involved in Height Growth Strategy

The studies mentioned previously have shown that sets of variables derived from mechanical models incorporating tree size, shape, and wood properties (available volume of wood, stem vs. crown biomass, shoot vs. root biomass, stem slenderness, taper and lean, and root and shoot architecture) are usually involved in the assessment of failure risk of trees. As often pointed out for any kind of tree functional trait [3], to describe a strategy, we must analyze those traits we estimated both at the individual and population levels. Thus, we must investigate how individual traits are influenced by both the environment (plasticity) and ontogeny. To define tree functional types, we assign greater importance to the trait variation or trajectory during ontogeny than to average values. For example, the decrease of the buckling safety factor during the early growth stages of saplings, and the maximal values reached in the most competitive environments, are more pertinent when comparing species' strategies than the mean value of risk per species. Certain size effects are physically obvious, and it is helpful to use modeling to define size-independent traits at the first order, for example buckling safety factors rather than critical height or structural mean modulus of elasticity rather than cross-sectional flexural stiffness (the product of the mean modulus of elasticity and the second moment of area of the cross section [113]). Because traits are potentially numerous, the minimal set able to describe a strategy for a given mechanical constraint in a given situation is a complex question. As far as we know, such a question is rarely considered, and traits are often chosen implicitly.

Lastly, height strategy involves not only selected morphological and anatomical features that are directly linked to tree failure, but two growth processes also exist, which allow a certain mechanical control over these features. One such process is thigmomorphogenesis [114], the phenomenon by which external mechanical loading can change (i) the biomass allocation between roots and shoots and also between their length and thickness, (ii) shoot and root architecture, (iii) organ cross-sectional shape, and (iv) internal plant structure. The second process is gravitropism [12], i.e., the phenomenon by which the verticality of a displaced stem or branch can be restored. A "hard" functional trait (see [115] for a discussion of the distinction between "hard" and "soft" traits) defining a species' strategy should be the species' sensitivity to these processes, i.e., its capacity to adapt functional growth. Such traits can also be measured experimentally by, for example, measuring the reorientation of artificially tilted stems or studying the growth response to applied mechanical loading (see Section 1.4).

1.4 THE GROWTH PROCESSES THAT CONTROL THE MECHANICAL STABILITY OF SLENDER TREE STEMS

1.4.1 THE MECHANICAL CONTROL OF GROWTH

Extensive research since the 1970s has demonstrated in many herbaceous and woody species that several growth parameters are affected in response to external mechanical loading (see [116] for a recent review). This general mechanoperceptive syndrome is known as thigmomorphogenesis [114] and is likely to be found in most erect plants, although variability exists in quantitative amounts between both species and genotypes (see, e.g., [117–119]). As shown in Figure 1.4, the major responses of trees to a mechanical stimulation of their aerial parts are (i) a decrease in biomass allocation to aerial parts, thereby favoring root growth (see Section 1.4.3 for more details on specific morphological and anatomical responses of roots), (ii) a clear stimulation of the secondary diametrical growth, (iii) a decrease in primary extension growth, and (iv) changes in the density and specific mechanical properties of the wood, usually toward a lower material stiffness and a higher strain at material failure (see, e.g., [73,119,120]). The effect of branching on tree sway characteristics has been recently investigated by Sellier and Fourcaud [121], who showed that tree adaptation to wind must also be considered with regards to tree architecture.

A better insight into the quantitative understanding of the signal perceived by a plant was shown by Coutand and co-workers [122,123]. These authors demonstrated that plants tend to perceive strain, not stress, and that the control of primary growth can only be explained assuming that a systemic signal is produced integrating the overall field of strain all over the living tissues in the stem. On this basis, they were able to produce a quantitative response curve for the mechanical control of stem extension growth. Although demonstrated on the apical growth of tomato (*Lycopersicon esculentum* Mill.), this analysis has now also been carried out for radial growth of poplar saplings (*Populus* sp.) [Coutand et al., unpublished data].

Even if the thigmomorphogenetic responses of trees to mechanical loading are now well documented, the ecological significance in terms of acclimation to wind or buckling largely remains to be addressed. Because of the difficulty in carrying out experiments on trees growing in natural conditions, few experiments have been conducted whereby the acclimation response of trees to wind has been measured in the field. Since the seminal studies by Jacobs [124], studies of thigmomorphogenesis in natural conditions involve comparisons between staked and free-standing trees. The effects of staking or guying on stem morphology are considerable. However, when trees are staked, a nonnatural situation is provoked whereby the stimulus is negligible in the trunk and an unnatural stimulus remains in the branches. Therefore, conditions are not realistic. Because growth response curves are highly nonlinear [122,125], the perception of a tree to a mechanical signal, e.g., wind, may actually occur at very low loading levels. It would be necessary to carry out more dose–response experiments, whereby trees are subjected to different levels of wind loading to determine better their thigmomorphogenetic responses. Moreover, most studies have been conducted on isolated trees in greenhouses or growth chambers, and recent evidence indicates that the photomorphogenetic response to shade in

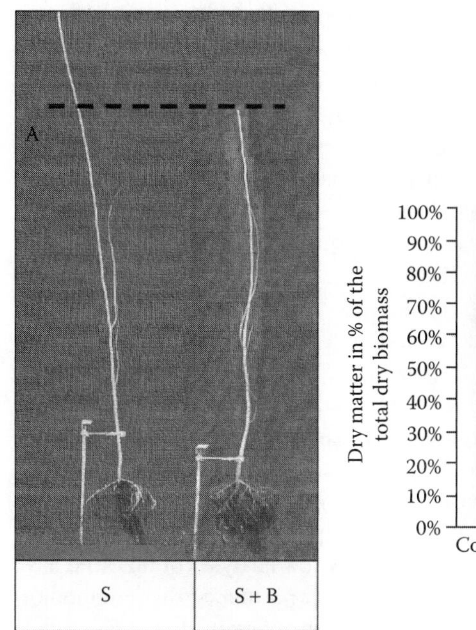

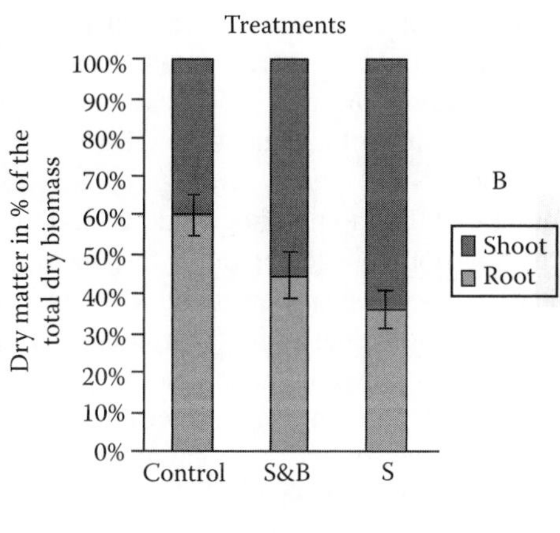

FIGURE 1.4 Thigmomorphogenetic responses in Wild Cherry saplings (*Prunus avium L.* cv. "monteil"). Three treatments were applied. Control: free standing submitted to natural wind sways; S: completely staked (trunk and branches); S+B: completely staked but with artificial bending of the trunk for 1 minute every 3 hours. (A) Typical morphologies of the stem and root system of plants subjected to treatments S and S+B. (B) Changes in dry matter partitioning between the shoot and root systems due to bending treatments. (Modified from Coutand et al., 2003.)

dense canopy may reduce thigmomorphogenetic responses (e.g., see [42,126], but see Mitchell [127]) or induce them when trees are grown in full sunlight [128]. The use of artificial fans to imitate wind loading on trees may be useful for identifying thigmomorphogenetic responses [90], but provides little information with regard to the natural significance of thigmomorphogenesis. Experiments where forest trees would be subjected to artificial loading by the use of fans, for example to simulate turbulence would require huge facilities (see, e.g., [129,130]). Recently, a new technique to demonstrate the occurrence of significant thigmomorphogenetic acclimation to wind in natural conditions has been proposed by Moulia and Combes [131]. These authors studied the variability in the difference between staked and free-standing plant canopies over several growing periods in alfalfa (*Medicago sativa* L.). Moulia and Combes [131] showed that the month-to-month variability in wind speed when winds were moderate was able to explain 65% of the reduction in aerial biomass and 41% of the reduction in total canopy height, thereby demonstrating highly significant thigmomophogenetic effects in dense canopies under natural conditions. However, similar studies on trees remain to be conducted and would take an extremely long time to carry out. A less cumbersome alternative is to study spatial changes in morphology associated with obvious natural gradients in wind conditions.

When applied to rain forest conditions where the evapotranspirative effects of wind can be neglected, this method has shown that thigmomorphogenetic acclimation is ongoing [132]. However, such an approach is only correlative and limited to very special conditions such as ridge crests and shelters in tropical rain forest. Of particular interest would be a study of the correlation between morphology and wind speed within a canopy using the natural spatial variability in wind speeds. A prerequisite for this would be to record wind-induced sways all over the canopy by using video recording and image correlation techniques for the kinematic tracking of wind-induced canopy movements [133]. However, long-term studies still remain to be conducted.

Assuming that thigmomorphogenetic acclimation does occur in nature, the second central question is whether these responses are adaptive or not, i.e., are the performance vs. mechanical constraints improved and are there consequences on the fitness of the individual in its environment? Because both the performance vs. the mechanical constraint and the thigmomorphogenetic "syndrome" involve several variables, qualitative inferences are uncertain, and only direct measurements of plant performance or the use of a mechanical model can help to determine the exact effect on plant performance vs. mechanical constraints. Very few analyses of this kind have been carried out. Concerning wind loading, it has been postulated that thigmomorphogenesis might be involved in allowing trees to reach a certain shape. This shape will permit a spatially homogeneous distribution of wind-induced stresses for wind conditions. Achieving such a constant stress is adaptive and even optimal in that all parts of the trees would display the same safety factor against material failure [47,52,134].

Mattheck [135] made a significant contribution to the old hypothesis of constant stress design [47,53] by providing a dynamic biomechanical model of stress equalization through growth. Mattheck and Bethge [136] also described a wide range of shapes that could be explained qualitatively through the constant stress hypothesis. However, no direct quantitative testing of the model's prediction has ever been produced. Moreover, subsequent studies that have attempted to verify the constant stress hypothesis have used fairly detailed modeling of the wind loads involved [137,138] and have even dismissed this hypothesis for wind loads on trees (but see [138,139]). More indirect tests comparing the height-to-diameter ratio have also been reviewed and not found convincing [51]. Although not optimal in terms of constant stress, thigmomorphogenesis is likely to improve the overall strength of a tree's structure but to an extent that remains to be quantified, and with strategies that still have to be studied.

Thigmomorphogenesis can also increase a tree's stability against buckling under self-weight. This phenomenon has been tested experimentally by Tateno [46] on mulberry trees (*Morus bombycis* Koidz).

1.4.2 The Control of Stem Orientation to Maintain or Restore the Tree Form, and Allow Vertical Growth

It may be logical to assume that trees submitted to gravitational or wind forces would lean more and more during their life span; however, stems can maintain or restore verticality more or less continuously using gravitropic reactions. Two "motors" are involved, both of which are relative to internally induced, asymmetric strains resulting in stem curvature. Initially, hydraulic turgor pressure in primary tissues associated with asymmetric growth can lead to reorientation of the apical, unlignified part of stems in just a few hours or days [140]. In other regions of the stem, where elongation has been achieved but radial growth is still active, the process of wood cell differentiation during maturation is modified, resulting in the formation of a particular type of wood called "reaction wood" [12,141,142]. This reaction wood is largely responsible for the control of the tree-righting mechanism. This second type of gravitropic reaction, which occurs during secondary growth, is much slower than the first, and the stem may take several months to several years, if at all, to straighten up, depending on its size.

The mechanism by which the stem rights itself occurs during cellular maturation and is associated with the shrinkage of fibers or tracheids [143,144]. In normal wood, as the newly created wood cells are "glued" onto the central core of existing stiff wood, shrinkage of these developing cells is impeded. Therefore, the peripheral wood is "stretched." Internal tensile forces along the grain are then generated and can be considered to act similar to a guy-rope system. When stems perceive the disequilibrium related to vertical growth, the wood structure and/or speed of cambial growth are modified on one side of the cross section, resulting in an asymmetry of these peripheral forces, and thus stem bending is induced (Figure 1.5). In gymnosperms, reaction wood formed in leaning stems is called "compression wood" [24] and tends to expand instead of shrink during maturation. Therefore, the peripheral force is compressive and is formed on the lower side of a leaning stem, on the opposite side to normal "stretched" wood [145]. In angiosperms, tension wood forms on the upper side of leaning stems [142]. During maturation, this wood shrinks with a higher resulting tensile force [143–145]. Both the origin of maturation strains and their variability in compression and tension wood can be explained by structural features at the cell wall level [145–148] and in particular, by the high MFA of compression wood [149].

At a macroscopic level, a mechanical model [37,150,151] allows a description of the induced curvature during a given stage of growth, which can be expressed in a general way as:

$$dC_R = \frac{\iint_{dS} \alpha \, y dy dz}{I} \tag{1.2}$$

where dC_R is the reaction curvature produced by the growth of the cross section from the area S to $S + dS$ (Figure 1.5), y and z are current coordinates in the cross-sectional plane from its geometric center, assuming that the cross section rotates around the z axis, I is the second moment of area of the cross section, and α is the maturation strain, which is usually heterogeneous within the cross section (a function of y and z). Heterogeneities in modulus of elasticity have been assumed as negligible with respect to α variations. Maturation strains can be estimated by measurements

FIGURE 1.5 Gravitropic reactions are induced by an asymmetry of peripheral tensile forces generated at the end of the wood cell differentiation process. Both the asymmetry of the wood structure in the case of (1a) compression wood formation on the lower side and (1b) tension wood formation on the upper side of a leaning stem, and (2) the asymmetry of growth, are involved (see Equation 1.2).

of released peripheral strains [143,152], which have also been used for studying growth stresses [145,153].

Equation 1.2 is useful when analyzing and comparing the effects of growth (*dS*), stem shape, wood heterogeneity, and values of maturation strains. For a circular, concentric cross section, assuming a sinusoidal variation of maturation strain around the circumference [37,150,151], Equation 1.2 becomes:

$$dC_R = 4\,\Delta\alpha \frac{dD}{D^2} \tag{1.3}$$

where $\Delta\alpha$ is the difference in maturation strain between the maximal value of shrinkage and the minimal value on the opposite side (in opposite wood for angiosperms or in compression wood for gymnosperms with a negative value because the wood does not shrink but rather expands), *D* is the diameter of the cross section, and *dD*, the increment of diameter (radial growth). Such an equation indicates that the initial size (*D*) is a constraint to the tree during stem reorientation. Young stems can thus be reoriented much more efficiently for a given amount of radial growth. In a general sense, Equation 1.2 gives relative scaling ideas about geometrical growth and anatomical effects. For instance, for a given diameter *D*, $\Delta\alpha = 0.005$ (which corresponds to the formation of very strong tension wood) associated with a relatively slow growth of $dD = 1$ mm/yr should have the same effect as $\Delta\alpha = 0.00025$, i.e., a slight asymmetry that does not require the formation of typical tension or compression wood, associated with a faster growth of $dD = 20$ mm/yr.

At this stage, the ecological significance of gravitropic curvatures induced by the asymmetry of tensile peripheral forces may not yet be obvious because reaction wood, usually opposed to "normal wood," could be supposed to be a quite rare and even a "pathological" phenomenon. Although the gravitropic function is well known [141,142,154], the formation of reaction wood is often assumed to be a response to mechanical stress (e.g., Turner [28]) associated with a lower technological wood quality [14] and not a motor process that counteracts the physical bending forces relative to growth and gravity, essential for the achievement of vertical growth in tree communities.

To demonstrate the ecological significance of gravitropic processes and because measurements of reaction and gravitational curvature in stems cannot be carried out separately over the long term because tree shape integrates both types, we use the approach of using a model (Equation 1.2 or Equation 1.3) to compare reaction curvatures to gravitational ones. The latter can be calculated using standard beam mechanics [8] through an incremental model [37] (see Section 1.3.1.2):

$$dC_g = \frac{1}{EI}\frac{dW\,H\sin\theta}{dS}dS \tag{1.4}$$

where

dC_g = the curvature physically induced by a weight increase during the growth of the cross section from S to $S + dS$
I = the second moment of area of the cross section
E = the modulus of elasticity
W = the weight
$H \sin\theta$ = the lever arm of the bending moment due to W (H is the height of the center of mass above the cross section and θ is the lean)

The derivative of the function $WH \sin\theta$ with respect to cross-sectional growth S can be obtained using allometric relations.

A practical application of Equations 1.3 and 1.4 on a community of saplings in the tropical rainforest of French Guiana provides evidence that reaction curvatures are usually greater than gravitational ones (Figure 1.6). This study demonstrates that the hypothesis previously inferred from Figure 1.2, whereby verticality is biologically controlled and plays a significant role under natural conditions, is sound. Use of the term "biologically controlled" raises the question of the signal perceived by plants that triggers the reorientation processes. This question is still a matter of debate, and the sensing system has not yet been clearly identified for either shoots or roots. The site of perception in roots has been shown to be located in the columnella cells of the root cap. Studies that aim to identify the perceived variable in shoots and to give quantitative relationships between the variation of the mechanical state induced by gravity and the plant's response are scarce [155,156]. At the

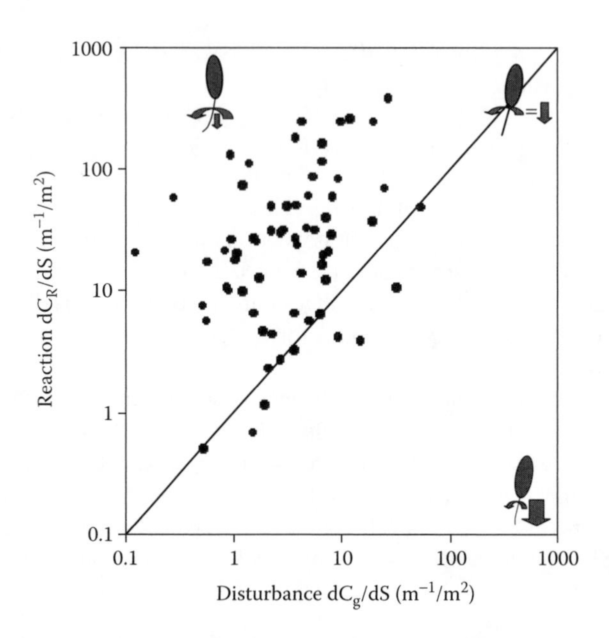

FIGURE 1.6 Comparison between two main bending processes in a community of saplings in a 750 m^2 plot of the tropical rainforest in French Guiana (DBH between 0.02 and 0.05 m): reaction and gravitational curvatures (see Equations 1.3 and 1.4) have been calculated for trunk basal cross sections (0.5 m height), using measurements of geometry (diameter, length, lean), weight (trees were cut and weighed), and maturation strains (Wap's method; see [152]) (Fournier, unpublished data).

organ scale, inclination of the tip (or local inclination) is still usually considered as the perceived variable, the Cholodny Went theory for elongating organs [157], and is used in the modeling of gravitropism [37,151]. However, the kinematics of the gravitropic response has revealed distinct phases: initially an upward curving occurs, i.e., a gravitropic response, followed by a phase of straightening out or autotropism, which is the automorphogenesis necessary to restore verticality of the whole organ. This straightening out has been observed in a wide range of plant organs and species [158]. Kinematical analyses of the gravitropic response demonstrated that autotropism starts well before the organ reaches the vertical [155,156]. This shows that the other variables are perceived to pilot the reorientation processes.

At the cellular scale, signal transduction has been much more studied with an emphasis on cloning genes involved in signal perception along with the role of the cytoskeleton. Through results from studies of gene expression, several gravi-sensing patterns have been hypothesized, but all are based on sedimentation of organites, such as statoliths and proteic globules. During their sedimentation, these organelles are thought to touch the rough endoplasmic reticulum and disrupt the cytoskeleton actin filament, thus resulting in an activation of mechanosensitive channels or stretch-activated channels. More recently, several authors have suggested that gravi-perception could rely on an activation of mechanosensitive channels of the plasmalemma induced by sedimentation of the entire cytoplasm. This mechanism suggests that strain could be the perceived variable, but no experimental study or mechanical model of gravi-sensing at the cellular scale has yet been carried out.

Further work should concern more studies in the field as well as in controlled environments whereby bending movements, such as stem longitudinal shape, curvature, and lean, at the whole trunk level are studied more carefully. The specific diversity of responses associated with the diversity of reaction wood structure and growth geometrical patterns with respect to reorientation efficiency and cost should also be investigated further.

1.4.3 The Control of Root Growth to Secure Anchorage

The knowledge from from recent studies on root biomechanics (see Section 1.3.2.2) is of particular interest when trying to understand the biomechanical adaptation of roots during plant growth. Several experiments have been carried out to quantify the root biomass investment according to the external loading on the plant [94,109]. Ennos and Fitter [92] showed that different root anchorage strategies can have an impact on this biomass allocation ratio between above and below ground parts during plant growth. Active biomechanical responses to external mechanical stresses have been studied and discussed by several authors, including the consequences of morphological modifications in root systems of plants regularly submitted to external forces, such as wind loading or shaking (see the review by Stokes and Guitard [159]). Such modifications include an increase in root branch number, cross-sectional area, and biomass along the axis of bending [73,90,109,159]. Responses may also occur at the cellular level, with consequences for wood mechanical properties [105]. In adult trees, uneven secondary growth occurs in the most highly stressed parts of individual roots, resulting in cross sections with shapes analogous to I or T beams, especially in the highly stressed region at the stem–root base. Such irregular radial growth results in an optimized root stiffness in the plane of bending for a given quantity of dry matter [103,160].

Although of importance for the anchorage of a plant, such responses of roots to external mechanical loading also have consequences for nutrient and water uptake. The optimal absorption of nutrients is significantly affected by root architecture and size [66,67], and a trade-off may exist between anchorage and nutrient uptake, particularly when the latter is limiting [159]. Nevertheless, it should be remembered that even though thigmomorphogenetic responses occur in the root system and can be even more pronounced than those in the stems of the same plants [90,159], such morphological changes are small when compared to the direct effects of differences in water and nutrient conditions on plant growth.

1.5 A Practical Application of Tree Biomechanics in Ecology

Research on plant biomechanics over the last decade has increased enormously, resulting in a large database of available knowledge. However, although certain fundamental aspects are understood in great detail, biologists, foresters, and mechanical engineers are not yet sure of how this knowledge can be applied in a practical way with regards to ecological sustainability. Nevertheless, several domains exist that benefit hugely from the recent collation of data typical of that presented in this chapter. Two examples of subject areas where biomechanics and ecology can be combined to produce highly effective ecological solutions to long-term sustainability issues are ground bio-engineering and eco-engineering [88].

Both eco-engineering and ground bio-engineering fall within the framework of "ecological engineering." Ecological engineering has been described as "the management of nature" [161] or as the proactive design of sustainable ecosystems that integrate human society with its natural environment for the benefit of both [162,163]. Ecological engineering has largely been devoted to the sustainability of wetlands, wastewater, and aquaculture [162], but can be applied to a larger range of environments. Focusing more on the restoration or protection of sites, the term "eco-engineering" has recently been defined as the *long-term* strategy to manage a site with regards to natural or man-made hazards [88]. Eco-engineering is not to be confused with ground bio-engineering. Ground or soil bioengineering methods integrate civil engineering techniques with natural materials to obtain *fast*, effective, and economic methods of protecting, restoring, and maintaining the environment [164,165]. The use of, for example, geotextiles or brush mattressing to arrest soil run off and the planting of fast-growing herbaceous species to fix soil are typical ground bio-engineering techniques. Although ground bio-engineering and eco-engineering can be used on all types of land from arid land prone to desertification to riverbanks subjected to flooding, these practices have largely been carried out on hill slopes [165,166]. Hill slopes are highly prone to soil loss once vegetation has been removed through deforestation, cropping, or natural hazards. For natural slopes, such hazards can involve the mass movement of soil, such as landslides, avalanches, and rockfall, or erosion, including sheet and gully erosion or river bank erosion. By combining ground bio-engineering techniques with long-term solutions, slopes can be managed effectively to minimize the risk of failure. The correct choice of plant material is difficult because bio-engineers need to know the ability of the plant to grow on a particular site as well as the efficiency of the root system in fixing and reinforcing soil on an unstable soil. This efficiency depends largely on the mechanical properties of individual roots as well as the overall anchorage and architecture of the root system [71,167]. Although such information may be available for a particular species, its performance in the long run also needs to be known: fo example, grasses often die back in summer and should be combined with shrubs so as to avoid slippage or erosion problems. Shade intolerant species will also decline as shrubs and trees grow taller over time. Long-term solutions therefore need to include the use of appropriate management strategies, the employment of decision support systems [168,169], and the integration of such tools into geographic information systems to predict future risks. Such management techniques are particularly effective in large-scale areas of Europe, e.g., ski resorts, mountain slopes, and forest stands [170].

Not only do a bio- or eco-engineers need to consider the suitability of a species to a particular site with regards to its ability to grow and to reinforce the soil through the fixing efficiency of the root system, but they also need to take into account the long-term effect of planting vegetation on the site, especially with regards to tree species. One example where the careful long-term planning of tree growth is crucial is in mountain forests. Mountain forests are more and more frequently used as protection forests, i.e., as natural barriers against the effect of snow avalanches, landslides, and rockfall [171–174]. Recent studies have shown that the spatial distribution of trees within a forest stand can have serious consequences on the ability of a mountain forest to withstand soil mass movement such as rockfall [10,175]. The choice of tree species, as well as the aerial and root architecture, is also of utmost importance because mechanical resistance to overturning and stem rupture will influence enormously a tree's ability to withstand abiotic loading [10,88,175]. Studies in the French Alps have shown that beech (*Fagus sylvatica* L.) is extremely resistant to rockfall, compared with fir (*Abies alba* Mill.) and Norway spruce (*Picea abies* L.), particularly if planted in a certain spatial pattern [10,88,175]. Nevertheless, even if a tree species is useful as a barrier against one particular type of hazard, the same species may not be suitable in protecting against a different type of hazard, for example, Norway spruce is not especially wind firm [68] nor resistant to rockfall [10,174]. However, in preventing snow movement, because of its aerial architecture, Norway spruce is highly effective in holding the snow mantle in place [174]. Therefore, it is always necessary to determine which species is best suited to a particular function. Similarly, in conifer forests subjected to frequent windstorms, appropriate long-term management of the upwind border of the stand could decrease the probability of damage during a storm; such borders could be planted with windfirm broadleaf species such as oak and pruned to create a "ramp," or shelter belt type structure. This type of structure would cost little to maintain and would allow the prevailing wind to pass over the plantation, rather than penetrate into the stand [176].

One further example of how effective planting and management can improve the stability of a slope was shown in a study by Roering et al. [79]. These authors calculated the additional cohesion in soil by carrying out tensile testing of tree roots in natural and planted forest stands found within a short distance of landslide scarps. It was shown that a mixture of species, even at a young age, proved more effective than a mature monospecific plantation in preventing against soil loss. If a forest is clear felled, roots decay within a period of 3 years, resulting in a decrease in tensile strength and the ability to fix soil, hence resulting in lost soil cohesion and compromised slope stability [77].

Examples of where eco-engineering techniques would be most useful are in situations whereby human safety is not an immediate issue or where protecting structures are already in place, such as safety nets or avalanche barriers. When deciding to carry out eco-engineering techniques on an unstable slope, the engineer must first determine the nature of the slope, type of soil, type of native or desired vegetation, and the likelihood of any catastrophic event occurring that would decrease slope stability during the restoration time. If the risk of danger to human life and infrastructures is low, the user must consider the size of the site and costs to be incurred throughout the life of the project. If the site is on a small scale and the cost of construction (e.g., fascines, gabions, and geotextiles), planting, and upkeep is equal to the economic, aesthetic, and safety gain at the end of the project, ground bio-engineering techniques can be considered. If the site is large scale, such as a mountain slope, the expenses incurred in carrying out certain bio-engineering techniques may be too high for the gain produced, and eco-engineering techniques may be used. However, it must be remembered that any gain as a result of an eco-engineering project will only be incurred in the long-term.

Still a relatively young subject, eco-engineering is beginning to emerge as a future research area in Europe, an area that engineers and ecologists should consider both in terms of education and application. Human activity over the last 100 years has been concerned with increasing productivity through technological progress at the cost of environmental degradation [162]. It is now necessary to repair this damage, although with limited resources, certain countries are unable to invest heavily in environmental restoration of degraded lands. Bio- and eco-engineering techniques can therefore provide a low-cost, long-term solution in many cases.

1.6 CONCLUSION

Previous studies in biomechanics should be considered when defining tree traits to be used to describe the mechanical aspects of height growth strategy. It is assumed that this strategy presents a competitive challenge with regards to physical constraints, especially in forest canopies where light is limited in the understory and increases with canopy height. Two growth processes have been suggested to play a major role in these strategies. The first is thigmomorphogenesis, i.e., the perception of mechanical loads and the plant growth response, which influences biomass allocation between roots and shoots, organ length and thickness, as well as stem and root system shape and their wood properties. The second is concerned with long-term gravitropic movements of lignified and stiff stems induced by secondary growth and maturation strains, which are continuously being generated to maintain verticality.

Studying the biomechanical aspect of a species' strategy is of general interest for forest ecologists for different reasons. Firstly, biomechanics can contribute to a better understanding of the relationships between diversity and functioning. For example, with regards to the ability of trees to coexist at different height strata, both competition and physical constraints can be interpreted as the use of different ecological niches with respect to light availability, which in turn could influence general ecosystem functioning and productivity. Biomechanics could therefore be useful for studying the relationships between tree functional types and natural ecosystem resilience or community dynamics. The efficiency of certain pioneer shade-intolerant species, which tend to dominate in gaps of disturbed forests and are able to survive under a closed canopy before becoming canopy trees in mature forests, should be linked to their ability to use support material at a low cost with a high mechanical efficiency, thus decreasing the risk of failure. Lastly, biomechanics can give valuable help when assessing species performance in engineering, such as stabilizing the soil on slopes through the use of vegetation or the choice of sylvicultural practices in a windy environment.

REFERENCES

1. Rowe, N.P. and Speck, T., Hydraulics and mechanics of plants: novelty, innovation and evolution, in *The Evolution of Plant Physiology*. Poole, I. and Hemsley, A.R., Eds., Elsevier Academic Press, Kew, 2004, chap. 16.
2. Niklas, K.J., The evolution of plant body plans — a biomechanical perspective, *Ann. Bot.*, 85, 411, 2000.
3. Westoby, M. et al., Plant ecological strategies: some leading dimensions of variation between species, *Ann. Rev. Ecol. Sys.*, 33, 125, 2002.
4. Cameron, T., The year of the "diversity-ecosystem function" debate, *Trends Ecol. Evol.*, 17, 495, 2002.
5. Diaz, S. and Cabido, M., Vive la différence: plant functional diversity matters to ecosystem processes, *Trends Ecol. Evol.*, 16, 646, 2001.
6. Wright, I.J. et al., The world-wide leaf economics spectrum, *Nature*, 428, 821, 2004.
7. Niklas, K.J., *Plant Biomechanics: An Engineering Approach to Plant Form and Function*, University of Chicago Press, Chicago, 1992.
8. Timoshenko, S.P., *Strength of Materials*, van Nostrand, Princeton, 1941.
9. Niklas, K.J. and Spatz, H.C., Methods for calculating factors of safety for plant stems, *J. Exp. Biol.*, 202, 3273, 1999.
10. Stokes, A. et al., Mechanical resistance of different tree species to rockfall in the French Alps, *Plant Soil*, in press. DOI: 10.1007/s11104-005-3899-3.2005.
11. Mosbrugger, V., *The tree habit in land plants: A functional comparison of trunk constructions with a brief introduction into the biomechanics of trees*, Lecture notes in earth sciences, Vol. 28, Springer-Verlag, Heidelberg, 1990, p. 161.
12. Wilson, B.F., *The Growing Tree*, The University of Massachusetts Press, Amherst, 1984.
13. Skaar, C., *Wood-Water Relations*, Springer-Verlag, Heidelberg, 1988.
14. Kollman, F.F.P. and Cote, W.A., *Principles of Wood Science and Technology*, I. *Solid Wood*, Springer-Verlag, Heidelberg, 1968.
15. Hogan, C.J. and Niklas, K.J., Temperature and water content effects on the viscoelastic behavior of *Tilia americana* (Tiliaceae) sapwood, *Trees-Struct. Funct.*, 18, 339, 2004.
16. Coutand, C. et al., Comparison of mechanical properties of tension and opposite wood in Populus, *Wood Sci. Technol.*, 38, 11, 2004.
17. Guitard, D., *Mécanique du Bois et Composites*, Cepadues, Toulouse, 1987.
18. Gibson, L.J. and Ashby, M.F., *Cellular Solids*, Pergamon Press, Oxford, 1988.
19. McMahon, T.A. and Kronauer, R.E., Tree structures: deducing the principle of mechanical design, *J. Theor. Biol.*, 59, 443, 1976.
20. Niklas, K.J., Influence of tissue density — specific mechanical properties on the scaling of the plant height, *Ann. Bot.*, 72, 173, 1993.
21. Sterck, F. and Bongers, F., Ontogenetic changes in size, allometry, and mechanical design of tropical rainforest trees, *Am. J. Bot.*, 85, 266, 1998.
22. Barnett, J.R. and Bonham, V.A., Cellulose microfibril angle in the cell wall of wood fibres, *Biol. Rev.*, 79, 2004.
23. Fratzl, P., Burgert, I., and Keckes, J., Mechanical model for the deformation of the wood cell wall, *Z. Metallkd*, 95, 579, 2004.
24. Timell, E., *Compression Wood in Conifers,* Springer-Verlag, Heidelberg, 1987.

25. Zobel, B.J. and Buijtenen, J.P.v., Wood variation: its causes and control, in *Wood Variation: Its Causes and Control*, Springer-Verlag, Berlin, 1989.
26. Baas, P., Jansen, S., and Wheeler, E., Ecological adpatations in wood microstructure and angiosperm phylogeny, *IAWA J.*, 23, 60, 2002.
27. Favrichon, V., Classification des espèces arborées en groupes fonctionnels en vue de la réalisation d'un modèle de dynamique de peuplement en Guyane française. (Classification of tree species in forests of French Guiana into functional groups based on a dynamic vegetation community matrix.), *Rev. Ecol. (Terre et Vie)*, 49, 379, 1994.
28. Turner, I.M., Ed. *The Ecology of Trees in the Tropical Rain Forest*, Cambridge Tropical Biology Series, Cambridge University Press, London, 2001, p. 298.
29. Zhang, S.Y. et al., Modelling wood density in European oak (*Quercus petraea* and *Quercus robur*) and simulating the silvicultural influence, *Can. J. Forest Res.*, 23, 2587, 1993.
30. Hillis, W.E., *Heartwood and Tree Exudates*, Springer-Verlag, Berlin, 1987.
31. Grzeskowiak, V., *Le bois juvénile de deux angiospermes à pores diffus (Populus euramericana cv I214, Dicorynia guianensis): variations radiales et avec la hauteur des caractères anatomiques, de l'infradensité et du retrait radial*, doctorate thesis, ENGREF: Nancy Paris Montpellier, 1997.
32. Schniewind, A.P., Horizontal specific gravity variation in tree stems in relation to their support function, *Forest Sci.*, 8, 111, 1962.
33. Wiemann, M.C. and Williamson, G.B., Wood specific gravity gradients in tropical and montane rain forest trees, *Am. J. Bot.*, 76, 924, 1989.
34. Woodcock, D.W. and Shier, A.D., Wood specific gravity and its radial variations: the many ways to make a tree, *Trees-Struct. Funct.*, 16, 437, 2002.
35. Laroze, S. and Barrau, J.J., *Calcul des structures en matériau composites. Mecanique des structures.* Masson, Paris, 1987.
36. Fournier, M. et al., Mécanique de l'arbre sur pied: modélisation d'une structure en croissance soumise à des chargements permanents et évolutifs. II. Analyse des contraintes de support, *Ann. Sci. Forest.*, 48, 513, 1991.
37. Fournier, M., Baillères, H., and Chanson, B., Tree biomechanics: growth, cumulative pre-stresses and reorientations, *Biomimetics*, 2, 229, 1994.
38. Falster, D.S. and Westoby, M., Plant height and game theory, *Trends Ecol. Evol.*, 18, 337, 2003.
39. Becker, P., Meinzer, F.C., and Wullschleger, S.D., Hydraulic limitation of tree height: a critique, *Funct. Ecol.*, 14, 4, 2000.
40. Ryan, M.G. and Yoder, B.J., Hydraulic limits to tree height and tree growth, *BioScience*, 47, 235, 1997.
41. Midgley, J.J., Is bigger better in plants? Hydraulic costs of increasing plant height, *Trends Ecol. Evol.*, 18, 5, 2003.
42. Holbrook, N.M. and Putz, F.E., Influence of neighbors on tree form: effects of lateral shade and prevention of sway on the allometry of *Liquidambar styraciflua*, *Am. J. Bot.*, 76, 1740, 1989.
43. Ancelin, P., *Modélisation du comportement biomécanique de l'arbre dans son environnement forestier.* Application au Pin maritime, in Ecole doctorale de sciences physiques et de l'ingénieur, doctorate thesis. Université Bordeaux I. Bordeaux, 2001, p. 182.
44. Ancelin, P., Courbaud, B., and Fourcaud, T., Developing an individual tree based mechanical model to predict wind damage within forest stands, *Forest Ecol. Manag.*, 203, 101, 2004.

45. Ancelin, P., Fourcaud, T., and Lac, P., Modelling the biomechanical behaviour of growing trees at the forest stand scale. Part I: Development of an incremental transfer matrix method and application on simplified tree structures, *Ann. Forest Sci.*, 61, 263, 2004.
46. Tateno, M., Increase in lodging safety factor of thigmomorphogenetically dwarfed shoots of mulberry tree, *Phys. Plant.*, 81, 239, 1991.
47. Metzger, K., Der Wind als massgebender Faktor für das Wachsthum der Baüme, *Mündener Forstl.*, 3, 35, 1893.
48. Greenhill, M.A., Determination of the greatest height consistent with stability that a vertical pole or mast can be made, and of the greatest height to which a tree of given proportions can grow, *Proc. Cambridge Philos. Soc.*, IV, PT II, 1881, 5.
49. King, D.A. and Loucks, O.L., The theory of tree bole and branch form, *Radiat. Environ. Bioph.*, 15, 141, 1978.
50. King, D.A., Tree dimensions: maximizing the height growth in dense stands, *Oecologia*, 51, 351, 1981.
51. Moulia, B. and Fournier-Djimbi, M., Optimal mechanical design of plant stems: the models behind the allometric power laws, in *Plant Biomechanics*, Centre for Biomimetics, The University of Reading, Reading, UK, 1997.
52. McMahon, T.A., The mechanical design of trees, *Sci. Am.*, 233, 92, 1975.
53. Esser, M.H.M., Tree trunks and branches as optimum mechanical supports of the crown: I. The trunk, *Bull. Math. Biophys.*, 8, 65, 1946.
54. King, D.A., Tree form, height growth, and susceptibility to wind damage in *Acer saccharum*, *Ecology*, 67, 980, 1986.
55. Mattheck, C., *Trees: The Mechanical Design*, Springer-Verlag, Heidelberg, 1991.
56. Niklas, K.J. and Spatz, H.C., Growth and hydraulic (not mechanical) constraints govern the scaling of tree height and mass, *Proc. Natl. Acad. Sci.*, 101, 15661, 2004.
57. Niklas, K.J., *Plant Allometry: The Scaling of Form and Process*, University of Chicago Press, Chicago, 1994.
58. Claussen, J.W. and Maycock, C.R., Stem allometry in a North Queensland tropical rainforest, *Biotropica*, 27, 1995.
59. Rich, P.M. et al., Height and diameter relationships for dicotyledonous trees and arborescent palms of Costa Rican tropical wet forest, *Bull. Torrey Bot. Club*, 113, 241, 1986.
60. King, D.A., Allometry and life history of tropical trees, *J. Trop. Ecol.*, 12, 25, 1996.
61. Kohyama, T. and Hotta, M., Significance of allometry in tropical saplings, *Funct. Ecol.*, 4, 515, 1990.
62. Almeras, T. and Gril, J., Bending of apricot tree branches under the weight of axillary growth: test of a mechanical model with experimental data, *Trees-Struct. Funct.*, 16, 5, 2002.
63. Schaeffer, B., Biomécanique. Forme d'équilibre d'une branche d'arbre, *C. R. Acad. Sci.*, 1991.
64. Castera, P. and Morlier, V., Growth patterns and bending mechanics of branches, *Trees-Struct. Funct.*, 5, 232, 1991.
65. Cruz, C. et al., Functional aspects of root architecture and mycorrhizal inoculation with respect to nutrient uptake capacity, *Mycorrhiza*, 14, 177, 2004.
66. Fitter, A.H. et al., Architectural analysis of plant root systems I. Architectural correlates of exploitation efficiency, *New Phytol.*, 118, 375, 1991.
67. Fitter, A.H. and Stickland, T.R., Architectural analysis of plant root systems II. Influence of nutrient supply on architecture in contrasting plant species, *New Phytol.*, 118, 383, 1991.

68. Stokes, A., Ed. The supporting roots of trees and woody plants: form, function and physiology, in *Developments in Plant and Soil Sciences,* Vol. 87, Kluwer Academic Publishers, Dordrecht, 2000, p. 426.
69. Ennos, A.R., The mechanics of root anchorage, *Adv. Bot. Res. Inc. Adv. Plant Path.*, 33, 133, 2000.
70. Ennos, A.R., The anchorage of leek seedlings: the effect of root length and soil strength, *Ann. Bot.*, 65, 409, 1990.
71. Waldron, L.J. and Dakessian, S., Soil reinforcement by roots: calculation of increased soil shear resistance from root properties, *Science*, 132, 427, 1981.
72. Stokes, A., Biomechanics of tree root anchorage, in *Plant Roots: The Hidden Part*, U. Kafkaki, Ed., Marcel Dekker, New York, 2002, p. 175.
73. Mickovski, S.B. and Ennos, A.R., The effect of unidirectional stem flexing on shoot and root morphology and architecture in young *Pinus sylvestris* trees, *Can. J. Forest Res.*, 33, 2202, 2003.
74. Ennos, A.R., The biomechanics of root anchorage, *Biomimetics*, 2, 129, 1994.
75. Crook, M.J. and Ennos, A.R., The anchorage mechanics of deep-rooted larch *Larix europea* × *L. japonica*, *J. Exp. Bot.*, 47, 1509, 1996.
76. Ennos, A.R., The mechanics of anchorage in seedlings of sunflower, *Helianthus annuus* L, *New Phytol.*, 113, 185, 1989.
77. Watson, A., Phillips, C., and Marden, M., Root strength, growth, and rates of decay: root reinforcement changes of two tree species and their contribution to slope stability, *Plant Soil*, 217, 39, 1999.
78. Schmidt, K.M. et al., The variability of root cohesion as an influence on shallow landslide susceptibility in the Oregon Coast Range, *Can. Geotechnol. J.*, 38, 995, 2001.
79. Roering, J.J. et al., Shallow landsliding, root reinforcement, and the spatial distribution of trees in the Oregon Coast Range, *Can. Geotechnol. J.*, 40, 237, 2003.
80. Hathaway, R.L. and Penny, D., Root strength in some Populus and Salix clones, *New Zeal. J. Bot.*, 13, 333, 1975.
81. Jonasson, S. and Callaghan, T.V., Root mechanical-properties related to disturbed and stressed habitats in the Arctic, *New Phytol.*, 22, 179, 992.
82. Makarova, O.V., Cofie, P. and Koolen, A.J., Axial stress-strain relationships of fine roots of beech and larch in loading to failure and in cyclic loading, *Soil Till. Res.*, 45, 175, 1998.
83. Genet, M. et al., The influence of cellulose content on tensile strength in tree roots, *Plant Soil*, in press, 2006.
84. Wu, W.M. and R.C. Sidle, A distributed slope stability model for steep forested basins, *Water Resour. Res.*, 31, 2097, 1995.
85. Cambell, K.A. and Hawkins, C.D.B., Paper birch and lodgepole pine root reinforcement in coarse-, medium-, and fine-textured soils, *Can. J. For. Res.*, 33, 1580, 2003.
86. Wu, T.H., McKinnel, W.P., and Swantson, D.N., Strength of tree roots and landslides on Prince of Wales Island, *Can. J. Geotechnol.*, 16, 19, 1979.
87. Nakamura, H., Nghiem, Q.M., and Iwasa, N., The influence of root reinforcement on slope stability: a case study from the Ozawa slope in Iwate Prefecture, Japan, in *Eco- and Ground Bio-Engineering: The Use of Vegetation to Improve Slope Stability*, Stokes, A., Norris, J.E., Spanos, I., and Cammeraat, L.H., Eds., Kluwer Academic Publishers, Dordrecht, 2006.

88. Stokes, A., Mickovski, S.B., and Thomas, B.R., Eco-engineering for the long-term protection of unstable slopes in Europe: developing management strategies for use in legislation, in *Landslides: Evaluation & Stabilization*, Lacerda, W.A., Erlich, M., Fontoura, S.A.B., and Sayão, A.S.F., Eds., Balkema, 1685.
89. Dupuy, L., Fourcaud, T., and Stokes, A., A numerical investigation into factors affecting the anchorage of roots in tension, *Eur. J. Soil Sci.*, 56, 319, 2005.
90. Stokes, A. et al., Responses of young trees to wind and shading: effects on root architecture, *J. Exp. Bot.*, 46, 21, 1995.
91. Stokes, A. et al., An experimental investigation of the resistance of model root systems to uprooting, *Ann. Bot.*, 78, 415, 1996.
92. Ennos, A.R. and Fitter, A.H., Comparative functional morphology of the anchorage systems of annual dicots, *Funct. Ecol.*, 6, 71, 1992.
93. Köstler, J.N., Brückner, E., and Bibelriether, H., *Die Wurzeln der Waldbäume*, Paul Parey, Hamburg, 1968.
94. Gartner, B.L., Root biomechanics and whole-plant allocation patterns — responses of tomato plants to stem flexure, *J. Exp. Bot.*, 45, 1647, 1994.
95. Fourcaud, T., Danjon, V., and Dupuy, V., Numerical analysis of the anchorage of maritime pine trees in connection with their root structure, in *International Conference "Wind Effects on Trees,"* University of Karlsruhe, Germany, September 16–18 2003, Ruck, B., Kottmeier, C., Mattheck, C., Quine, G., Wilhelm, G., pp. 323–330.
96. Dupuy, L., Fourcaud, T., and Stokes, A., A numerical investigation into the influence of soil type and root architecture on tree anchorage, *Plant Soil*, in press, 2006.
97. Coutts, M.P., Root architecture and tree stability, in *Tree Root Systems and Their Mycorrhizas*, *Plant and Soil*, 71, 171, 1983.
98. Coutts, M.P., Components of tree stability in Sitka spruce on peaty gley soil, *Forestry*, 59, 173, 1986.
99. Richards, P.W., *The Tropical Rainforest*, Second Edition, Cambridge University Press, Cambridge, 1996, p. 453.
100. Kaufman, L., The role of developmental crises in the formation of buttresses: a unified hypothesis, *Evol. Trend. Plant.*, 2, 39, 1988.
101. Crook, M.J., Ennos, A.R., and Banks, J.R., The function of buttress roots: a comparative study of the anchorage systems of buttressed (Aglaia and *Nephelium ramboutan* species) and not-buttressed (*Mallotus wrayi*) tropical trees, *J. Exp. Bot.*, 48, 1703, 1997.
102. Clair, B. et al., Biomechanics of buttressed trees: bending strains and stresses, *Am. J. Bot.*, 90, 1345, 2003.
103. Nicoll, B.C. and Ray, D., Adaptive growth of tree root systems in response to wind action and site conditions, *Tree Physiol.*, 16, 891, 1996.
104. Ennos, A.R., The function and formation of buttresses, *Trends Ecol. Evol.*, 8, 350, 1993.
105. Stokes, A., Strain distribution during anchorage failure of *Pinus pinaster* Ait. at different ages and tree growth response to wind-induced root movement, *Plant Soil*, 217, 17, 1999.
106. Hintikka, V., Wind-induced movements in forest trees, *Metsäntutkimuslitoksen Julkaisuja*, 76, 1, 1972.
107. Mickovski, S.B. and Ennos, A.R., A morphological and mechanical study of the root systems of suppressed crown Scots pine *Pinus sylvestris*, *Trees-Struct. Funct.*, 16, 274, 2002.
108. Niklas, K.J. et al., The biomechanics of *Pachycereus pringlei* root systems, *Am. J. Bot.*, 89, 12, 2002.

109. Tamasi, E. et al., Influence of wind stress on root system development and architecture in oak seedlings (*Quercus robur* L.), *Trees-Struct. Funct.*, 19, 374, 2005.
110. Moore, J.R., Differences in maximum resistive bending moments of *Pinus radiata* trees grown on a range of soil types, *Forest Ecol. Manag.*, 135, 63, 2000.
111. Fourcaud, T. et al. Application of plant architectural models to biomechanics, in *PMA03 – Plant growth modeling and applications*, Hu, B. and Jaeger, M. Eds., Tsinghua University Press-Springer, Beijing, China, Springer, 2003.
112. Dupuy, L., Fourcand, T., Lac, P., and Stokes, A., Modelling the influence of morphological and mechanical properties on the anchorage of root systems, in International conference "Wind Effect on Trees," September 16–18, 2003, University of Karlsruhe, Germany, Ruck, B., Koltmeier, C., Mattheck, C., Quine, G., and Wilhelm, G., pp. 315–322.
113. Speck, T., Bending stability of plant stems: ontogenetical, ecological and phylogenetical aspects, *Biomimetics*, 2, 109, 1994.
114. Jaffe, M.J., Morphogenetic responses of plants to mechanical stimuli or stress, *Bioscience*, 30, 239, 1980.
115. Weiher, E. et al., Challenging Theophrastus: a common core list of plant traits for functional ecology, *J. Veg. Sci.*, 10, 609, 1999.
116. Jaffe, M.J., Leopold, A.C., and Staples, R.C., Thigmoresponses in plants and fungi, *Am. J. Bot.*, 89, 382, 2002.
117. Telewski, F.W. and Jaffe, M.J., Thigmomorphogenesis: anatomical, morphological and mechanical analysis of genetically different sibs of *Pinus taeda* in response to mechanical perturbation, *Physiol. Plantarum*, 66, 219, 1986.
118. Pruyn, M., Ewers, B.J., and Telewski, F.W., Thigmomorphogenesis: changes in the morphology and mechanical properties of two Populus hybrids in response to mechanical perturbation, *Tree Physiol.*, 20, 535, 2000.
119. Coutand, C. et al., *Les sollicitations mécaniques, des régulateurs de la répartition de la biomasse chez les ligneux*, Conference "Réseau d'Ecophysiologie de l'Arbre," France, Cruiziat, P. and Dreyer, E., Eds., INRA, 2003.
120. Telewski, F.W., Structure and function of flexure wood in *Abies fraseri*, *Tree Physiol.*, 5, 123, 1989.
121. Sellier, D. and Fourcaud, T., Relationship between the oscillations of young pines (*Pinus pinaster* Aït.) and their aerial architecture, *J. Exp. Bot.*, 56, 1563, 2005.
122. Coutand, C. and Moulia, B., Biomechanical study of the effect of a controlled bending on tomato stem elongation: local strain sensing and spatial integration of the signal, *J. Exp. Bot.*, 51, 1825, 2000.
123. Coutand, C. et al., Biomechanical study of the effect of a controlled bending on tomato stem elongation: global mechanical analysis, *J. Exp. Bot.*, 51, 1813, 2000.
124. Jacobs M.R., The effect of wind sway on the form and development of *Pinus radiata* D. Don., *Aust. J. Bot.*, 2, 35, 1954.
125. Telewski, F.W. and Pruyn, M.L., Thigmomorphogenesis: a dose response to flexing in *Ulmus americana* seedlings, *Tree Physiol.*, 18, 65, 1998.
126. Hal, H. and Thomas, S.C., Interactive effects of lateral shade and wind on stem allometry, biomass allocation, and mechanical stability in *Abutilon theophrasti* (Malvaceae), *Am. J. Bot.*, 89, 1609, 2002.
127. Mitchell, S.J., Effects of mechanical stimulus, shade, and nitrogen fertilization on morphology and bending resistance in Douglas-fir seedlings, *Can. J. Forest Res.*, 33, 1602, 2003.
128. Berthier, S. and Stokes, A., Phototropic response induced by wind loading in maritime pine seedlings (*Pinus pinaster* Ait.), *J. Exp. Bot.*, 56, 851, 2005.

129. Stacey, G.R. et al., Wind flows and forces in a model spruce forest, *Bound.-Lay. Meteorol.*, 69, 311, 1994.
130. Rudnicki, M., Mitchell, S.J., and Novak, M.D., Wind tunnel measurements of crown streamlining and drag relationships for three conifer species, *Can. J. Forest Res.*, 34, 666, 2004.
131. Moulia, B. and Combes, D., *Thigmomorphogenetic Acclimation of Plants to Moderate Winds Greatly Affects Height Structure in Field-Gown Alfalfa (Medicago sativa L.), an Indeterminate Herb*, Society for Experimental Biology, General Congress, Edinburgh, 2004.
132. Lawton, R.O., Wind stresses and elfin stature in a montane rain forest tree: an adaptive explanation, *Am. J. Bot.*, 69, 1224, 1982.
133. Py, C. et al., A new technique for the measurement of the wind induced motion of a plant canopy, in *Flow Induced Vibration*, Congress, Paris, 2004.
134. Mattheck, C., Bethge, K., and Schafer, J., Safety factors in trees, *J. Theoret. Biol.*, 165, 185, 1993.
135. Mattheck, C., Why they grow, how they grow: the mechanics of trees, *Arboricultural J.*, 14, 1, 1990.
136. Mattheck, C. and Bethge, K., The structural optimization of trees, *Naturwissenschaften*, 85, 1, 1998.
137. West, P.W., Jackett, D.R., and Sykes, S.J., Stresses in, and the shape of, tree stems in forest monoculture, *J. Theoret. Biol.*, 140, 327, 1989.
138. Niklas, K.J. and Spatz, H.C., Wind-induced stresses in cherry trees: evidence against the hypothesis of constant stress levels, *Trees-Struct. Funct.*, 14, 230, 2000.
139. Mattheck, C., Comments on "Wind-induced stresses in cherry trees: evidence against the hypothesis of constant stress levels" by Niklas, K.J., Spatz, H.-C., *Trees-Struct. Funct.,* 15, 2000.
140. Salisbury, F.B. and Ross, C.W., The power of movement in plants, in *Plant Physiology*, 4th ed., Wadsworth, Belmont, CA, 1992, chap. 19.
141. Wilson, B.F. and Archer, R.R., Tree design: some biological solutions to mechanical problems, *Bioscience*, 29, 293, 1979.
142. Scurfield, G., Reaction wood: its structure and function, *Science*, 179, 647, 1973.
143. Fournier, M. et al., Mécanique de l'arbre sur pied: modélisation d'une structure en croissance soumise à des chargements permanents et évolutifs. II. Analyse tridimensionnelle des contraintes de maturation — cas du feuillu standard, *Ann. Sci. Forest.*, 48, 527, 1991.
144. Archer, R.R. and Byrnes, F.E., On the distribution of tree growth stresses. Part I. An anisotropic plane strain theory, *Wood Sci. Technol.*, 8, 184, 1973.
145. Archer, R.R., Growth stresses and strains in trees. Springer Series in Wood Sciences, Springer Verlag, Heidelberg, 1987.
146. Bamber, R.K., A general theory for the origin of growth stresses in reaction wood: how trees stay upright, *IAWA J.*, 22, 205, 2001.
147. Yamamoto, H. and Okuyama, T., Analysis of the generation process of growth stresses in cell walls, *Mokuzai Gakkaishi*, 34, 788, 1988.
148. Yamamoto, H. et al., Origin of the biomechanical properties of wood related to the fine structure of the multilayered cell wall, *Trans. ASME*, 124, 432, 2002.
149. Boyd, J., Compression wood force generation and functional mechanics, *New Zeal. J. Forest Sci.*, 3, 240, 1973.
150. Fourcaud, T. et al., Numerical modelling of shape regulation and growth stresses in trees. II. Implementation in the AMAPpara software and simulation of growth, *Trees-Struct. Funct.*, 17, 31, 2003.

151. Fourcaud, T. and Lac, P., Numerical modelling of shape regulation and growth stresses in trees. I. An incremental static finite element formulation, *Trees-Struct. Funct.*, 17, 23, 2003.
152. Fournier, M. et al., Mesures des déformations résiduelles de croissance à la surface des arbres en relation avec leur morphologie. Observations sur différentes espèces, *Ann. Sci. Forest.*, 51, 249, 1994.
153. Yamamoto, H., Okuyama, T., and Iguchi, M., Measurement of growth stresses on the surface of a leaning stem, *Mokuzai Gakkaishi*, 35, 595, 1989.
154. Sinnott, E.W., Reaction wood and the regulation of tree form, *Am. J. Bot.*, 39, 69, 1952.
155. Meskausas, A., Jurkoniene, S., and Moore, D., Spatial organization of the gravitropic response in plants: applicability of the revised local curvature distribution model to *Triticum aestivum* coleoptiles, *New Phytol.*, 143, 401, 1999.
156. Meskausas, A., Moore, D., and Novak Frazer, L., Mathematical modelling of morphogenesis in fungi: spatial organization of the gravitropic response in the mushroom stem of *Coprinus cinereus*, *New Phytol.*, 140, 111, 1998.
157. Salisbury, F.B., Gravitropism: changing ideas, *Hortic. Rev.*, 15, 233, 1993.
158. Hart, J.W., *Plant Tropisms and Other Growth Movements*, Unwin Hyman, London, 1989.
159. Stokes, A. and Guitard, D., Tree root response to mechanical stress, in *Biology of Root Formation and Development*, Waisel, Y., Ed., Plenum Press: New York, 227, 1997.
160. Niklas, K.J., Variations of the mechanical properties of *Acer saccharum* roots, *J. Exp. Bot.*, 50, 193, 1999.
161. Odum, H.T., *Environment, Power and Society*, Wiley Interscience, New York, 1971.
162. Painter, D.J., Forty-nine shades of green: ecology and sustainability in the academic formation of engineers, *Ecol. Eng.*, 20, 267, 2003.
163. Mitsch, W.J., Ecological engineering: a new paradigm for engineers and ecologists, in *Engineering within Ecological Constraints*, S. PC, Ed., National Academy Press: Washington DC, 1996, p. 111.
164. Coppin, N.J. and Richards, I.G., *Use of Vegetation in Civil Engineering*, Butterworths, London, 1990.
165. Schiechtl, H.M., *Bioengineering for Land Reclamation and Conservation,* University of Alberta Press, Edmonton (Alberta), 1980, p. 404.
166. Barker, D.H., *Vegetation and Slopes,* Thomas Telford, London, 1995.
167. Wu, T.H., Watson, A., and El-Khouly, M.A., Soil-root interaction and slope stability, in *Ground and Water Bioengineering for Erosion Control and Slope Stabilization*, Barker, D.H. et al., Eds., Science Publishers, Enfield, NH, 2004, p. 183.
168. Mickovski, S.B., Decision support systems in eco-engineering: The case of the SDSS, in *Eco- and Ground Bio-Engineering: The Use of Vegetation to Improve Slope Stability,* Cammeraat, L.H., Ed., Kluwer Academic Publishers, Dordrecht, 2005.
169. Mickovski, S.B., Stokes, A., and van Beek, L.P.H., A decision support tool for windthrow hazard assessment and prevention, *Forest Ecol. Manag.*, 216, 64, 2005.
170. Dorren, L.K.A. and Seijmonsbergen, A.C., Comparison of three GIS-based models for predicting rockfall runout zones at a regional scale, *Geomorphology*, 56, 49, 2003.
171. Ott, E., Guidelines for the protective role of forests in avalanche formation, *Forstwissenssch. Centralblatt*, 115, 223, 1996.
172. Motta, R. and Haudemand, J.C., Protective forests and sylvicultural stability: an example of planning in the Aosta Valley, *Mountain. Res. Dev.*, 20, 180, 2000.

173. Brang, P., Resistance and elasticity: promising concepts for the management of protection forests in the European Alps, *Forest Ecol. Manag.*, 145, 107, 2001.
174. Hurand, A. and Berger, F., Forêts et risques naturels. Protection contre l'érosion, les mouvements de terrain et les avalanches, *La Houille Blanche*, 3, 64, 2002.
175. Dorren, L.K.A. et al., Mechanisms, effects and management implications of rockfall in forests, *Forest Ecol. Manag.*, 215, 183, 2005.
176. Quine, C.P. et al., *Forests and Wind: Management to Minimise Damage*, Bulletin 114, Her Majesty's Stationery Office, London, 1995.

2 Diversity of Mechanical Architectures in Climbing Plants: An Ecological Perspective

Nicholas P. Rowe, Sandrine Isnard, Friederike Gallenmüller, and Thomas Speck

CONTENTS

2.1 INTRODUCTION

Many vines and lianas show extreme variations in mechanical properties during development from young to adult growth. These developmental trends differ from those observed in most self-supporting plants and have a significant effect on how climbing plants exploit the environment. Furthermore, different climbers show a wide range of attachment methods and may differ considerably in size as well as mechanical properties according to how they attach to host plants. To better understand how climbing habits differ, we investigate the mechanical properties of 43 species of tropical and temperate climbing plants according to their mode of attachment. The study includes stem twiners, tendril climbers, hook climbers, branch-angle climbers, and leaning climbers. Mechanical trends during growth are discussed in terms of mode of attachment, taxonomic affinity, and inherent developmental constraints (i.e., existence or lack of secondary growth) as well as approximate size and position of the climbing plant in the surrounding vegetation.

2.1.1 Importance of Climbers

Climbing plants can differ considerably from self-supporting plants in terms of growth strategy and stem mechanics. Self-supporting plants can be understood to make up the overall primary mechanical structure of many terrestrial ecosystems, whereas climbers fill vacant space on and around their self-supporting hosts. Despite a number of commentators noting that the ecology of lianas is more poorly known compared with other plant growth forms such as trees, recent studies over the last decade, including biomechanical studies, have unequivocally demonstrated that lianas are extremely important constituents of many tropical ecosystems [1,2]. Furthermore, recent reports based on long-term censuses, particularly from the neotropics, have demonstrated that liana abundance is apparently increasing relative to tree species, possibly as a result of anthropogenic disturbance and/or climatic change [3,4]. Because climbing plants represent important elements as well as potential indicators of regional and possibly global vegetation change, there is much need for developing techniques and approaches to better understand their growth, development, and ecological significance, especially under natural conditions.

2.1.2 Mechanical Structure and Development of Climbers

Lianas are currently understood to have a range of important ecological functions in forest dynamics, of which a number are mechanical (Figure 2.1). Mature lianas climbing to the forest canopy can literally tie neighboring trees together (Figures 2.1c and 2.1d); they can weigh down and distort tree growth or even cause main branches or entire trees to fall. Young stages of liana growth can be particularly well adapted to spanning long distances even though they are composed of only narrow stems (Figures 2.1a and 2.1b), bridging host supports and exploring and filling open and disturbed areas, often to the detriment of self-supporting species. These specialized strategies are made possible in many climbing species by highly adapted mechanical properties of the stem. A number of recent studies have explored the developmental, ecological, and evolutionary significance of the transition from relatively stiff young shoots to older more flexible stems [5–10]. This developmental trend differs from most self-supporting plants, which are normally composed of relatively compliant tissues when young and relatively stiff tissues when old. Many vine and liana stems show extreme changes in mechanical properties during development from young to adult plants (Figure 2.1), and these changes have a significant effect on how they exploit the environment. Climbers are particularly active in early successional and secondary forests as well as forest margins and disturbances within forests, including tree falls and tree fall gaps.

2.1.3 Attachment Modes of Climbers

Climbing plants can differ considerably in size, maximal diameter, mechanical properties of young and old stems, and exact ecological preference. Furthermore, they show a wide range of attachment organs connecting the climbing plant to its host plant or plants [11]. These attachment devices vary widely in mechanical function. Climbers, which attach themselves most firmly to host supports, are represented by twining stems, which can literally form knots around host trunks and branches. The main stem of the climber itself constitutes the connection with the host and such stems can be extremely firmly attached (Figure 2.1d). Climbers can also attach via tendrils and sensitive petioles or other touch-sensitive appendicular organs, which extend out and twine around relatively slender supports (Figure 2.6e and Figure 2.6f). These provide a relatively firm attachment for small-diameter searchers and for initial attachment to the host but are less strong than main twining axes that can provide extremely secure attachment after further woody secondary growth. Hooks, grapnels, recurved spines, and straight spines provide attachments, which are less secure than either twining leading stems or tendrils. Hooks and spines can be jolted loose from their attachment. Some hooks, such as the appendicular organs of the genus *Strychnos*, maintain woody growth after attachment and can eventually close the hook around the host support, providing a firmer attachment than when open (Figure 2.6d and 2.6e). Other climbers attain mechanical dependence on other host plants by even less secure and possibly less specialized attachment devices, including wide-angled, stiff branches, which can interlock with branches of neighboring plants (Figure 2.6a to 2.6c). Some climbing plants have roughened surfaces such as hairs, scales, or a roughened cuticular surface of the leaf that can increase frictional contact with potential host supports. Other climbers can generate friction or some degree of anchorage with the host plants via lobed or reflexed leaves.

2.1.4 Mechanical Constraints and Types of Attachment

The attachment of a climber to its support is a crucial aspect of its survival and ecology, and it can also mechanically influence its host plant. Climbers that twine via the main stem on host trees are less likely to fall than plants that deploy recurved hooks. An interesting question is whether the mechanical properties of the climbing plant stem differ according to the type of attachment device [5]. Climbers that deploy “weak” types of attachment might be expected to produce relatively stiffer stem materials than plants that are irreversibly twined onto the host support. If hook

FIGURE 2.1 (a) Juvenile stage of a twining liana. The plant is self-supporting and the stem below the circumnutating apex is relatively stiff and has a high E_{str} (French Guiana). This stage of development often has mechanical properties that are similar to those of the stiff terminal "searcher" twigs of adult climbing plants [see (b)]. (b) Apical, young stage of development of a mature twining liana (*Condylocarpon guianense*) (French Guiana). Such axes are relatively stiff and span long distances to reach host supports. (c) Old stages of growth of a woody twining liana. The stems are firmly attached to the branches of the canopy, and the older basal portions are swinging freely from the canopy (French Guiana). Such stages have often very low values of E_{str}. (d) Old stage of growth of a twining liana stem traversing from one tree to another and forming a very firm attachment between the two (French Guiana). The stem was observed to be under significant tensile stress when the wind blew and the trees swayed violently in the wind. (e) Old stage of growth of *Bauhinia guianensis* (monkey ladder) (French Guiana). The mechanical properties of the old ribbonlike stem are characterized by low values of E_{str} because this species attaches to the host tree by numerous tendrilar organs. The branch to the left is a reiterative branch with "young" development and circular cross section. These reiterative branches are very stiff.

attachment fails, a relatively stiff stem would prevent the stem from buckling and falling to the ground like a coil of rope; the relatively stiff stem would allow the plant to sway and perhaps secure attachment to another host. In this case, a good strategy would be to retain relatively stiff mechanical properties. Because climbing stems employing hooks and spines are more loosely attached to the host plant, swaying or movement of the host would also probably detach the climber from the support rather than being deformed and possibly becoming damaged during movement, especially if the host supports were massive. Climbers that are firmly attached, such as stem twiners possibly have rather different constraints between attachment type and stem mechanics. Although mature to aged twining stems can sometimes be observed to have loosened and fallen in coils from host trunks, many firmly secured twining connections that are reinforced by wood production will probably never disconnect from the stem unless by failure under extreme loading. In such cases, it is of no advantage to the climbing plant to invest in mechanically stiffening the older basal stem. In fact, with such types of connections, the host or climber will actually fail, possibly under the weight of the growing liana, before the twining attachment does. Failure of the host tree would probably result in catastrophic movement and stresses to the climbing plant stem; under such circumstances, a relatively slender stem is less likely to fail if it is flexible rather than relatively stiff. Furthermore, movement or swaying of host trees could exert extreme stresses and strains on the connected climbing stem, which if firmly twined or knotted to the host, would require high compliance to avoid failure [12]. Extreme movement of host trees might not occur often, but infrequent storm conditions with strong winds and large host movements could excessively load the stems of firmly attached climbers. Furthermore, tree falls and the formation of forest gaps are important elements of forest dynamics, particularly in the tropics. Plants that are firmly attached to trees and branches that fail are more likely to survive if compliant. Alternatively, more weakly attached climbers might avoid mechanical stresses by becoming unattached from the host.

In the following account, we investigate the bending elastic mechanical properties of 43 species of tropical and temperate climbing plants; we explore how mechanical properties of the stem vary with the type of attachment, overall height, size, and diameter of climbers ranging from woody lianas, which can extend to the canopy of the tropical rain forest, to more diminutive climbers or vines barely reaching a meter in height. The study has particular significance for understanding the likely ecological constraints on different climbing habits of different species. We have listed the species tested along with their type of attachment, taxonomic affinity, and approximate size and position of the mature climbing plant in the surrounding vegetation. We have differentiated between five types of attachment modes: stem twiners, tendril climbers, hook climbers, branch-angle climbers, and leaning climbers. The categories of climber are somewhat generalized because, for example, both tendril and hook climbers include plants with varying types of hooks and tendrils. An important omission in this study is root climbers, plants that climb on the host plant via roots produced along the stem. While these are an important element in many ecosystems, they are usually attached firmly along much of their entire length apart from the apex and as such differ from the majority of vines and lianas. In this study, we compare plants that attach themselves to the host plant but do not "grow" directly on the host plant as do most root climbers.

2.2 METHODS AND MATERIALS

A wide range of mechanical properties is relevant to the understanding of the extremely specialized stem structures found in some climbing stems. They include the stiffness in the elastic range of bending, torsion, and tension as well as other properties in the nonelastic range up to failure of the plant stem as well as toughness, extensibility, and critical strain — to mention a few [13,14]. Additional properties of interest and inextricably linked to stem mechanical properties are hydraulic conductance of the stem and the extent of mechanical deformation that can occur before the essential hydraulic supply of the stem is compromised. In the following account, we confine our attention to bending stiffness in the elastic range. We have chosen this single parameter because it can be measured relatively easily under field or laboratory conditions and because it gives a measure of the bending mechanical properties of the stem before irrereversible deformation and damage outside the normal properties of living healthy plant tissues. Our objective is to compare a wide range of tested species with different climbing habits and, where possible, from plants sampled in their natural habitats. The inclusion of a wide range of mechanical parameters is not feasible with this view in mind. Bending mechanical properties can be measured in a number of ways and via a number of types of bending tests. In the following study, we include data from previous and ongoing research projects in which we use similar bending experimental protocols for a wide range of climbing species, in the laboratory, in relatively sophisticated "field laboratories," and in remote field conditions.

2.2.1 EXPERIMENTAL PROTOCOLS

While bending tests are relatively straightforward in theory, the heterogeneity, geometry, combination of tissues contributing to the plant stem, and abrupt changes in development and anatomy provide a variety of practical and theoretical difficulties. A range of bending tests can be set up for the majority of climbing plant species, but the type of test depends on the diameter and resistance to bending of the material. A bending apparatus that we have used in many laboratory and field situations (Figure 2.2a) consists of a steel frame fixed to a solid base and equipped with adjustable holders that can be positioned at variable distances. A stereoscope binocular microscope equipped with an eyepiece graticule is attached to the base and can be adjusted vertically and focused in a horizontal plane to a fixed point on the plant specimen to critically observe deflection [7,15]. The apparatus can be configured for two types of bending tests: three-point (Figure 2.2b) and four-point (Figure 2.2c) bending. This kind of apparatus can be modified into more portable versions for making measurements in the field (Figure 2.2d). The principal aim of the bending experiment is to measure flexural stiffness, *EI,* of the stem, and from this quantity calculate the structural Young's modulus, E_{str}, of the plant stem [7].

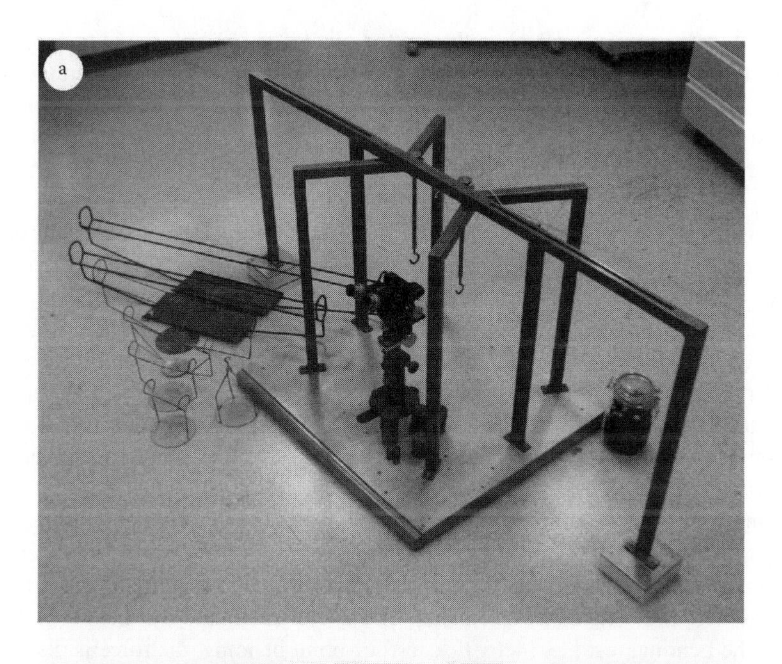

FIGURE 2.2 (a) Mechanical bending apparatus for use in field camps and field laboratories. The methods are entirely mechanical and do not require electricity. Plant stems are placed on supports suspended from a horizontal beam. Span distances can be adjusted to the size and bending resistance of the material. A range of panniers are used for tests in three- and four-point bending (left) as are a series of weights from 10 to 500 g (center). A series of up to six to eight weight increments is placed on the pannier, and the deflections are observed via a dissecting microscope equipped with an eyepiece graticule. The central part of each segment tested is stored in alcohol or FAA (1 part acetic acid 80%, 25 parts formalin 40%, 218 parts ethanol 50%) (right). (b) Configuration for three-point bending. The plant stem is placed on the two supports at a known distance, and the pannier is suspended from the exact center via a plastic ring, which can be turned obliquely to allow observation of the stem through the microscope. (c) Configuration for four-point bending. The plant stem is placed on the two supports at a known distance, and the two-armed pannier is placed on the specimen at equal distances from each central support. With this type of four-point bending, the weights applied to the pannier produce a deflection of the stem in an upward direction, and the influence of shear on the experiment is insignificant. (d) Portable, field, bending apparatus. The plant specimen is placed on the two supports, and the span distance can be adjusted easily. Both three- and four-point bending tests are possible.

Plant stem segments are pruned from the plant body and placed in the apparatus configured for either three- or four-point bending. Stems are first checked for splits, fractures, or other damage that could affect their mechanical properties. Stems that taper significantly (more than 10%) and stems that include significantly thickened nodes or branch points are excluded from the study. Apical parts of some plant stems such as the leading searchers of lianas can show relatively abrupt changes in mechanical properties from immature tissue near the very apex to the mature, differentiated stem lower down. Tests of young axes should avoid this transition, which is not always obvious from external inspection.

There are a number of advantages and disadvantages to both three- and four-point bending. Three-point bending is relatively easy to apply because the weight is added centrally (Figure 2.2b) and the stem does not tend to move about and spoil the deflection reading when the weights are added. In four-point bending, however, the stem is loaded via a two-armed pannier (Figure 2.2c), and slightly irregular stems can twist or flip over after one or more weights are added. This means that it is sometimes difficult to study all of the stem segments of a sample. In three-point bending, because the force is applied and concentrated at one point at the center of the stem, the measurement can include deflection of the axis caused by shear of the plant tissues, thus such measurements would underestimate the flexural stiffness during the bending test [13,14,16,17]. In four-point bending, the force is distributed equally over the central portion of the tested stem, and the effect of shear is insignificant. Tests in three-point bending should be verified to find out the minimum span-to-depth ratio above which the effect of shear will be insignificant. A long length of plant stem is selected, and a series of tests at increasing span lengths is carried out until a span-to-depth ratio is reached at which the flexural stiffness becomes a constant value [15,18]. In a plant stem of, say, a liana, ideally such tests should be carried out prior to measuring stems with widely differing mechanical properties such as young and old developmental stages. Stiff combinations of plant tissues are less liable to shear than more compliant ones. Many references refer to a span-to-depth ratio of 20 as being suitable for three-point bending based on the properties of many industrially tested woods and timbers. This value, however, can change significantly. Some compliant plant materials we have observed require span-to-depth ratios of 40 or more. Four-point bending can also require the addition of quite substantial weight increments to obtain a reasonable deflection in the region of 1 to 3 mm for a range of stem diameters of 1 to 30 mm. Adding excessive weight in a four-point bending test can cause problems because the outer tissue of the plant stem resting on the fixed supports can indent and affect the deflection distance observed; this is especially the case when the deflection is small. Such problems can be addressed by carefully trimming soft bark tissue from around the of the stem's point of contact with the supports so that the support contacts harder wood beneath. But if the wood itself is compressible, the problem might be difficult to solve. In summary, three-point bending is somewhat easier and quicker to carry out, but the need for carrying out span tests can make the approach time-consuming, especially under field conditions. Four-point bending can be trickier to set up and carry out because of movement of the stem, and it might not be possible to measure all of the stem segments in a sample, resulting in a loss in continuity in the analysis. In general, it is best to stick with one or the other types of test for a given analysis on a given plant. Having said that, it is also informative to first test a specimen in three-point bending and then repeat the test in four-point bending, combining this kind of verification with span tests; however, this takes time.

Successive weights are added during the bending experiment in three- or four-point bending at fixed time intervals of normally 1 min, which are sufficient for the plant stem to achieve its maximum deflection for a given weight. Creep is sometimes observed when the plant stem continues to deflect and is an indication that sufficient force is being applied to exceed the elastic range of the stem material. Some stems react extremely quickly to the force applied, while others react more slowly. The time increment should take this into account, but 1 min is normally enough to reach maximum deflection.

2.2.2 Calculation of Structural Young's Modulus

The bending test results are plotted as the deflection (in millimeters) on the *y*-axis against the force applied (in Newtons) on the *x*-axis. The bending test was successful if the value of r^2 for the force-deflection curve is greater than 0.98 (i.e., if the entire measurement is in the linear elastic range), the specimen has not moved or slipped during the test accidentally, and the experiment has remained within the elastic range. Immediately after the experiment, the diameter of the stem segment is measured in three to six positions along its length in the direction of the force applied and perpendicular to it. The mean axial second moment of area (I) (in mm^4) of the stem is then calculated from these measurements where the stem cross section is approximated as an ellipse:

$$I = (\pi/4) \cdot (r_1{}^3 \cdot r_2)$$

where r_1 is the radial thickness of the stem in the direction of the applied force and r_2 is the radial thickness in the perpendicular direction.

In three-point bending, the flexural stiffness (EI) (Nmm^2) of the tested stem is calculated via:

$$EI = l^3/(48 \cdot b)$$

where l is the distance (mm) between the two supports (Figure 2b) and b is the slope of the force-deflection curve (deflection/force) (mm/N).

In four-point bending, the flexural stiffness (EI) (Nmm^2) of the tested stem is calculated via:

$$EI = l^2 \cdot a/(16 \cdot b)$$

where l is the distance between the two internal supports, a is the distance between the outside pannier and the internal support (Figure 2c), and b is again the slope of the force–deflection curve (mm/N).

The structural Young's modulus of the stem (E_{str}) (MNm^{-2}) is calculated via:

$$E_{str} = EI/I$$

We use the term "structural Young's modulus" (E_{str}) as the parameter to describe a structure consisting of different tissues, i.e., a plant stem [7,15].

2.2.3 Sampling

We present values of structural Young's modulus E_{str} for a wide taxonomic and morphological range of climbers; the results are presented as means and standard deviations for young and old developmental stages of climbing plants (Figure 2.3 and Figure 2.4; Table 2.1). Despite problems in assigning specific ages for young and old plant segments, young axes of the plant are generally positioned toward the apex and, often in climbing plants, represent "searchers" for reaching new hosts (Figures 2.1a and 2.1b). In general, older stages of development represent more basal and generally, but not always, more flexible parts of the climbing plant (Figures 2.1c and 2.1e) [10,19]. The object is to compare how mechanical properties vary with type of attachment among different plant groups. Attachment modes were summarized as:

1. Twining or winding of the main leading axis and/or principal branches.
2. Attachment to the host plant via sensitive tendrils or tendril-like organs. There are a range of such types of organs derived from stem, leaf, or fertile axis. For simplicity, we have grouped these together. From a mechanical perspective and in terms of the growth form of the whole plant, we suspect that these types of attachment act similarly.
3. Attachment to host plants via hooks.
4. Attachment to host plants via wide-angled branches acting as anchors.
5. Leaning against the surrounding vegetation.

The analysis includes 43 species from 18 vascular plant families, including 1 lycopsid, 3 gymnosperms, 8 monocotyledons, and 31 dicotyledons (Table 2.1).

2.3 RESULTS: MECHANICAL PROPERTIES AND TYPE OF ATTACHMENT

2.3.1 Twining Climbers

All twining climbers show a marked decrease in E_{str} from young to older stages of development (Table 2.1; Figure 2.3 and Figure 2.4). All but five twiners show relatively high values of E_{str} in young stages of development of 2000 to 6000 MNm^{-2}. The highest value for young stages occurs in *Doliocarpus* sp. at about 6900 MNm^{-2}. Except for the species of *Lonicera* and two species of *Secamone* (*S. bosserii*, *S. geayii*), values of older stages (Figure 2.4) are below 1200 MNm^{-2} and about half of these are less than 600 MNm^{-2}. All stem twiners tested are dicotyledonous angiosperms (Table 2.1) with the exception of the gymnosperm genus *Gnetum*. Interestingly, the two African species of *Gnetum* show markedly lower values of E_{str}

in the young searcher shoots compared with all the other angiosperm twiners tested (Table 2.1; Figure 2.3 and Figure 2.4).

2.3.2 Tendril Climbers

All dicotyledonous tendril climbers tested show a marked drop in E_{str} from young to older stages with final values of E_{str} in the range of below 1000 MNm^{-2} (Table 2.1; Figure 2.3 and Figure 2.4). The species of *Bauhinia* (monkey ladder) shows an extreme drop in mechanical properties from young stages of over 8000 MNm^{-2} (Table 2.1). This varies from one of the highest tested of all categories (8480 MNm^{-2}) to one of the lowest (350 MNm^{-2}). The species shows a marked change in both material properties of the stem (Figure 2.3 and Figure 2.4) as well as a profound change in stem geometry leading to ribbonlike, highly flexible stems (Figure 2.1e). In contrast to the dicotyledonous twiners, the small monocotyledonous climber *Smilax aspera*, which generally grows on herbs, shrubs, and small trees, shows an increase in E_{str} from younger parts of the plant to older basal parts (Table 2.1).

2.3.3 Hook Climbers

The hook climbers show a more variable range of shifts in mechanical properties compared with twiners and tendril climbers (Table 2.1; Figure 2.3 and Figure 2.4). The woody dicotyledonous liana *Strychnos* (Figure 2.6d and 2.6e) shows the highest value of E_{str} of all plants tested in young stages of development, i.e., over 12,000 MNm^{-2}, followed by a large reduction in E_{str} (Table 2.1). In older stages, its value of E_{str} is also rather higher than most twining and tendril-climbing species. The two woody hook-climbers *Bougainvillea* and *Rosa* do not show a noticeable drop in E_{str} during development; rather, they retain relatively high values in old stages or increase them.

Monocotyledonous hook climbers (Figure 2.5) also show variable trends, depending on the taxonomic group they belong to. Species of *Calamus* (Figure 2.5b) show decreases in E_{str} to values of 1830 and 2300 MNm^{-2}, equivalent to the highest values of old twining species (*Lonicera* and *Secamone*) (Table 2.1; Figure 2.3 and Figure 2.4). Old stages of these *Calamus* species are canelike and have lost the leaf sheath that contributed to relatively high values of E_{str} in young stages of growth. In *Daemonorops jenkinsiana*, old stages included stems, which are still surrounded by leaf sheaths. For this species, very old and long stems were not available in the studied area. In *Plectocomia himalayana*, the older stages measured show no changes in mechanical properties compared with young stages (Table 2.1; Figure 2.3 and Figure 2.4). For this species, values of E_{str} remain high, around 3000 to 4000 MNm^{-2} (Table 2.1). The South American climbing palm *Desmoncus* (Arecoideae) (Figure 2.5a and 2.5f) shows a quite different trend. Interestingly, both species show a marked increase in E_{str} in older stages of development, for which the leaf sheath has senesced and, in the case of *D. orthacanthos*, was finally lost. Values in young stages of around 4000 to 5000 MNm^{-2} are in the similar range, with higher values found among twiners and tendril climbers, whereas values in older stages of 8000 to 9000

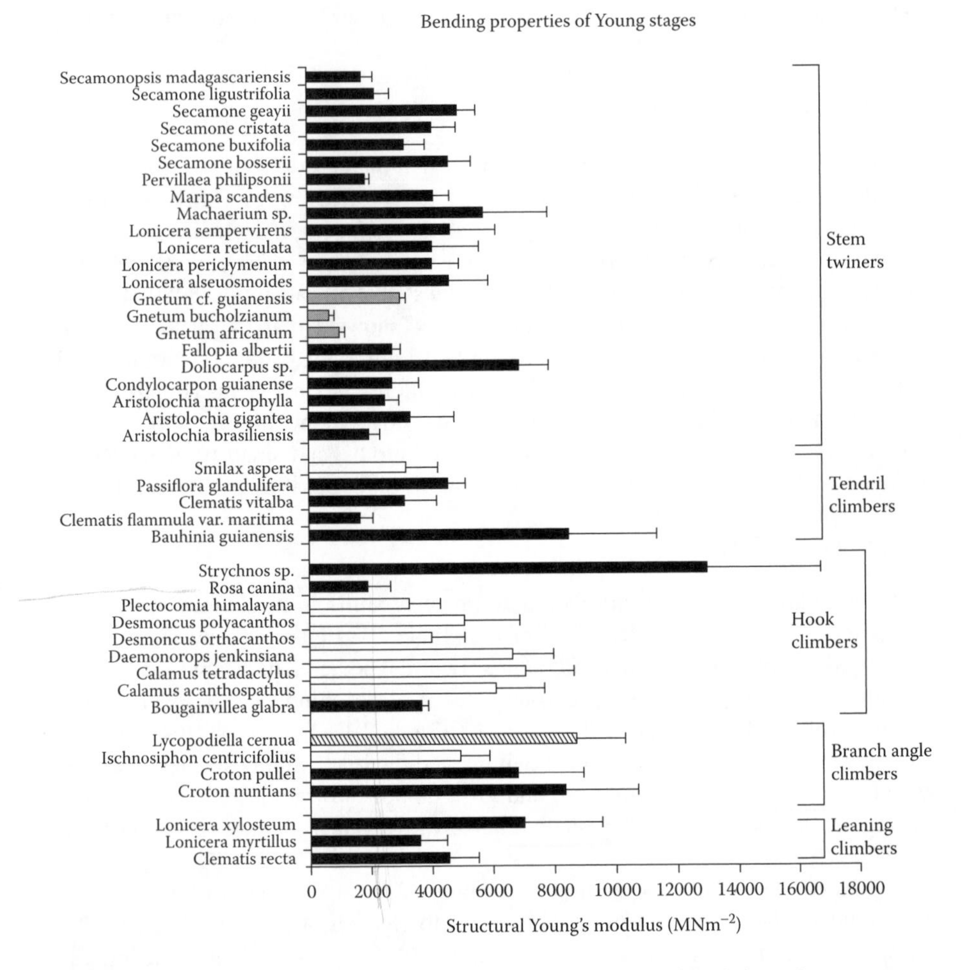

FIGURE 2.3 Mean structural Young's modulus and one standard deviation of young stages of development of climbing plants with different modes of attachment. Black bars = dicotyledons, grey bars = gymnosperms, white bars = monocotyledons, striped bar = lycopod.

MNm^{-2} are higher than all other plants tested with the exception of the monocotyledon *Ischnosiphon* and the leaning species *Lonicera xylosteum* (Table 2.1).

2.3.4 Branch-Angle Climbers

Plants forming attachments via wide-angled branching also show widely differing patterns of change in E_{str}. The species of *Croton*, representing a woody dicotyledon (Figure 2.6a), shows a similar trend as the majority of twiners and tendril-climbers (Table 2.1; Figure 2.3 and Figure 2.4). The young parts of climbing stems have high values of E_{str} and above the general range of nearly all twiners and most of the hook climbers and tendril climbers. The older parts show a marked drop in E_{str} with values

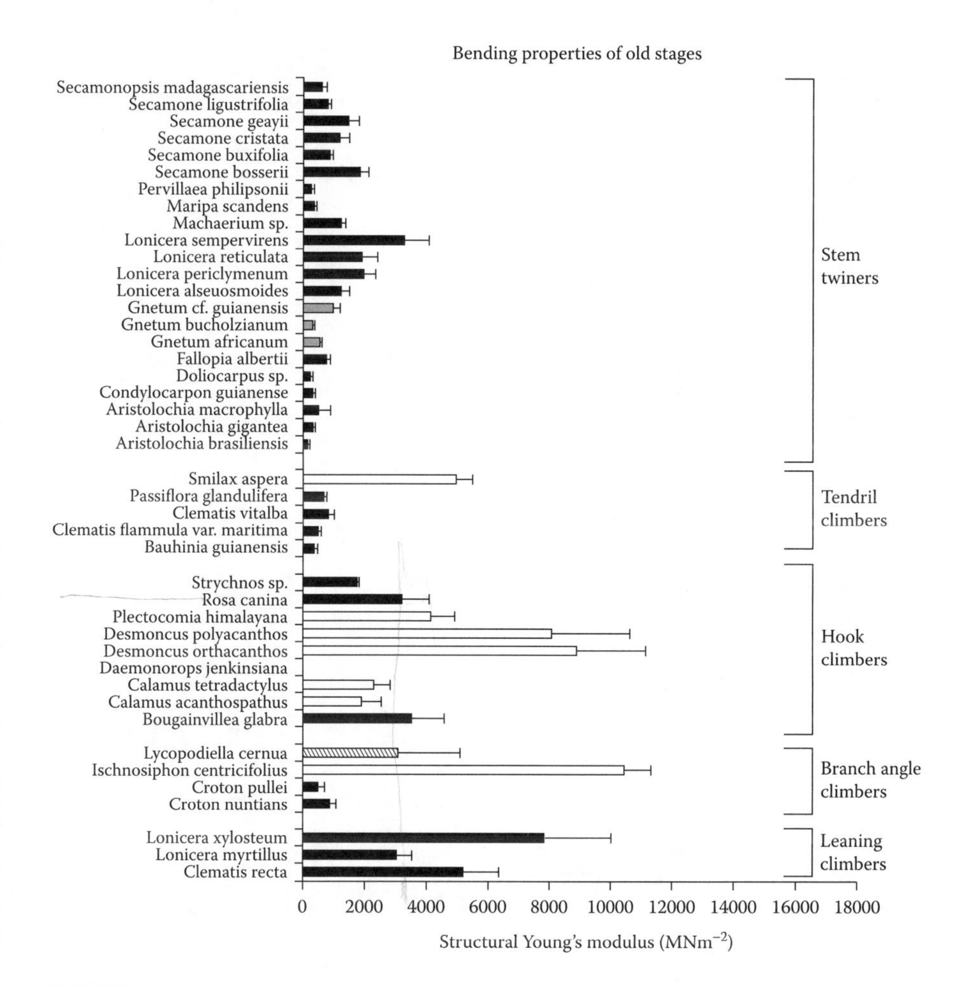

FIGURE 2.4 Mean structural Young's modulus and one standard deviation of old stages of development of climbing plants with different modes of attachment. Black bars = dicotyledons, stippled bars = gymnosperms, white bars = monocotyledons, striped bar = lycopod.

of less than 1000 MNm^{-2}, approaching that of many tendril climbers and twiners and, interestingly, well below that of all of the other hook climbers.

Stems of *Lycopodiella cernua* interlock with the surrounding vegetation via lateral branches. The younger parts of the plant show relatively stiff mechanical properties of over 8000 MNm^{-2}, in the range of woody rigid species like *Bauhinia* and *Strychnos* (Table 2.1; Figure 2.3 and Figure 2.4). The older stage at the base of the plant is more flexible but still retains high values of around 3000 MNm^{-2}.

Finally, the monocotyledonous branch climber, *Ischnosiphon centricifolius* (Marantaceae) has highly characteristic "pencil-like" stems, which produce rosettes of leaves at nodes along the stem (Figure 2.6b and 2.6c). Both the leaves and the angle formed by the insertion of the next internode can anchor the plant stem onto surrounding host plants in the forest understory. Younger distal parts of the climbing

TABLE 2.1
Structural Young's Modulus, Attachment Type, Size, Diameter, and Taxonomic Group of 43 Tested Climbing Plants from 18 Different Vascular Plant Families, Arranged according to Attachment Type

Species	Notes		E_{str} [MNm^{-2}] Young Stages	E_{str} [MNm^{-2}] Old Stages	Climbing to: Forest Canopy Forest Understory Herb to Shrub Level Herb Level	Diameter Of Widest Stem Tested [Mm]	No. of Samples (Young Stages)	No. of Samples (Old Stages)	Family	Group
Leaning										
Clematis recta	**	b	4521 ± 980	5132 ± 1200	Herb to shrub	6	20	33	Ranunculaceae	Dicot
Lonicera myrtillus	**	f	3560 ± 939	2978 ± 560	Forest understory	10	3	7	Caprifoliaceae	Dicot
Lonicera xylosteum	*	f	7022 ± 2497	7790 ± 2222	Forest understory	18	4	5	Caprifoliaceae	Dicot
Angled Branch										
Croton nuntians	*	d	8318 ± 2401	849 ± 211	Forest canopy	49	14	5	Euphorbiaceae	Dicot
Croton pullei	*	e	6741 ± 2206	490 ± 203	Forest canopy	102	38	8	Euphorbiaceae	Dicot
Ischnosiphon centricifolius	*		4956 ± 921	10427 ± 921	Forest understory	15	5	8	Marantaceae	Monocot
Lycopodiella cernua	*	e	8719 ± 1571	3031 ± 2054	Herb to shrub	4	10	10	Lycopodiaceae	Lycopsid
Hook										
Bougainvillea glabra	**	a	3640 ± 219	3460 ± 1113	Forest canopy	11	41	20	Nyctaginaceae	Dicot
Calamus acanthospathus	*	i	6110 ± 1560	1830 ± 670	Forest canopy	15	3	2	Arecaceae	Monocot
Calamus tetradactylus	**	i	7070 ± 1560	2300 ± 500	Forest canopy	13	13	40	Arecaceae	Monocot
Daemonorops jenkinsiana	**		6630 ± 1330	-	Forest canopy	33	8	–	Arecaceae	Monocot
Desmoncus orthacanthos	*	i	4000 ± 1070	8906 ± 2240	Forest canopy	30	17	9	Arecaceae	Monocot
Desmoncus polyacanthos	*		5074 ± 1782	8091 ± 2546	Forest understory	10	16	11	Arecaceae	Monocot
Plectocomia himalayana	*		3250 ± 1030	4120 ± 780	Forest canopy	40	3	3	Arecaceae	Monocot
Rosa canina	*	h	1848 ± 787	3155 ± 954	Herb to shrub	13	12	23	Rosaceae	Dicot
Strychnos sp.	*	a	12920 ± 3790	1750 ± 40	Forest canopy	49	14	2	Loganiaceae	Dicot

Tendril										
Bauhinia guianensis	*	a,g	8480 ± 2820	350 ± 100	Forest canopy	18	3	3	Fabaceae	Dicot
Clematis flammula var. *maritima*	*	b	1619 ± 492	470 ± 113	Herb to shrub	10	21	10	Ranunculaceae	Dicot
Clematis vitalba	*	b	3074 ± 1083	772 ± 261	Forest canopy	18	18	17	Ranunculaceae	Dicot
Passiflora glandulifera	*	a	4540 ± 620	630 ± 130	Forest canopy	21	3	3	Passifloraceae	Dicot
Smilax aspera	*		3178 ± 1057	4937 ± 529	Herb to shrub	6	7	7	Smilacaceae	Monocot
Stem Twiner										
Aristolochia brasiliensis	**		1955 ± 411	104 ± 72	Forest canopy	35	6	8	Aristolochiaceae	Dicot
Aristolochia gigantea	**		3343 ± 1405	294 ± 89	Forest canopy	23	6	5	Aristolochiaceae	Dicot
Aristolochia macrophylla	**	b	2465 ± 479	497 ± 385	Forest understory	28	40	45	Aristolochiaceae	Dicot
Condylocarpon guianense	*	c	2720 ± 900	310 ± 50	Forest canopy	38	36	11	Apocynaceae	Dicot
Doliocarpus sp.	*	a	6870 ± 980	290 ± 30	Forest canopy	33	3	3	Dilleniaceae	Dicot
Fallopia albertii	**	a	2740 ± 290	730 ± 150	Forest canopy	13	10	3	Polygonaceae	Dicot
Gnetum africanum	*		1036 ± 176	560 ± 64	Forest canopy	24	2	4	Gnetaceae	Gymnosperm
Gnetum bucholzianum	*		683 ± 180	293 ± 90	Forest canopy	17	11	7	Gnetaceae	Gymnosperm
Gnetum cf. *guianensis*	*		3037 ± 126	961 ± 222	Forest canopy	17	2	3	Gnetaceae	Gymnosperm
Lonicera alseuosmoides	**	f	4561 ± 1330	1170 ± 342	Forest understory	7	10	2	Caprifoliaceae	Dicot
Lonicera periclymenum	*	f	4050 ± 894	1912 ± 444	Forest understory	24	3	6	Caprifoliaceae	Dicot
Lonicera reticulata	**	f	4028 ± 1589	1876 ± 528	Forest understory	9	8	5	Caprifoliaceae	Dicot
Lonicera sempervirens	**	f	4666 ± 1516	3262 ± 796	Forest understory	12	8	3	Caprifoliaceae	Dicot
Machaerium sp.	*	a	5750 ± 2110	1200 ± 150	Forest canopy	29	3	3	Fabaceae	Dicot
Maripa scandens	*	a	4111 ± 510	390 ± 30	Forest canopy	32	5	3	Convolvulaceae	Dicot
Pervillaea philipsonii	*		1851 ± 232	266 ± 79	Forest canopy	15	12	11	Apocynaceae	Dicot
Secamone bosserii	*		4573 ± 818	1793 ± 305	Herb to shrub	12	12	6	Apocynaceae	Dicot
Secamone buxifolia	*		3145 ± 693	850 ± 85	Forest canopy	17	12	6	Apocynaceae	Dicot
Secamone cristata	*		4057 ± 851	1158 ± 357	Forest to shrub	11	9	2	Apocynaceae	Dicot
Secamone geayii	*		4909 ± 647	1415 ± 396	Herb to shrub	11	8	6	Apocynaceae	Dicot

(continued)

TABLE 2.1 (CONTINUED)
Structural Young's Modulus, Attachment Type, Size, Diameter, and Taxonomic Group of 43 Tested Climbing Plants from 18 Different Vascular Plant Families, Arranged according to Attachment Type

Species	Notes	E_{str} [MNm^{-2}] Young Stages	E_{str} [MNm^{-2}] Old Stages	Climbing to: Forest Canopy Forest Understory Herb to Shrub Level Herb Level	Diameter Of Widest Stem Tested [Mm]	No. of Samples (Young Stages)	No. of Samples (Old Stages)	Family	Group
Stem Twiner (continued)									
Secamone ligustrifolia	*	1976 ± 544	767 ± 105	Forest canopy	29	12	7	Apocynaceae	Dicot
Secamonopsis madagascariensis	*	1734 ± 413	570 ± 192	Forest canopy	23	12	4	Apocynaceae	Dicot

Note: Values of E_{str} are based on means and 1 standard deviation.

* Plant measured in natural habitat
** Plant measured in botanical gardens or greenhouses.
[a] See ref. [8]
[b] See ref. [27]
[c] See ref. [7]
[d] See ref. [21]
[e] See ref. [15]
[f] See ref. [25]
[g] Old stems of *Bauhinia guianensis* are ribbon-shaped; the maximum diameter given here corresponds to the diameter of a circular surface equal in area to the ribbon-shaped stem in cross section.
[h] Measurements of *Rosa canina* are confined to leaning or climbing stems up to 1 year old.
[i] Data for these climbing palms include measurements based on old stages in which the outer leaf sheath has been lost via natural aging. The other species are all based on older portions of climbing stems in which the leaf sheath is still attached but usually, at least, partially senescent.

FIGURE 2.5 (a) Young phase of growth of the understory climbing palm *Desmoncus polyacanthos* (French Guiana). When the plant becomes mechanically unstable, it produces leaves equipped with distal hooks and grapnel-bearing cirri (left). (b) Young phase of growth of *Calamus acanthospathus* (Yunnan Province, China). When the plant becomes unstable, long modified inflorescences bearing recurved hooks (flagellae) are produced, and the plant can reach the canopy, producing stems over 70 m in length. (c) The calamoid rattan palm *Plectocomia himalayana* (Yunnan Province, China), old stage of development in which the young part of the plant is firmly attached in the canopy and the older part of the axis is spanning a wide space between supports. The leaves and leaf sheaths of this part of the stem are senescing and lose their attachment with the original supports. (d) Apical parts of *P. himalayana* emerge from the forest canopy (Yunnan Province, China) and bear many cirrus-bearing leaves and many potential attachment points with the surrounding canopy vegetation. (e) The small-bodied *Smilax aspera* (southern France) uses both apendicular tendrils and recurved hooks for attaching to host plants. Tendrils firmly connect the plant to the host. (f) Apex of the attachment organ (cirrus) of *Desmoncus polyacanthos* (French Guiana) produce laterally deployed grapnel-like modified leaflets (acanthophylls) and abaxially placed recurved spines.

FIGURE 2.6 (a) The branch-angle climber *Croton nuntians* (French Guiana) shows a well-attached apex and flexible older stems descending to ground level. (b) The climbing monocotyledon *Ischnosiphon centricifolius* (French Guiana). Its narrow pencil-like stems ascend into the understory. (c) The angles formed between successive nodes and internodes in stems of *Ischnosiphon centricifolius* provide effective points of attachment. (d) Stiff distal branches and sensitive hooks of *Strychnos* sp. (French Guiana). (e) Unattached hook of *Strychnos* sp. When hooks locate a host branch, they can develop a more secure attachment by tendril-like growth and woody development (French Guiana). (f) Appendicular woody tendrils produced from *Bauhinia guianensis*. Such structures can form firm attachments to narrow diameter host branches (French Guiana).

stem show relatively high values of E_{str} of around 5000 MNm^{-2}, which is in the upper range of values for twiners and hook climbers. Interestingly older, more basal parts of the plant show very high values of E_{str} of over 10,000 MNm^{-2} (Table 2.1; Figure 2.3 and Figure 2.4).

2.3.5 Leaning Climbers

This final category includes plants that simply lean against the surrounding vegetation without any specialized attachment, often showing a limited range of development and small body size. Species tested do not show significant changes in mechanical properties from young to old stages (Table 2.1; Figure 2.3 and Figure 2.4). The two woody dicotyledons, *Lonicera xylosteum* and *L. myrtillus*, which resemble "unstable shrubs" show high values of E_{str} in old stages with values of *L. xylosteum* among the four stiffest stems tested of older stages. The small statured *Clematis recta* also shows very similar values of E_{str} in both younger apical parts of the plant and older basal stages of development.

2.4 DISCUSSION

2.4.1 Mechanical Properties and Attachment of Dicotyledonous Climbers

All twiners, tendril climbers, and branch-angle climbers of dicotyledons show a drop of E_{str} from young to old stages of growth. These changes in mechanical properties probably reflect a need for twining and tendril plants to be protected against extreme forces generated by movement of the host trees. Climbers that twine from one tree to another risk experiencing extremely high forces during high winds and after failure of branches and trunks. The twining strategy therefore requires highly compliant older stems that can withstand severe stresses and strains. Being strongly attached to a support also means that once the stem is attached, the part of the stem that is below the point of attachment has little need of stem stiffness to partially support the stem or lean against surrounding vegetation to retain its position. Mature trunks of large-bodied twining species are often so compliant that they hang in tension from the point of attachment above. Such stems can have very low structural Young's moduli as reflected in some of the values presented among twiners.

Tendrils can also form an extremely firm attachment to the host supports, but the attachment is usually weaker than for stem twiners because the tendrilar organs are often derived from modified stipules, leaves, or fertile axes, which have narrow diameters. Like stem twiners, tendril climbers also have little need of stem stiffness in basal parts of the plant once they are firmly attached at many points.

Hook climbers are generally less firmly attached to the host plant than either stem twiners or tendril climbers. Hooks and curved spines can be effective means of attaching to host supports, but attachment is constrained by the internal angle and length of the hook. They can also be dislodged by movement of the host branches. Compared with twiners, hook climbers are considerably less firmly attached to the host plant, particularly in young stages. Movement and swaying of the host plant could potentially dislodge the climber from the support rather than exert extreme stresses on the main stem as in twiners. Furthermore, relatively stiff, older parts would ensure that the plant stem would maintain its position in the vegetation even though one or more hooks had become unattached. After detachment, the young apical part of the plant would remain in the vicinity of potential supports. This is quite different from the effect of removing the attachment from a mature or aging stem twiner because such stems can buckle and collapse in coils because of their own weight. A hook-climbing strategy with relatively stiff older stems is probably adopted by many different plant groups and with different developmental constraints [20]. Among the plants tested, two of the three woody plants, *Bougainvillea* and *Rosa*, retain relatively high Young's moduli in older stages of development, with even higher values found in the old stages of *Rosa*. Interestingly, both climb via less specialized curved spines rather than acutely angled hooks. In species of *Strychnos* tested from French Guiana, early growth of apical branches bear open hooks that are green and sensitive to touch (Figure 2.6e). They can engage small-diameter supports, and the limb of the hook eventually thickens and closes around the supporting host branch. *Strychnos*, like all the dicotyledons that produce active attachment organs, shows a large reduction in E_{str} to just below 2000 MNm^{-2} from relatively large values of over 12,000 MNm^{-2} found in young stages.

Plants climbing via wide-angled branches show high values of E_{str} in young stages, generally higher than those of twiners and consistent with the idea that young stages of growth require stem stiffness for attachment. Wide-angled branches can interlock with the branches of neighboring trees and provide very firm anchorage. It is important that young stages of growth be stiff and not deflect when the plant has established its climbing phase. The branch-angle climbers of *Croton* also show marked drops in E_{str} toward the base of the plant. Even though the attachment is not as active as in tendril and twining species, branch anchorage can be very effective and can be coupled with very low values of E_{str} in old stems [21].

For the three plants grouped as leaning climbers, general observations indicate that branches and leaves do not anchor the plant in the vegetation. We saw previously that the adult stems of *Croton* that were tested can be suspended from the host canopy via their wide-angled attachment and finally produce highly compliant wood. Unlike climbing species of *Clematis*, the petioles of *C. recta* do not twine extensively around supports. Similarly the two leaning species of *Lonicera* tested differ from the more typical climbing species in that they do not twine around neighboring stems. All three plants produce little change in E_{str} from young to old stages of development and from near the apex to the base. The values do remain relatively high, and this kind of pattern has been termed "semi-self-supporting" [10]. Such growth forms retain relatively stiff mechanical properties through their growth trajectory. They form little attachment with host vegetation apart from simply leaning.

Most dicotyledonous species that show a marked decrease in E_{str} during development, such as stem twiners and tendril climbers, have active attachment organs. The data also indicate that some species (e.g., *Croton*) that climb by wide-angled branches are also capable of producing compliant wood. The production of branches with many potential points of anchorage might finally permit a firm connection with the surrounding vegetation, but twining stems are always more firmly bound.

Interestingly, both *Strychnos* and *Croton* eventually produce older stages of growth with relatively low values of E_{str}. Recent studies have shown that these plants have relatively extended "stiff" phases of development, and this is possibly linked to their type of attachment as well as other potential environmental factors such as light availability [9,20]. The type of attachment shown by these plants via many potential points of attachment and the degree of security this affords to the climbing plant differs considerably from the mechanical constraints imposed by firmly twining or attaching via tendrils.

2.4.2 Climbing Growth Strategies in Monocots and Other Plants without Secondary Growth

In dicotyledonous species, the secondary growth leads to an increase in the second moment of area (I) and stem flexural stiffness (EI). The reduction of E_{str} during development can be interpreted as a mechanical strategy that "compensates" for this increase in I and confers relative stem compliance. Despite the diversity of attachment modes, all the monocots tested, except for species of the rattan *Calamus*, show high values of E_{str} in old stages and otherwise little change in mechanical properties during development. The latter can be explained by the structural constrains inherent to monocotyledons as a result of the absence of secondary growth. By retaining relatively small stem diameters (i.e., a small I) in old stems, the monocots escape this constraint and can, even with a high value of E_{str} remain compliant in older stems, i.e., have a relatively small EI in old developmental stages [20,22]. The tendril-climbing species *Smilax aspera* also lacks secondary growth and shows an opposite trend to other dicotyledonous tendril climbers, with an increase in structural Young's modulus. *Smilax aspera* is a relatively lightly built, small-bodied plant rarely exceeding shrub level in its natural habitat (Figure 2.5e), and may therefore not be exposed to high stresses caused by the swaying of the host plants as would be the case for large tendril climbers. The branch-angle climber *Ischnosiphon* with its pencil-like stem has a peculiar growth form that relies largely on stem stiffness to maintain the plant in position between points at which leaf-bearing nodes catch onto branches of host plants. High stiffness from top to bottom is used throughout the life history of the plant to ensure a weakly attached, partially leaning growth form. A similar strategy is observed in the more diminutive lycopsid *Lycopodiella cernua,* which has very high values of E_{str} in near apical portions of their small stems that are only loosely supported by the wide-angled branches [9,23]. This plant also lacks secondary tissues, but the primary tissues toward the base of the stem are more compliant. The developmental and ecological significance of the basal compliancy in *Lycopodiella* is not clear [9,20,23]. This herbaceous plant is self-supporting with climbing and horizontal or trailing axes. It is possible that the basal compliancy observed may permit the plant to occupy a procumbent orientation by bending at the base if it becomes detached from the surrounding herbs and shrubs. Many formerly upright or leaning individuals end up as procumbent stems. If true, this is an interesting variant of how herbaceous plants might use stem compliancy to avoid damage via mechanical stresses after movement and falling from host supports.

The palms tested climb via hook-like organs placed on modified frond tips (cirri) or modified inflorescence stalks (flagellae) that emerge from the leaf sheath. Two species of the South American genus *Desmoncus* show an actual increase in E_{str} toward the base of the plant [22]. In the larger species, *D. orthacanthos*, old stages have shed the outer leaf sheath and resemble canes. This stage of development does nevertheless show relatively high values of E_{str}. With the exception of the two tested *Calamus* species, the other climbing palms in the Calamoideae, *P. himalayana* (Plectocominae) [24] also show high values of E_{str} in old stages. This is consistent with the idea that hook climbers that are not firmly bound to host supports do not need to develop high compliancy to survive movement of host branches. Instead, the relatively high stiffness of the stem ensures that the stems can retain their orientation in the surrounding vegetation when movement of the host branches moves or detaches any hooks and grapnels from their points of attachment. Rattans are wonderful examples of how monocots can escape the structural constraints to develop a comparable mechanical architecture as dicotyledonous lianas. Species of *Calamus* can show a decrease in E_{str} toward the base, a trend typical of many dicotyledonous lianas. However, the values of old stages of *Calamus* measured are in the range of the highest values of old stages of dicotyledonous twiners but do not approach the highly compliant properties of many other twiners.

Finally, future studies should incorporate studies based on hydraulic and mechanical properties of the stem. Many climbers show a reduction in structural Young's modulus during ontogeny, and this might be consistent with a relative increase in large-diameter vessels and increased hydraulic conductivity at the expense of relatively stiff tissues in the wood. The developmental and ecological trade-offs between mechanical functioning (stem stiffness and support) and hydraulic conductivity are complex [29]. While increased vessel diameter might influence stem stiffness in some climbers, many dicotyledonous climbers show relative increases in other compliant tissues during ontogeny such as rays and development of thinner-walled wood fibers. Such tissues are more easily interpreted as structural modifications that increase compliancy, resistance to fracture, and stem repair in vines and lianas.

2.4.3 Ecological Diversity of Climbers among Different Groups

Climbing growth forms are common among woody families of angiosperms, relatively common among monocots, but much rarer in other groups, including gymnosperms and ferns. The mechanical properties of climbing stems as well as the type of attachment are undoubtedly coupled to a wide range of evolutionary constraints among different plant groups [19,20,26–29]. Most monocot climbers tested so far retain the relatively stiff properties of host location and initial stiff phase of growth. Such plants without secondary growth are unable to make the developmentally complex, though ecologically important, transition from high stiffness to high compliance of tissues. However, because of their small axial second moment of area, flexural stiffness also remains relatively low in old stems. Such climbing forms nevertheless occupy both small-bodied and large-bodied climbing niches and can show important diverse modes of climbing attachment. Only species of *Calamus*, arguably the most diverse and specialized group of climbing monocots, show relatively low values of E_{str} of the inner stem, which produces flexible canes after the loss of the leaf sheath. This particular mechanical architecture in *Calamus* could have been an important innovation in the group and might explain its ecological success in Southeast Asia.

2.5 CONCLUSIONS

Twining climbers that attach firmly to the host support eventually require substantial compliance to escape stresses. Most of the climbing strategies require a stiff initial development to (a) get off the ground, (b) locate host supports, and (c) span distances between host supports. A twining strategy would therefore require an initial phase of stiff development followed by a later stage of compliant development. This requires a relatively sophisticated change in stem development and is mostly observed among dicotyledonous angiosperms that possess a bifacial vascular cambium and produce a specialized type of highly compliant wood during later growth. Many of these wood types have been termed "anomalous wood" (or more recently, variant cambia), and produce a wide range of vascular configurations that often incorporate enhanced hydraulic conductance, wound repair, and mechanical compliance properties. Is high compliance necessary for all climbing habits? For firmly attached twining plants that produce large diameter climbing axes, the answer is probably yes. However, some kinds of hook climbing may require relatively stiff stems so that the plant stem will be more likely to sway onto another neighboring support rather than buckle and collapse.

In summary, climbing plants can show many types of attachment from firmly attached twiners to less well attached hook climbers and to leaning plants. Bending mechanical properties of the stem and its change during development can differ between these categories: Twining climbers tend to produce lower values of E_{str} in older stages of development than other more loosely attached climbers. This is probably because firm attachment to host plants requires high compliancy to survive movement of the host plant. Other niches linked to more loosely attached climbing modes are open to a wider range of plant groups and do not require high compliancy in older growth stages. This type of climbing strategy has been adopted by a wide range of plant groups, notably monocotyledons.

REFERENCES

1. Schnitzer, S.A. and Bongers, F., The ecology of lianas and their role in forests, *TREE*, 17, 223, 2002.
2. Rowe, N.P. and Speck T. Plant growth forms: an ecological and evolutionary perspective, *New Phytol.*, 166, 61, 2005.
3. Phillips, O.L. et al., Increasing dominance of large lianas in Amazonian forests, *Nature*, 418, 770, 2002.
4. Wright, S.J., Calderón, O., Hernandéz, A., and Paton, S., Are lianas increasing in importance in tropical forests? A 17 year record from Panama, *Ecology*, 85, 484, 2004.

5. Putz, F.E. and Mooney, H.A. (Eds.), *The Biology of Vines*, Cambridge University Press, Cambridge, 1991.
6. Speck, T., Bending stability of plant stems: ontogenetical, ecological, and phylogenetical aspects, *Biomimetics* 2, 109, 1994.
7. Rowe, N.P. and Speck T., Biomechanical characteristics of the ontogeny and growth habit of the tropical liana *Condylocarpon guianense* (Apocynaceae), *Int. J. Plant Sci.* 157, 406, 1996.
8. Speck T. et al., How plants adjust the "material properties" of their stems according to differing mechanical constraints during growth: an example of smart design in nature, in *Bioengineering. Proceedings of the 1996 Engineering Systems Design and Analysis Conference*, Engin, A.E., Ed., PD-Vol. 77, Vol. 5. The American Society of Mechanical Engineers, New York, 1996, p. 233,.
9. Rowe, N.P. and Speck T., Biomechanics of plant growth forms: the trouble with fossil plants, *Rev. Palaeobot. Palynol.*, 102, 43, 1998.
10. Speck, T. and Rowe, N.P., A quantitative approach to analytically defining size, form and habit in living and fossil plants, in *The Evolution of Plant Architecture*, Hemsley, A.R. and Kurmann, M., Eds., Royal Botanical Gardens, Kew, UK, 1999, p. 447.
11. Hegarty E.E., Vine-host interactions, in *The Biology of Vines*, Putz, F.E. and Mooney, H.A., Eds., Cambridge University Press, Cambridge, 73, 1991.
12. Fisher, B. and Ewers, F.W., Structural responses to stem injury in vines, in *The Biology of Vines*, in Putz, F.E. and Mooney, H.A., Eds., Cambridge University Press, Cambridge, 99, 1991.
13. Wainwright, S.A., Biggs, W.D., Currey, J.D., and Gosline, J.M., *Mechanical Design in Organisms*, John Wiley, New York. 1976.
14. Niklas, K.J., *Plant Biomechanics, an Engineering Approach to Plant Form and Function*, The University of Chicago Press, Chicago, 1992.
15. Gallenmüller, F., Müller, U., Rowe, N.P., and Speck, T., The growth form of *Croton pullei* (Euphorbiaceae) — Functional morphology and biomechanics of a neotropical liana, *Pl. Biol.*, 3, 50, 2001.
16. Vincent, J.F.V., *Structural Biomaterials*, Princeton University Press, London, 1990.
17. Vincent, J.F.V., Plants, in *Biomechanics — Material: A Practical Approach*, Vincent, J.F.V., ed., IRL Press, Oxford, 1992, p. 165.
18. Bodig, J. and Jayne, B.A., *Mechanics of Wood and Wood Composites,* Krieger Publishing Company, Malabar, FL, 1993.
19. Speck, T. et al., The potential of plant biomechanics in functional biology and systematics, in *Deep Morphology: Toward a Renaissance of Morphology in Plant Systematics*, Stuessy, T.F., Mayer, V.,and Hörandl, E, Eds., Koeltz, Königstein, 2003, p. 241,.
20. Rowe, N.P., Isnard, S., and Speck, T., Diversity of mechanical architectures in climbing plants: an evolutionary perspective, *J. Plant Growth Regul.*, 23, 108, 2004.
21. Gallenmüller, F., Rowe, N.P. and Speck, T., Development and growth form of the neotropical liana *Croton nuntians*: the effect of light and mode of attachment on the biomechanics of the stem, *J. Plant Growth Regul.*, 23, 83, 2004.
22. Isnard, S., Speck, T., and Rowe, N.P., Biomechanics and development of the climbing habit in two species of the South American palm genus *Desmoncus* (Arecaceae). *Am. J. Bot.*, 92, 1444, 2005
23. Rowe, N.P. and Speck, T., Biomechanics of *Lycopodiella cernua* and *Huperzia squarrosa*: implications for inferring growth habits of fossil small-bodied Lycopsids, *Med. Ned. Ins. Toeg. Geo. TNO.*, 58, 290, 1997.

24. Isnard, S., Biomechanics and development of rattans: what is special about *P. himalayana*?, *Bot. J. Linn. Soc.*, (in press) 2006.
25. Traiser, C., Reidelstrüz, P., and Speck, T., Biomechanische, anatomische und morphologische Untersuchungen verschiedener Wuchsformtypen der Gattung Lonicera L. *Mitt. Bad. Landesver. Naturkunde u. Naturschutz N.F.*, 17, 123, 1998.
26. Civeyrel, L. and Rowe, N.P., Phylogenetic relationships of Secamonoideae based on the plastid gene matK, morphology and biomechanics, *Ann. Missouri Bot. Gard.,* 88, 583, 2001.
27. Isnard, S., Speck, T., and Rowe, N.P. Mechanical architecture and development in *Clematis*: implications for canalised evolution of growth forms, *New Phytol.,* 158, 543, 2003.
28. Lahaye, R., Civeyrel, L., Speck, T., and Rowe, N.P., Evolution of shrub-like growth forms in the lianoid sub family Secamonoideae (Apocynaceae s.l.) of Madagascar: phylogeny, biomechanics and development. *Am. J. Bot.,* 92, 1381, 2005.
29. Rowe, N.P. and Speck, T., Hydraulics and mechanics of plants: novelty, innovation and evolution, in *The Evolution of Plant Physiology*, Poole, I. and Hemsley, A.R., Eds., Elsevier Academic Press, Kew, 2004, p. 297.

3 The Role of Blade Buoyancy and Reconfiguration in the Mechanical Adaptation of the Southern Bullkelp *Durvillaea*

Deane L. Harder, Craig L. Stevens, Thomas Speck, and Catriona L. Hurd

CONTENTS

3.1 INTRODUCTION

3.1.1 THE INTERTIDAL ZONE

The intertidal habitat is mechanically very demanding [1]. High flow rates (greater than 25 m s^{-1}) and accelerations (greater than 500 m s^{-2}) require special mechanical adaptations by intertidal organisms [2–8]. In general, it is advantageous to minimize the overall size to avoid excessive wave-induced forces [9]. Intertidal seaweeds, however, deviate from this pattern. Based on common presumptions of how forces scale with size, this group seems to be oversized [9].

Seaweeds can adapt their mechanical properties in response to ambient wave climates [2,4,7]. Possibly even more important, seaweeds are very flexible and can change their overall shape [3,5,6,8]. By streamlining, seaweeds are able to reduce the magnitude of acting forces that can potentially be generated at high velocities [10–12]. The overall goal of this study was to quantify the process of streamlining and reconfiguration and to assess the importance of the positively buoyant lamina in the large intertidal seaweed *Durvillaea*.

3.1.2 THE SOUTHERN BULLKELPS *DURVILLAEA ANTARCTICA* AND *D. WILLANA*

The southern bull kelp *Durvillaea* is a member of the Fucales [13]. Its morphology is typical for large brown seaweeds with a holdfast, a stalklike stipe, a transitionary palm zone at the apical end of the stipe, and a large blade. Unlike other members of the Fucales, growth in *Durvillaea* is not restricted to a small apical meristematic zone but is diffuse [14]. The distribution of *Durvillaea* is confined to the Southern hemisphere where it grows on temperate rocky shores [15].

Durvillaea is the largest intertidal seaweed in the world. Individuals with a length of greater than 13 m [16] and a mass of more than 80 kg (C. Hurd, unpublished data) have been recorded. This genus can thrive even in the harsh conditions of the wave-swept surf zone. Moreover, it needs at least a moderate wave exposure for the successful establishment at a particular site [14].

Durvillaea antarctica occurs along the coasts of New Zealand, Chile, and some sub-Antarctic islands [15]. Its size and morphology are highly dependent on the ambient wave climate [15,17]. Three morphotypes can be identified [15]

FIGURE 3.1 The morphology of *Durvillaea antarctica* is highly dependent on wave exposure. (A) At comparatively sheltered sites, the blade becomes broad and undulating. (B) If wave exposure is more severe, the blade is subdivided into many whip-like thongs. The overall length of the blade is approximately 5 to 7 m in both photographs.

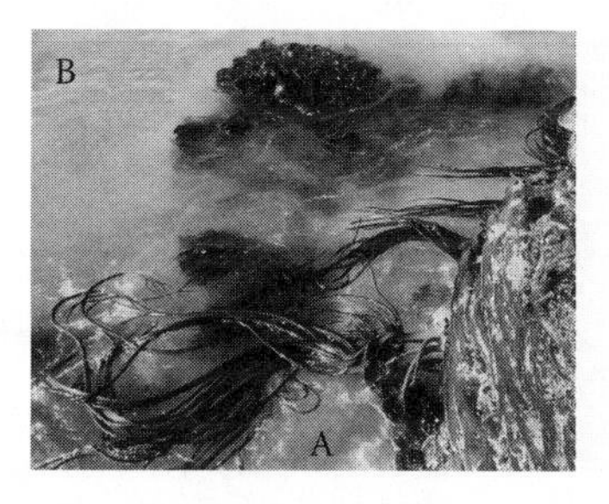

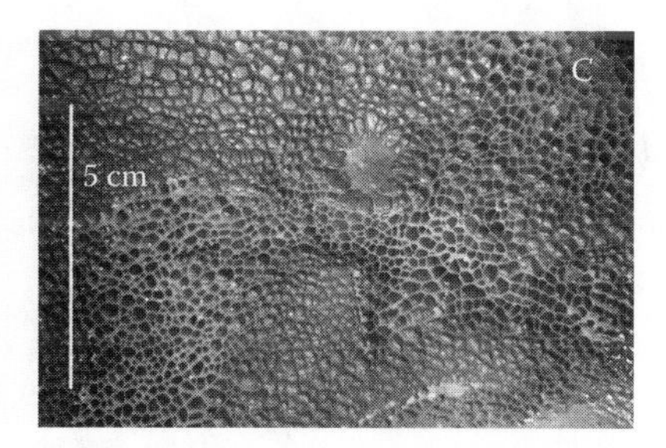

FIGURE 3.2 (A) The blade of *D. antarctica* is positively buoyant so the lamina is floating at the water surface, whereas (B) the blade of *D. willana* is neutrally buoyant, so that the lamina is upright in the water column. (C) The medulla of *D. antarctica* contains honeycomb-shaped, gas-filled sacs.

(Figure 3.1). At wave-sheltered sites, the overall morphology of the blade is broad and cape-like, with undulating edges (Figure 3.1A, left). At more wave-exposed sites, the blade becomes flatter and subdivided into many whip-like thongs (Figure 3.1B, right). At extremely wave-exposed sites, the stipe becomes longer, the blade shorter, and the overall morphology is stunted [15]. The morphology of *D. antarctica* is therefore a qualitative measure of the predominant wave exposure at a particular site.

The medulla of the blade of *D. antarctica* consists of gas-filled sacs [14], which make the whole blade positively buoyant (Figure 3.2C). At low tide, the photosynthetically active area can therefore be maximized as the blade floats at the surface while minimizing self-shading [18]. The thickness of the medulla is not uniform but is dependent on a variety of factors such as wave exposure, age, and overall morphology (C. Hurd, unpublished data). The thallus of *D. antarctica* can consequently be very voluminous at a comparatively low weight.

The congeneric species *D. willana* is endemic to New Zealand. In general, the stipe is larger and stiffer and bears lateral secondary blades of smaller size in addition to the apical main blade [19]. If the main blade is lost as a result of failure, one of the lateral blades can increase in size considerably. *Durvillaea willana* commonly form a belt in the intertidal–subtidal zone just below the belt of *D. antarctica.* Sometimes stands of the two *Durvillaea* species will be mixed. The ecological range of *D. willana*, however, seems to be more restricted than for *D. antarctica*, because this species is absent at sites of very severe wave exposure and also at sites of moderate wave exposure, where populations of *D. antarctica* can still exist.

The main morphological–anatomical difference between the two species of *Durvillaea* is the makeup of the blade. With *D. antarctica*, the blade is positively buoyant and has the tendency to float at the water's surface (Figure 3.2A). Unlike many other seaweeds, e.g., *Macrocystis pyrifera* or *Ascophyllum nodosum*, the entire medulla of the blade of *D. antarctica* is gas-filled rather than only the pneumatocysts. The blade of *D. willana* lacks the honeycomb-shaped, gas-filled sacks of the medulla. As a consequence, the blade of *D. willana* is neutrally buoyant and floats upright in the water column if no wave action or currents are present (Figure 3.2B) and is generally not as bulky as the blade of *D. antarctica*. A difference in the way these two species react to flow-induced loading can therefore be expected.

3.1.3 Drag and Streamlining

Commonly, drag is determined by [20]:

$$F_d = \tfrac{1}{2}\rho A_c C_d u_r^2 \tag{3.1}$$

where

F_d	= drag force (N)
ρ	= density of the fluid (kg m^{-3})
A_c	= characteristic area of the drag-producing body [m^2]
C_d	= drag coefficient
u_r	= fluid's velocity relative to an object [m s^{-1}] (cf. Figure 3.3)

With flexible organisms, it is commonly observed that the drag coefficient is not constant but changes with increasing velocity as the body reconfigures itself [10,21,22]. Consequently, comparisons between different individuals or different species often are restricted to a certain velocity [6,11]. Additionally, a constant drag coefficient typically does not yield the expected increase of drag with the velocity squared [23]. The process of reconfiguration, which leads to a lower increase of drag than would be expected, is described by Vogel [24,25]. The deviation from a second-power relation between drag and velocity is maintained by the introduction of a "figure of merit" as an addend in the power function. Since the shape is not constant, a more general shape factor can be introduced, leading to the following extended equation for drag [6]:

$$F_d = \tfrac{1}{2}\rho A_c S_d u_r^{2+B} \tag{3.2}$$

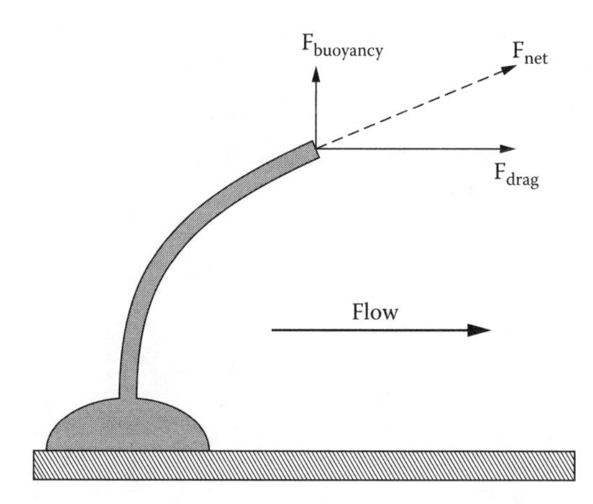

FIGURE 3.3 A simple model of the resulting net force on a seaweed stipe if force due to drag and buoyancy are superimposed.

where S_d is the shape coefficient and B is the figure of merit. For clarity and simplicity, Gaylord et al. [10] have introduced the term "Vogel number" for this figure of merit, which is used henceforth in this study.

The more negative the Vogel number, the lower is the increase in drag with increasing velocity. It is therefore a means of quantifying the effect of reconfiguration.

3.1.4 Objectives

The aim of this study was to examine how *Durvillaea* spp. are adapted to the surf zone with its various degrees of wave exposure. This was mainly done by measuring drag forces on entire thalli in a flume and in the field and by quantifying the process of reconfiguration of the blade. Accompanying tests yielded information on the buoyancy, acceleration, and the way different forces act together in *D. antarctica* and *D. willana*. These findings were then related to a field survey of several morphological parameters.

3.2 MATERIAL AND METHODS

3.2.1 Tested Seaweeds

For flume experiments, a total of eight individual specimens of *D. antarctica* and two individual specimens of *D. willana* were haphazardly collected from Brighton Beach, New Zealand (46° S, 170° E), during low tide on June 25, 2002 and July 26, 2002. They were transported to a nearby laboratory in Dunedin, New Zealand, and tested within 24 hr. Prior to the tests in a flume, the morphometrical parameters of length, mass, volume, and planform area of the blade of the harvested seaweeds were recorded (Table 3.1). The overall length was measured with a tape measure to the nearest centimeter. The mass was measured to the nearest 0.1 kg by placing the seaweeds in a basket and attaching a spring balance. The volume was determined by immersing the seaweeds in a barrel of seawater and weighing the displaced

TABLE 3.1
Morphometrical Data of the Eight Individuals of *Durvillaea antarctica* (Specimens I to VIII) and the Two Individuals of *D. willana* (Specimens IX and X) Tested in the Flume

Individual	Morphology	Length (m)	Area (m^2)	Mass (kg)	Volume (10^{-3} m^3)
I	Exposed	4.97	1.70	23.5	38.0
II	Exposed	7.15	1.52	17.5	38.0
III	Exposed	3.10	1.25	8.5	22.5
IV	Exposed	4.25	1.65	22.0	23.3
V	Intermediate	7.49	2.52	51.0	92.8
VI	Intermediate	2.18	1.07	2.5	7.0
VII	Intermediate	3.80	1.26	9.0	13.5
VIII	Sheltered	6.40	1.46	15.5	27.0
IX	Intermediate	6.03	1.55	18.5	17.8
X	Intermediate	0.35	1.01	0.4	0.4

amount of water to the closest 0.1 kg. The mass of the displaced water was then divided by the density of seawater (1024 kg m^{-3}), giving the volume. The planform area of the seaweeds was determined by photographing the fully extended blade. Because there was no suitable point of elevation for taking an orthographic image from exactly above the spread out individuals, photographs were taken at an angle. The images were then photogrammatically rectified with a vector-based program routine (MatLab version 12, The Mathworks) to account for and correct the distortions introduced by photographing at an angle. Subsequently, the planform area was analyzed with an image analysis program (Optimas version 6.5, Media Cybernetics). The recorded morphometrical parameters were then correlated with the drag forces on the seaweeds.

3.2.2 Drag Forces

Drag forces were tested in a flume at the Human Performance Centre, Dunedin. The dimensions of the flume — length, width, and depth — were 10, 2.5, and 1.4 m, respectively. The tests were conducted at flow velocities of 0.5, 1.0, 2.0, and 2.8 m s^{-1}, the latter being the maximum velocity of the flume. The forces and concurrent flow velocities of each test run were logged by an online data recorder for at least 2 min at a logging frequency of 10 Hz. To see if high-frequency events occurred, three individuals were logged at a frequency of 1000 Hz. As the flume at the "Human Performance Centre" could not be run with highly corrosive sea water, the drag tests were conducted in freshwater. Since *Durvillaea* is an intertidal seaweed and frequently experiences rain water, a temporary exposure to freshwater of 10–15 minutes was not considered to change the seaweed's mechanical performance, and no obvious signs of changes in appearance were observed.

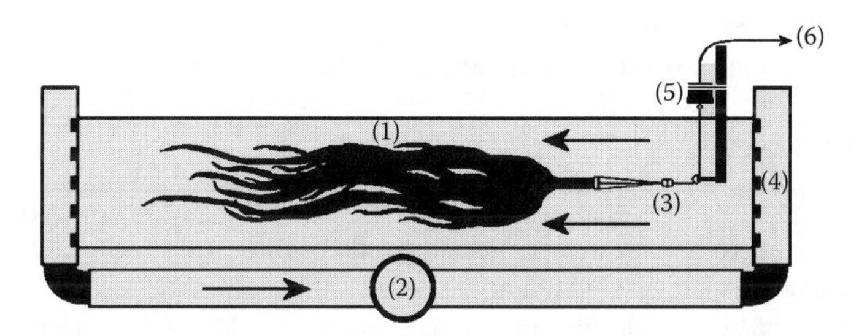

FIGURE 3.4 Schematic drawing of the experimental setup of the flume experiments: (1) test specimen of *Durvillaea antarctica*, (2) pump, (3) attachment, (4) homogenizer, (5) force transducer, and (6) connection to online PC. Not to scale. For details, see text.

Prior to testing, the seaweeds were cut just above the holdfast and prepared for testing as shown in Figure 3.4. The stipe was fastened with a hose clamp (also called "jubilee clip"), which was fixed to a swivel by four pieces of low-strain yachting rope of 4 mm diameter. The swivel was connected to another piece of low-strain rope, which was redirected via a pulley and attached to a force transducer (RDP Group, Model 41, maximum load 250 lb) outside the water. The pulley was screwed to a wing spar, which had only a small influence on the flow in the flume and was therefore considered negligible.

3.2.3 Shortening Experiments

To test the importance of the overall shape and the length on drag and reconfiguration, shortening experiments were conducted. Two individuals of an intermediate morphology were tested at a velocity of 2.0 m s^{-1}. The blades had initial lengths of L_{IV} = 4.25 m (specimen IV; Table 3.1) and L_{VIII} = 3.80 m (specimen VII; Table 3.1) and were then both shortened twice by cutting off 1 m from the distal end and tested again. By cutting of the ends of the blades, the stream-optimized shapes of the kelp were disturbed. The resulting flow-induced forces on the kelp can be expected to reflect the changes in size but also in shape.

3.2.4 Drag Coefficients and Reconfiguration

Based on the overall morphology, the eight individuals of *D. antarctica* were grouped as "wave exposed" or "intermediate/wave sheltered." Drag coefficients were calculated using Equation 3.1, and the planform area of the seaweeds was used for A_c, which is common for long flexible organisms, rather than the projected area [11]. The process of passive reconfiguration was examined by the Vogel number. Considering the factor $\frac{1}{2}\rho A_c S_d$ of Equation 3.2 as constant gives the following proportionality:

$$F_d \sim u_r^{2+B} \tag{3.3}$$

The Vogel number, B, can therefore be written as the slope of a double-logarithmic plot of the velocity-specific drag $[\log(F_d / u_r^2)]$ as a function of velocity $[\log(u_r)]$. The greater the absolute value of the negative slope, the better the reconfiguration process was considered to be. The Vogel number was subsequently correlated with the previously recorded morphometrical parameters.

3.2.5 Buoyancy

To measure the buoyancy forces generated by the gas-filled medulla of *D. antarctica*, 10 individuals were haphazardly collected from Brighton Beach on July 23, 2002. All measurements were carried out at the beach so that all replicates were fresh and weight reduction due to desiccation effects could be ruled out. To test the forces exerted by the buoyancy of the blades, thalli cut at the stipe were submerged by placing a neutrally buoyant plastic mesh container upside down over the kelp in a seawater-filled barrel. The force necessary to keep the container with the kelp at water level was measured with a spring scale attached to a metal rod, which was used to push the container with the kelp down, and taken as the buoyancy of the tested individual. To analyze the correlation of exerted buoyancy forces with morphometrical parameters, the overall length, planform area, and fresh weight of the tested kelp were also determined.

3.2.6 Field Studies

Because of their morphological differences, the mechanical behavior *in situ* of *D. antarctica* and *D. willana* can be expected to differ. The effect of the buoyancy of the blade can be gauged by examining the simultaneous response of *D. antarctica* and *D. willana* to waves. Field experiments studying *D. antarctica* and *D. willana* under natural conditions were conducted at St. Clair, a suburban beach near Dunedin, during the period January 18 to 28, 2000 [26]. The sampling all took place at St. Clair seawall. This site is characterized by a rocky shoaling platform backed by a seawall. The beach boulders were in the range of 0.2 to 0.6 m in diameter. It is not directly exposed to open ocean surf, and waves occasionally broke directly in this region; more often, the waves broke slightly offshore and then would rush in as a bore. A local *D. antarctica* population was located some 10 m offshore from the site of the experiments, whereas *D. willana* did not occur there.

Samples of *D. antarctica* and *D. willana* of intermediate morphology were taken from Lawyers Head, a rocky outcrop about 3 km away, using a chisel to remove the thalli from the substratum. The harvested individuals were then mounted in small concrete blocks, which were then attached to a region of flat substratum using eight self-fastening metal bolts (dynabolts) and four webbing belts with ratchet locks. Equipment used included three-dimensional accelerometry (Figure 3.5) and wave gauges (see [26] for methodological details). The tidal range during the experiments was 2 m.

The accelerometers were calibrated before and after each experiment. This was necessary because the long cables (greater than 40 m) affected nominal factory calibration. The wave gauge data can only be considered representative of wave height, and the arrival time of the waves depended on the relative position to the plants. The wave gauge was guyed to dynabolts to hold it securely in position. The wave gauge data were logged using a Tattletale® logger (Onset Computer Corporation) running at 32 Hz.

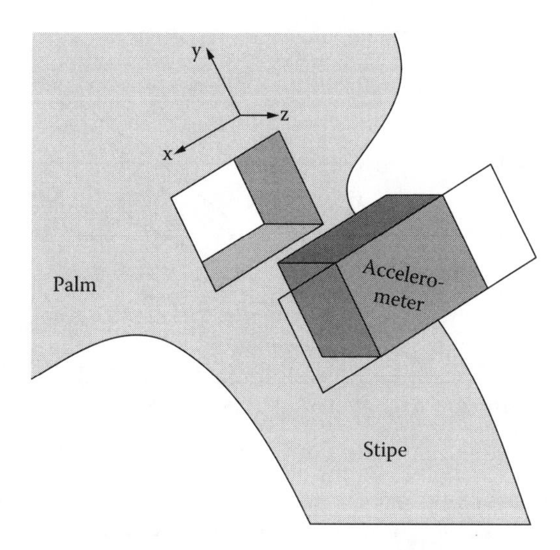

FIGURE 3.5 A three-dimensional accelerometer was mounted within a cut section in the palm of the *Durvillaea* blade. A second accelerometer was attached at the distal end of the lamina.

3.2.7 Morphological Survey

To compare the morphology of individuals of *D. antarctica* and *D. willana* with different degrees of wave exposure, we conducted a field survey at St. Kilda Beach, a suburban beach near Dunedin, in February 1999. Quadrats (1 × 1 m) were randomly placed within stands of kelp of both *D. antarctica* and *D. willana*. The wave exposure typical of any particular quadrat was qualitatively determined by the predominant blade morphology of the kelp growing within the quadrat. Thus, individuals of both species were categorized as either "wave exposed" or "wave sheltered." Factored by species and wave exposure, four random quadrats were used to sample each of the four groups, giving a total of 16 quadrats (*D. antarctica*: sheltered/exposed; *D. willana*: sheltered/exposed). All individuals of either *Durvillaea* species growing within a quadrat were harvested and four morphological parameters were recorded. Measurements of the blade length, stipe length, and maximum stipe diameter were used to examine possible correlations between these three morphological parameters and the species or wave exposure as indicated by the forth parameter, blade morphology.

3.2.8 Statistical Analysis

Statistical tests were performed with SPSS, version 12.0, and SigmaPlot, SPSS, version 8. Differences between two groups were determined by Welch's *t*-test, adapted to unequal variances. Statistical tests were considered significant at a level of $p < 0.05$. The results are either presented with ±0.1 standard deviation (SD) or the 95% confidence interval (CI) as indicated. Results of correlation tests are presented with Pearson's adjusted R^2.

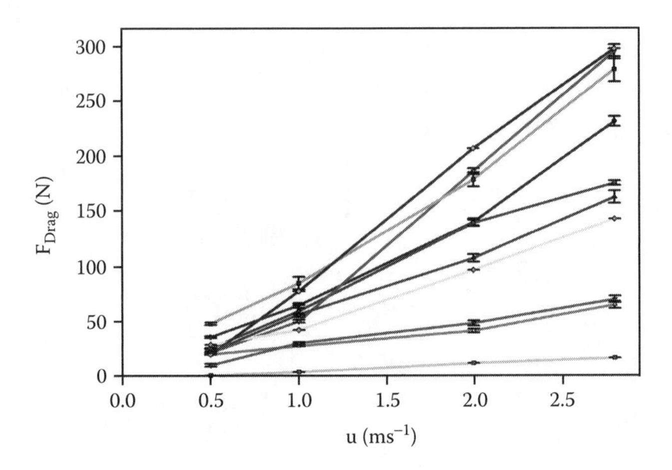

FIGURE 3.6 The relation between force and velocity for the eight individuals of *Durvillaea antarctica* tested in a flume. F_{Drag} is the drag force, and u is the velocity. Error bars indicate standard deviations of 60 s of data, recorded at 10 Hz (i.e., 600 data points).

3.3 RESULTS

3.3.1 DRAG FORCES

In general, the drag increased with increasing velocity (Figure 3.6). The variation in data also increased with increasing velocity. No transient drag peaks were observed at the higher recording frequency of 1000 Hz, and so the lower recording frequency of 10 Hz was sufficient for capturing all relevant velocity-dependent changes in drag forces. The highest recorded forces during the flume tests were almost 300 N for the two largest individuals (i.e., individuals II and V in Table 3.1). The increase, however, often deviated from the second power of the velocity as predicted by the standard equation for drag (Equation 3.1) and was nearly linear.

Correlation tests of drag and the four measured morphometrical parameters for flume specimens yielded only low correlation coefficients (Figure 3.7). The best correlation with drag was found with length ($R^2_{length} = 0.63$). Planform area and mass both showed a slightly lower correlation with drag ($R^2_{area} = 0.58$ and $R^2_{mass} = 0.58$), whereas only a poor correlation was found between drag and volume ($R^2_{volume} = 0.36$). The correlations, however, improved considerably by taking the wave-dependent morphology as an additional independent variable into account so that the combined information on length and wave-dependent morphology of individuals (exposed or intermediate/sheltered) gave the best correlation with the measured drag forces ($R^2_{length\ +\ wave\ exposure} = 0.71$).

3.3.2 SHORTENING EXPERIMENTS

The shortening experiments for *D. antarctica* in the flume yielded a nonlinear relation between drag and each of the four measured morphometrical parameters (Figure 3.8). A linear reduction in length caused a reduction in drag that was less

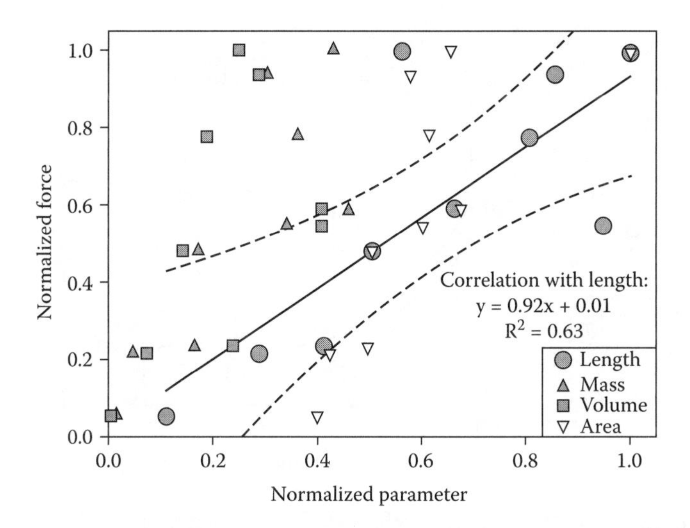

FIGURE 3.7 There was a significant correlation between drag and length ($R^2 = 0.63$), whereas only weak or nonsignificant correlations were found between drag and blade mass, volume, or area. The morphometric parameters and the forces were normalized by the values for the largest individual, which was also the heaviest and most voluminous one of the test sample. The regression is only for the normalized length data, while the dashed lines represent the 95% CI.

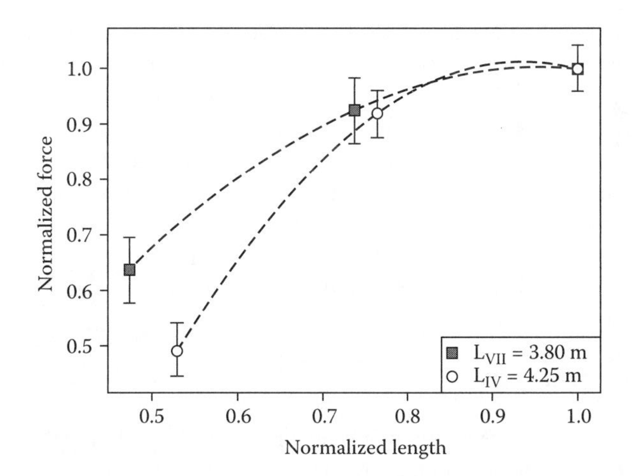

FIGURE 3.8 Shortening experiment with two individual *D. antarctica* (specimens VII and IV; Table 3.1) tested at a velocity of 2.0 m s^{-1}. The nonlinear trend between drag and velocity indicates that a simple cut prevents the thallus body from reconfiguring into a more streamlined shape. The parameters are normalized to the maximum forces during the individual test runs and the individual lengths for ease of comparison. Error bars indicate standard deviations of 60 s of data, recorded at 10 Hz (i.e., 600 data points).

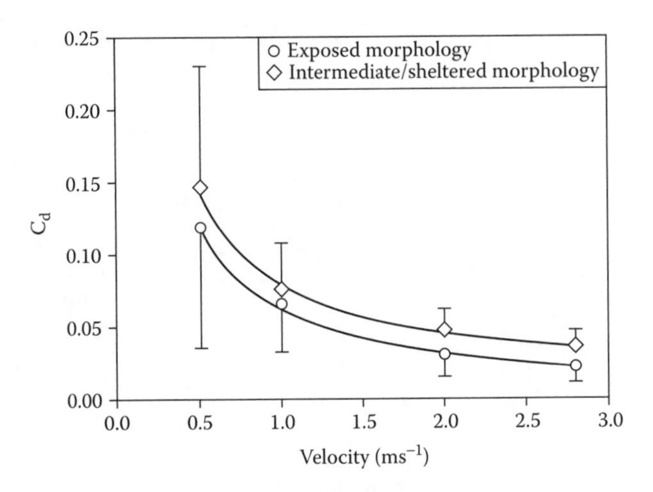

FIGURE 3.9 The change of the drag coefficient, C_d, with increasing velocity of the fluid (u) for *Durvillaea antarctica*, grouped by wave-exposed and intermediate/wave-sheltered morphology. Error bars indicate one standard deviation.

than would be predicted if the relation between a morphological parameter and drag was linear (see previous paragraph) or squared (Equation 3.1).

3.3.3 Drag Coefficients and Reconfiguration

Drag coefficients of the tested seaweeds were highly dependent on velocity (Figure 3.9). The mean of the drag coefficients decreased hyperbolically with increasing velocity, using Equation 3.1 and keeping A_c constant. At all tested velocities, the mean drag coefficients of the group with wave-exposed morphology were always lower than the ones of the group with intermediate/wave-sheltered morphology. The minimum mean drag coefficient was found for the wave-exposed group at a velocity of 2.8 m s^{-1} at $C_d = 0.023$. The maximum mean drag coefficient was recorded for the intermediate/wave-sheltered group at a velocity of 0.5 m s^{-1} at $C_d = 0.147$. The variation in C_d expressed by the standard deviation decreased for both groups with increasing velocity.

3.3.4 Vogel Number

The efficiency of passive reconfiguration processes of individual seaweeds was characterized by the Vogel number. All tested individuals exhibited an increase in drag with increasing velocity that was less than could be expected from Equation 3.1. Vogel numbers ranged from a maximum of $B = -0.25$ to a minimum of $B = -1.21$, with an average of $B = -0.86 \pm 0.31$ (mean ± 1 SD). Grouped by morphology, the wave-exposed individuals averaged a lower Vogel number than the intermediate/wave-sheltered individuals ($B = -1.08 \pm 0.15$ and $B = -0.65 \pm 0.28$, respectively), i.e., the wave-exposed individuals performed with a significantly more efficient mode of streamlining (Welch's t-test, $p < 0.05$).

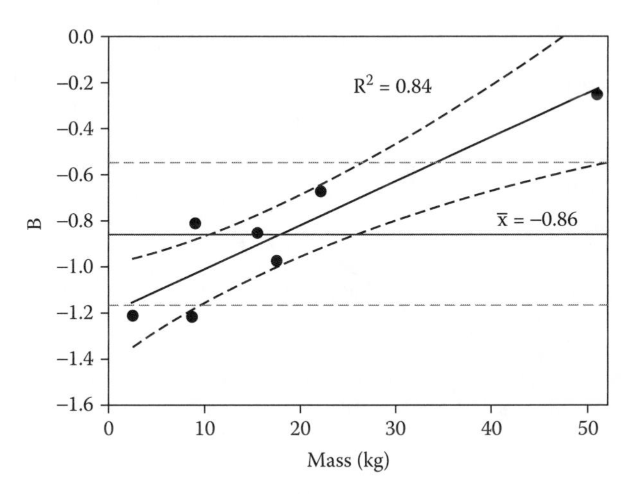

FIGURE 3.10 Correlation of the Vogel number, *B*, with the mass of the tested seaweeds. The black oblique line indicates the regression, while the dashed black lines are the 95% CI for the linear regression. The gray horizontal line indicates the mean at $B = 0.86$ for all tested individuals, while the dashed gray horizontal lines represent the 95% CI.

The Vogel number could be correlated best with the mass of an individual ($R^2_{mass} = 0.84$) (Figure 3.10). The other tested morphometrical parameters — volume, length, and planform area — yielded lower correlation coefficients ($R^2_{volume} = 0.56$, $R^2_{length} = 0.32$, and $R^2_{area} = 0.77$). The correlation could be improved by considering the overall morphology as an additional independent variable (e.g., $R^2_{mass + wave\ exposure} = 0.97$).

3.3.5 Buoyancy

For the sample of 10 individuals of *D. antarctica*, the highest recorded force due to buoyancy was about 150 N. Buoyancy could be correlated best with blade mass ($R^2_{mass} = 0.94$) (Figure 3.11) and correlated well with blade area ($R^2_{area} = 0.79$), whereas the blade length yielded a low correlation coefficient ($R^2_{length} = 0.09$).

3.3.6 Field Studies

Accelerometers provided a proxy for frond motion [26]. Time series of water elevation data and accelerometer data (Figure 3.12) showed marked differences. The forcing of the tested individuals by the waves was obviously complex (Figure 3.12a). Despite the wave field being visually well ordered with contiguous crests arriving at regular intervals, the direct observations revealed a complicated distribution of peaks and troughs. There were substantial differences in response of the two species to waves as well as differences in how the regions of the frond reacted. In particular, *D. antarctica* (Figure 3.12, samples 1 and 2) exhibited far sharper "shock-like" responses, lasting only a fraction of a second (sampling at 10 Hz) than those exhibited by *D. willana* (Figure 3.12, samples 3 and 4). The *D. antarctica* response was not particularly different on the palm compared with the blade. In contrast, the accelerometers attached to *D. willana* on the palm (Figure 3.12, samples 3) and blade

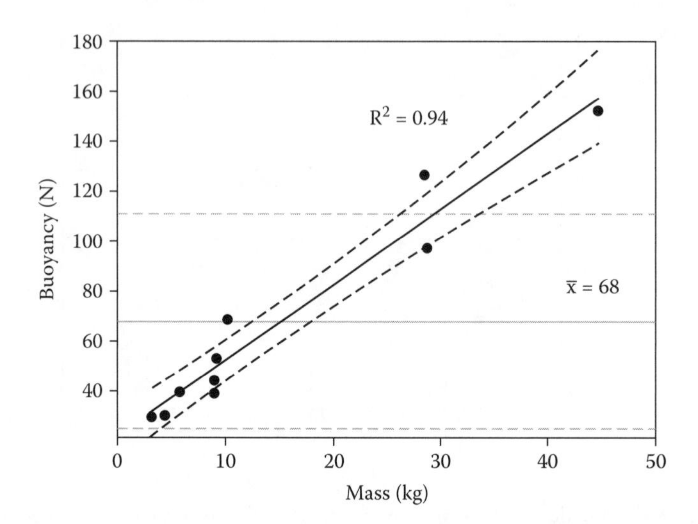

FIGURE 3.11 Correlation between the mass and buoyancy for 10 individuals of *D. antarctica*. The black oblique line indicates the regression, while the black dashed lines are the 95% CI for the linear regression. The gray horizontal line indicates the mean for all tested individuals, while the dashed horizontal lines represent the 95% CI.

(Figure 3.12, samples 4) showed a slower response than *D. antarctica* and differed from one another. The palm response of the *D. willana* sample could only weakly be visually correlated with the wave gauge record.

3.3.7 Morphological Survey

For *D. antarctica* (N = 131) from wave-exposed (N = 76) and wave-sheltered sites (N = 55), the blade length was weakly correlated with stipe diameter (Figure 3.13), and no correlation was found between blade length and stipe length (Figure 3.14). For *D. willana* (N = 102) from wave-exposed (N = 38) and wave-sheltered sites (N = 64), there were clear correlations between blade length and both stipe diameter (Figure 3.13) and stipe length (Figure 3.14).

3.4 DISCUSSION

3.4.1 Drag Forces

The way drag acts on an intertidal seaweed like *Durvillaea* changes as the algal body is moved by a wave. A passing wave agitates water particles in a circular orbital motion. As the water depth decreases near shore, the circular motion becomes more and more elliptical [23]. In shallow water, waves can thus cause a simple back and forth swaying of seaweeds. This swaying can be described for both *Durvillaea* species by two extreme states and a series of intermediate ("midsway") positions.

1. The algal body is stretched normal to the oncoming wave and against the direction of the upstream flow. The characteristic area is then the projected

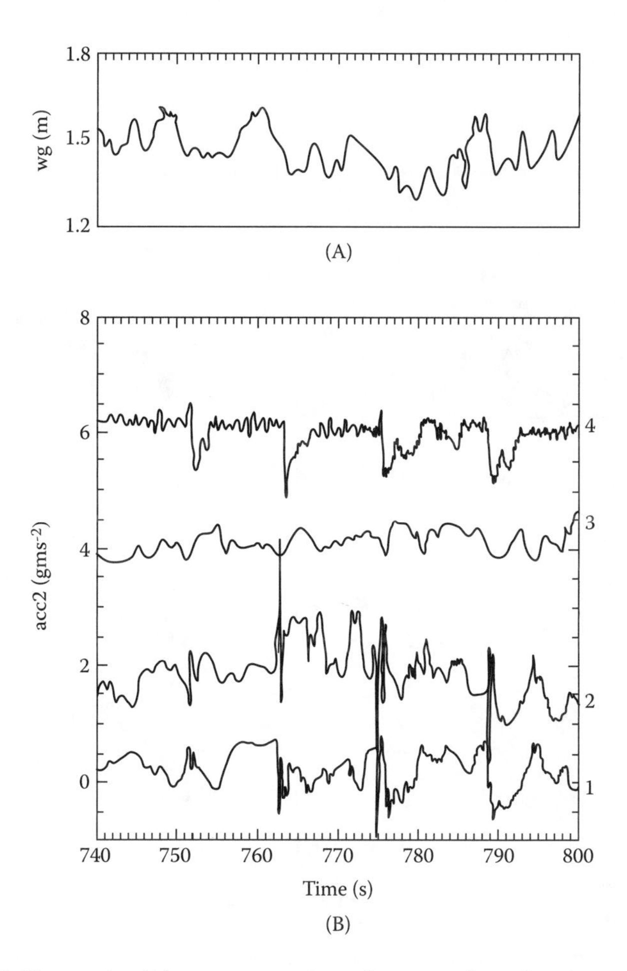

FIGURE 3.12 Time series (A) wave gauge data of water surface elevation and (B) along-blade accelerations offset by 1 unit from their mean value. Accelerometers 1 (palm) and 2 (blade) are from the *D. antarctica* sample and accelerometers 3 (palm) and 4 (blade) are from the *D. willana* sample. The water surface elevation is contaminated with foam during extreme nearby breaking events (e.g., 772 s). In addition, the wave gauge and samples were not exactly colocated in space, and the fronds themselves have a finite scale so that the peaks in elevation do not exactly correspond with the accelerometer responses.

area. Mainly pressure drag is acting on the seaweed, pushing its body downstream. As *Durvillaea* is reacting to the drag, it starts to reconfigure its blade relative to the flow, exposing more and more area normal to the flow direction.

2. Midway between the extremes, the algal body is comparatively upright in the water column. The main area of the blade is thus normal to the flow. Subsequently, pressure drag is acting on a large surface. Although drag could be expected to be high, the resulting tensional forces will be low as long as the blade can deflect further ("going with the flow") [27].

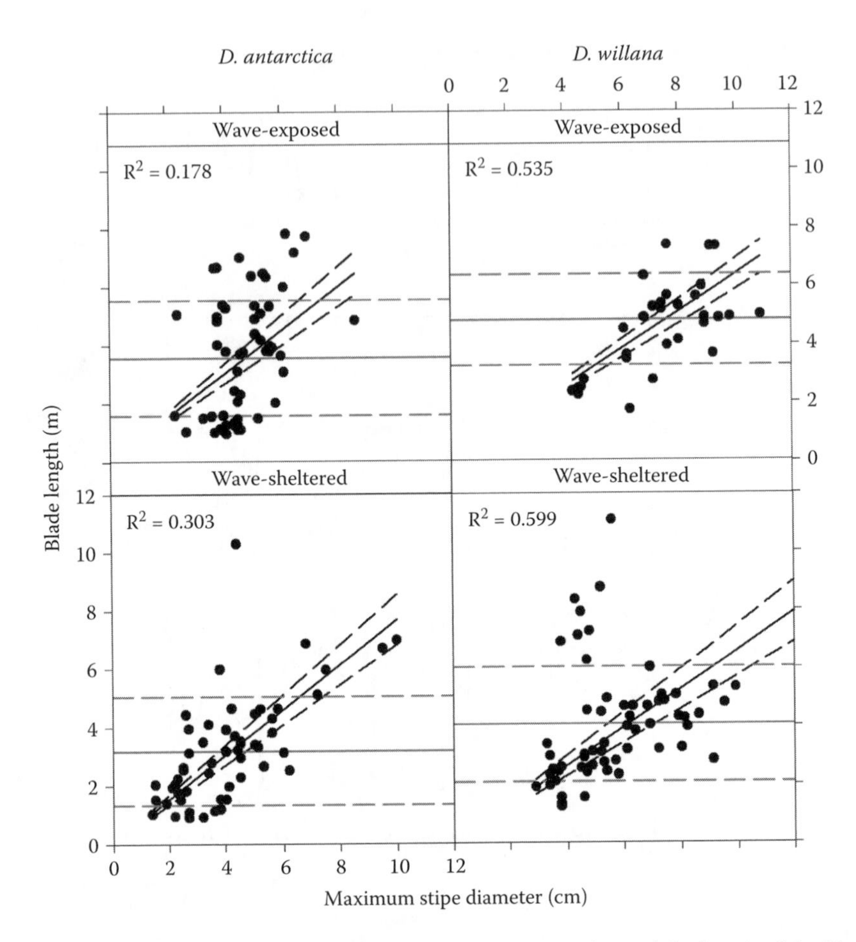

FIGURE 3.13 Lattice plot of the maximum diameter of the stipe and the length of the blade for *D. antarctica* and *D. willana*, grouped by species and wave exposure. Dashed lines represent the 95% CI.

3. The algal body is outstretched along the direction of the flow. The area normal to the flow is comparatively small again (projected area) so that the pressure drag is low. The friction drag, however, is comparatively high as water is pushed along the surface of the thongs of the blade so that the planform area is the relevant factor. The overall drag will often be less in this position than in the midsway position (2), however, because the blade cannot deflect any further, the reactive tensional forces will increase.

Because the shape of an individual *Durvillaea* changes and because the drag acting on the body changes, a constant drag coefficient cannot be expected. Although the real situation in the intertidal zone is more complex, in particular with breaking waves, these assumptions seem sound as a first-order approximation, and the morphological parameters, which can most reliably predict drag, need to be identified.

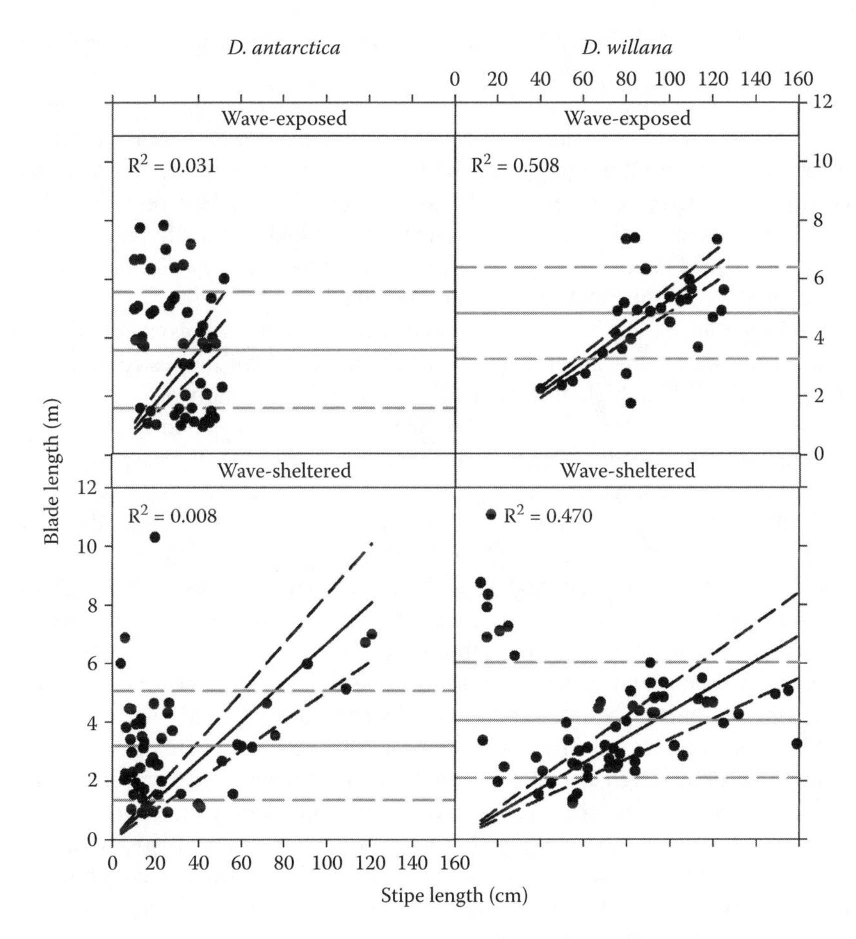

FIGURE 3.14 Lattice plot of the length of the stipe and the length of the blade for *D. antarctica* and *D. willana*, grouped by species and wave exposure. Dashed lines represent the 95% CI.

The maximum forces obtained from drag tests in a flume were comparable to forces recorded during field experiments with transplants of *Durvillaea* at 300 N [26]. Based on Equation 3.1, area is expected to correlate well with drag. Unfortunately, the common approach of taking the projected area is of little use if the study organisms change their shapes by reconfiguring into streamlined bundles with increasing flow velocities. Therefore, the planform area is often taken instead as the characteristic area, but this does not account for undulations or corrugations typical of many seaweeds [28,29]. A third approach is to take the wetted area as the characteristic area [6]. In the case of *Durvillaea*, however, the determination of the wetted area is extremely difficult for individuals with wave-exposed morphologies because the blade is reconfigured in such a way that parts of the surface areas of neighboring thongs will be in close contact on the "inside" of the streamlined body. The contact area will subsequently not act as a friction surface for the surrounding medium, and drag forces will be lower than could be expected from correlations with the wetted area.

The highest correlation for *D. antarctica* between drag and morphological parameters was with blade length combined with information about the type of wave-dependent overall morphology. The highest correlation of the Vogel number was with mass. It seems therefore justified to expect that the best predictor of the way a given individual will behave in different flow regimes and at different velocities will be a combined factor of length, mass, and overall morphology. The factor would be an expression of the "bulkiness" of an individual.

The importance of an adapted morphology and its bulkiness is clearly demonstrated by the shortening experiments. The original shapes of the tested specimens were well adapted to high flow velocities. The experimentally shortened blade prevented a reconfiguration into a streamlined bundle and resulted in drag forces that were higher than could be expected from the results of the flume tests on intact individuals. The importance of the bulkiness factor is further supported by the two outliers in Figure 3.7. The individual above the 95% CI was previously damaged, probably in a storm. The shortened blade was very "bulky" with many thongs, resulting in disproportionately high drag. The other outlier below the 95% CI also had fractured tips. Although missing large parts of its blade, a few thongs had remained unbroken. The blade was subsequently very long, whereas the bulkiness was very low. These two outliers demonstrate the shortcomings of simple morphometrical measurements for correlations with drag because a similar line of argument could be applied to area or mass as morphologically relevant parameters.

Although intuitively easy to understand, the quantification of such a bulkiness factor is complex. Approaches that involve the determination of the "branchiness" of a lamina, similar to a fractal analysis used in computer models of algae, seem plausible but not very practical for seaweeds of the size of *Durvillaea*. Other and simpler methods still need to be developed to describe the intricate shape variations of these seaweeds to predict drag based on morphology.

3.4.2 Drag Coefficients, Reconfiguration, and the Vogel Number

The decrease in the standard deviations of the drag coefficients as the velocity increased, as found in this study, was probably due to the multifactorial optimization of the seaweed blade with respect to physiological and mechanical boundary conditions [30]. The variability of blade shapes is high at low velocities, which are mechanically harmless. Under these conditions, the shape of a blade can be optimized with respect to other, nonmechanical requirements, e.g., light interception or nutrient uptake [31,32]. At higher velocities, the different shapes all reconfigure into streamlined bundles with similar overall shapes. This seems to be the case for seaweeds with highly variable morphotypes such as *Durvillaea* and also for a whole range of flexible seaweeds in general [33].

The Vogel numbers found for *Durvillaea* are similar to those reported from other studies on flexible seaweeds [11]. Because the mean was close to $B = -1$, an almost linear increase of drag with velocity can thus be explained. The least negative value

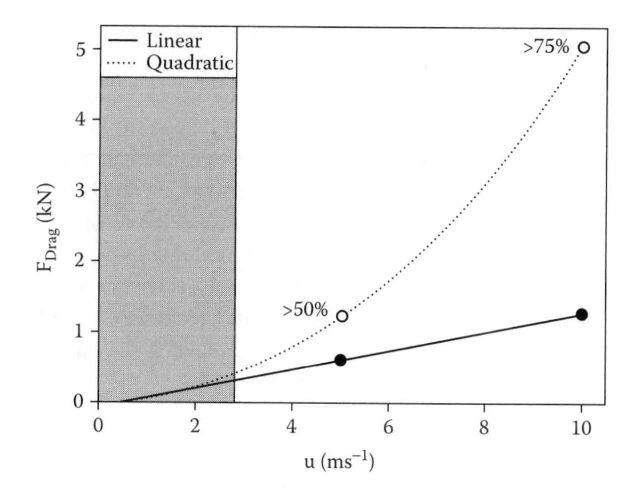

FIGURE 3.15 Hypothetical drag forces, assuming a linear increase and a quadratic increase with velocity, respectively. For the individual represented in the graph, the reduction in drag due to streamlining and subsequent linear (rather than a quadratic) increase is about 52% at a velocity of 5 m s^{-1} and 75% at 10 m s^{-1}. The experimental range is shaded in gray. F is the drag force, and u is the velocity of the fluid.

was found for the same individual that was an outlier below the 95% CI in the correlation of drag and length (Figure 3.7). Reduced in bulkiness, the shape of the blade could hardly be further optimized, making it comparable to a rope. The most effective reconfiguration process and subsequent reduction in actual drag compared with the drag predicted by Equation 3.1 will be achieved for a limited range of aspect ratios. The most negative Vogel number was found for an individual with wave-exposed morphology and a very massive blade with no apparent damage, the second outlier above the 95% CI (Figure 3.7). This type of morphology seems to be the optimized morphology for reconfiguration under very unsteady, rapidly changing flow conditions.

Assuming a Vogel number of $B = -1$, the theoretical reduction in drag due to reconfiguration can be calculated. The result of this simple extrapolation can be seen in Figure 3.15. At a velocity of 5 m s^{-1}, the reduction in drag due to reconfiguration is already about 50%. At a velocity of 10 m s^{-1} — common in stormy conditions at exposed sites — the reduction is even more than 75%. It is noteworthy that similar findings have been reported for terrestrial plants, e.g., the giant reed *Arundo donax* [22,34]. Reconfiguration is therefore an effective general process of plants for adapting on a small temporal scale to variable flow conditions, which does not require any further mechanical changes at the "material" level of the organism.

How can *Durvillaea* grow to a size an order of magnitude larger than other intertidal seaweeds at that position on the shore given that its biomechanical properties are rather typical for a large spectrum of seaweeds [35]. The answer may lie in the way *Durvillaea* grows. Unlike other members of the Fucales, *Durvillaea* lacks an apical meristem but has diffuse growth. This allows two mechanisms to interact. First, broken tips can regrow, regardless of previous damage. Second, a strip of a blade fractured longitudinally at its distal end can change its growth form. The tip of the crack will become blunt to avoid crack propagation [36]. The strip of blade will then have two termini, which will both be able to grow in length. The tip of the crack, however, remains a permanent point of separation of the two newly generated termini, forming two only loosely connected mechanical subunits that are differentially agitated by wave action. It is thus possible for *Durvillaea* to change its morphology as it grows, adapting to the ambient wave exposure. The potentially endangered large unit blade is divided into many smaller subunits, possibly reducing the physiological efficiency of the photosynthetically active area, but more importantly, reducing the risk of total blade loss. The indeterminate morphology of *Durvillaea*, therefore, seems to be a key factor for the successful establishment of this seaweed in the wave-swept intertidal environment.

3.4.3 Buoyancy and Field Studies

In *D. antarctica*, the whole blade is buoyant, whereas many other large brown seaweeds have only distinct floating organs, e.g., pneumatocysts. The recorded buoyancy forces of up to 150 N are high, e.g., 15 times higher than the buoyancy forces recorded for an 8 m long individual of *Nereocystis luetkeana* [37]. It can therefore be expected that the buoyancy of the blade of *D. antarctica* has a considerable influence on the overall mechanical behavior of this species in the surf zone.

The finding that the acceleration response of *D. antarctica* was not particularly different on the palm compared with the frond was initially surprising as one might expect the blades to respond more strongly because buoyancy holds them up high during the passage of breaking waves. This suggests that the drag force imparted to the blades was transferred along the stipe and made itself apparent even near the holdfast. Contemporary load cell data quantifying the actual load transferred to the substrate connection supports this [26].

The frond response of the *D. willana* sample was probably the best track of the water elevation, clearly marking the passage of the wave crest and the gradual decay (recall that the wave gauge elevation did not exactly reflect the velocity of a passing wave). It is unclear why the *D. willana* structure did not pass the response in accelerometer on to the stipe. A possible explanation exists in the load driving realignment of the frond that was not constrained by buoyancy.

An alternate viewpoint for the same data is provided by examining the domain of along-blade vs. across-blade accelerations (Figure 3.16). Stevens et al. [26] studied all three axes combinations (x–y, x–z, and y–z); however, here we consider only x–y for simplicity. Clearly, the *D. willana* stipe palm is more constrained than the *D. antarctica* stipe (Figure 3.16C vs. 3.16A). The scatter in the data points of *D. antarctica* indicates that the blade has many degrees of freedom to attain a certain position in the water column (Figure 3.16B). In contrast, the acceleration data of *D. willana* shows less scatter (Figure 3.16D). This might be indicative of a higher constraint of movement of the blade in the water column. Thus, the positive buoyancy of *D. antarctica* increases the "movability" of the blade. This could increase rates of nutrient uptake as well as lower the fatigue strains due to repetitive bending in a preferred direction predefined by the predominant wave direction.

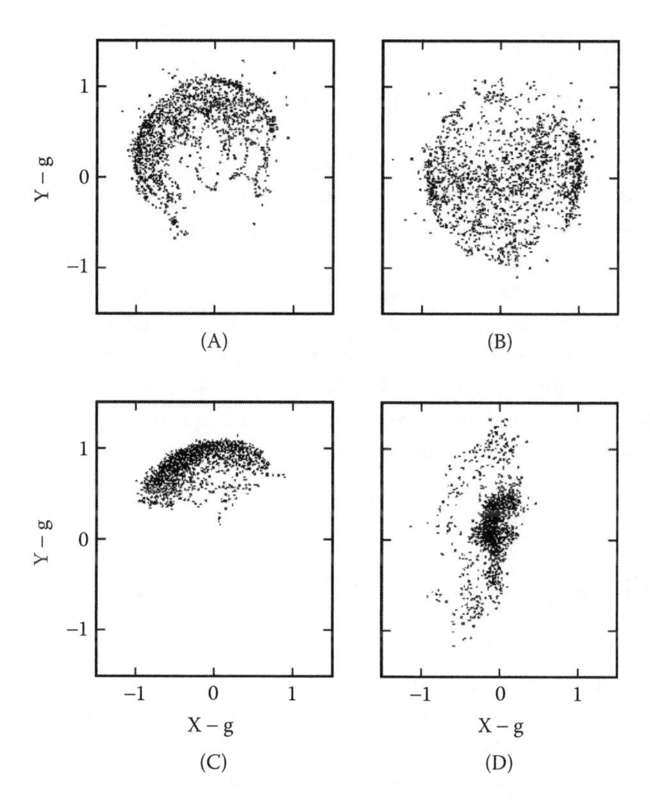

FIGURE 3.16 Comparison of accelerometer response (in units of g, acceleration due to gravity, where 1 g = 9.81 m s^{-2}) for *D. antarctica* (A) palm and (B) frond and *D. willana* (C) palm and (D) frond. The results show along-frond acceleration (X) vs. across-frond acceleration (Y). Around 170 s of data subsectioned from a 2000s time series are shown.

3.4.4 Morphological Survey

The morphological survey yielded only weak correlations between morphological parameters. One reason for this finding seems to be the distribution of data in two overlapping "clouds" as can be seen in particular with the correlation between blade length and stipe length (Figure 3.14, bottom panels). The two overlapping data sets in Figure 3.13 and Figure 3.14 possibly represent older, well-established individuals and young individuals with overproportioned large blades. It can be hypothesized that some of these fast-growing individuals have not yet experienced severe winter storms that have the potential to prune the blades to a size that can be easily correlated to the dimensions of the stipe or, alternatively, will cause the dislodgment of these individuals. As the determination of age is difficult with *Durvillaea*, long-term studies with tagged populations are necessary to establish a clearer view on correlations between typical wave exposure and subsequent morphological adaptations.

3.5 CONCLUSION

This study demonstrates how organisms can adjust to severe physical conditions by predominantly passive processes. Further adaptations on the tissue level or the biochemical composition of cell walls might be important factors for the mechanical fine-tuning of an individual to its habitat [19,22]. However, the main competitive factors, reconfiguration and morphological plasticity, are both directly linked to the indeterminate growth of *Durvillaea*. A main process of rapid adaptation to severe flow conditions is the passive reconfiguration of the flexible blade. In conjunction with the positive buoyancy of the blade, *D. antarctica* seems to possess a very high degree of adaptability to a great variety of flow conditions, which allows this species to occupy a larger range of habitats than *D. willana*. These adaptations act on a short to intermediate time scale (few seconds to one growth season), which is advantageous in a highly unsteady and unpredictable habitat like the intertidal. It can therefore be concluded that seemingly primitive organisms like *Durvillaea* are actually very well-adapted to their habitat by being less specialized.

ACKNOWLEDGMENTS

The authors thank Dave Pease and George Neill, who were in charge of running the flume during the experiments; Rob Daly and Murray Smith, who helped with harvesting and setting up the thalli for testing; and Louise Kregting and José Derraik for field assistance. This study was supported by a Marsden Grant to CLS, a University of Otago scholarship, and a DAAD scholarship to DLH, and a University of Otago research grant to CLH.

REFERENCES

1. Denny, M.W., Life in the maelstrom — the biomechanics of wave-swept rocky shores, *Trends Ecol. Evol.,* 2, 61, 1987.
2. Armstrong, S.L., Mechanical properties of the tissues of the brown alga *Hedophyllum sessile* (C. Ag.) Setchell: variability with habitat, *J. Exp. Mar. Biol. Ecol.,* 114, 143, 1987.
3. Charters, A.C., Neushul, M., and Barilotti, C., The functional morphology of *Eisenia arborea*, *Proc. Int. Seaweed Symp.,* 6, 89, 1969.
4. Delf, E.M., Experiments with the stipes of *Fucus* and *Laminaria*, *J. Exp. Biol.,* 9, 300, 1932.
5. Denny, M.W., Gaylord, B.P., and Cowen, E.A., Flow and flexibility — II — The roles of size and shape in determining wave-forces on the bull kelp *Nereocystis luetkeana*, *J. Exp. Biol.,* 200, 3165, 1997.
6. Gaylord, B. and Denny, M.W., Flow and flexibility — I — Effects of size, shape and stiffness in determining wave-forces on the stipitate kelps *Eisenia arborea* and *Pterygophora californica*, *J. Exp. Biol.,* 200, 3141, 1997.
7. Koehl, M.A.R. and Wainwright, B., Mechanical adaptations of a giant kelp, *Limnol. Oceanogr.,* 22, 1067, 1977.
8. Koehl, M.A.R., When does morphology matter, *Annu. Rev. Ecol. Syst.,* 27, 501, 1996.

9. Denny, M.W., Daniel, T.L., and Koehl, M.A.R., Mechanical limits to size in wave-swept organisms, *Ecol. Monogr.,* 55, 69, 1985.
10. Gaylord, B., Blanchette, C.A., and Denny, M.W., Mechanical consequences of size in wave-swept algae, *Ecol. Monogr.,* 64, 287, 1994.
11. Koehl, M.A.R., Mechanical design and hydrodynamics of blade-like algae, in *3rd Plant Biomechanics Conference*, Spatz, H.C. and Speck, T., Eds., Georg Thieme Verlag, Stuttgart, 299, 2000.
12. Koehl, M.A.R., How do benthic organisms withstand moving water? *Am. Zool.,* 24, 57, 1984.
13. Rousseau, F. and De Reviers, B., Phylogenetic relationships within the Fucales (Phaeophyceae) based on combined partial SSU+LSU rDNA sequence data, *Eur. J. Phycol.,* 34, 53, 1999.
14. Naylor, M., The New Zealand species of *Durvillea*, *Trans. Roy. Soc. New Zealand,* 80, 277, 1953.
15. Hay, C.H., *Durvillaea*, in *Biology of Economic Algae I,* Akatsuka, I., Ed., SBS Academic Publishing, the Hague, 1994, p. 353.
16. Smith, J.M.B. and Bayliss-Smith, T.P., Kelp-plucking — coastal erosion facilitated by bull-kelp *Durvillaea antarctica* at sub-Antarctic Macquarie-Island, *Antarct. Sci.,* 10, 431, 1998.
17. Hay, C.H., Growth mortality, longevity and standing crop of *Durvillaea antarctica* (Phaeophyceae) in New Zealand, *Proc. Int. Seaweed Symp.,* 9, 97, 1979.
18. Wing, S.R., Leichter, J.J., and Denny, M.W., A dynamic model for wave-induced light fluctuations in a kelp forest, *Limnol. Oceanogr.,* 38, 396, 1993.
19. Harder, D.L., Hurd, C.L., and Speck, T., Biomechanics of sympatric macroalgae in the surf zone of New Zealand and Helgoland, Germany, in *3rd Plant Biomechanics Conference*, Spatz, H.C. and Speck, T., Eds. Georg Thieme Verlag, Stuttgart, 287, 2000.
20. Niklas, K.J., *Plant biomechanics: An Engineering Approach to Plant Form and Function*, The University of Chicago Press, Chicago, 1992.
21. Hoerner, S.F., *Fluid-Dynamic Drag*, Hoerner, S.F., Brick Town, NJ, 1965.
22. Harder, D.L. et al., Reconfiguration as a prerequisite for survival in highly unstable flow-dominated habitats, *J. Plant Growth Reg.,* 23, 98, 2004.
23. Denny, M.W., *Biology and the Mechanics of the Wave-Swept Environment*, Princeton University Press, Princeton, NJ, 1988.
24. Vogel, S., Drag and reconfiguration of broad leaves in high winds, *J. Exp. Bot.,* 40, 941, 1989.
25. Vogel, S., Drag and flexibility in sessile organisms, *Am. Zool.,* 24, 37 1984.
26. Stevens, C.L., Hurd, C.L., and Smith, M.J., Field measurement of the dynamics of the bull kelp *Durvillaea antarctica* (Chamisso) Heriot, *J. Exp. Mar. Biol. Ecol.,* 269, 147, 2002.
27. Koehl, M.A.R., Jumars, P.A., and Karp-Boss, L., Algal biophysics, in *Out of the Past: Collected Reviews To Celebrate the Jubilee of the British Phycological Society,* Norton, T.A., Ed., The British Phycological Society, Belfast, 115, 2003.
28. Hurd, C.L. et al., Visualization of seawater flow around morphologically distinct forms of the giant kelp *Macrocystis integrifolia* from wave-sheltered and exposed sites, *Limnol. Oceanogr.,* 42, 156, 1997.
29. Koehl, M.A.R. and Alberte, R.S., Flow, flapping, and photosynthesis of *Nereocystis luetkeana:* a functional comparison of undulate and flat blade morphology, *Mar. Biol.,* 99, 435, 1988.

30. Koehl, M.A.R., Seaweeds in moving water: form and mechanical function, in *On the Economy of Plant Form and Function,* Givinish, T.J.L., Ed., Cambridge University Press, Cambridge, 1986, p. 603.
31. Hurd, C.L., Water motion, marine macroalgal physiology, and production, *J. Phycol.,* 36, 453, 2000.
32. Niklas, K.J., Petiole mechanics, light interception by lamina, and economy in design, *Oecologia,* 90, 518, 1992.
33. Carrington, E., Drag and dislodgment of an intertidal macroalga: consequences of morphological variation in *Mastocarpus papillatus* Kützing, *J. Exp. Mar. Biol. Ecol.,* 139, 185, 1990.
34. Speck, O., Field measurements of wind speed and reconfiguration in *Arundo donax* (Poaceae) with estimates of drag forces, *Am. J. Bot.,* 90, 1253, 2003.
35. Harder, D.L. et al., Comparison of mechanical properties of four large, wave-exposed seaweeds, in preparation.
36. Denny, M.W. et al., Fracture mechanics and the survival of wave-swept macroalgae, *J. Exp. Mar. Biol. Ecol.,* 127, 211, 1989.
37. Denny, M.W. et al., The menace of momentum — dynamic forces on flexible organisms, *Limnol. Oceanogr.,* 43, 955, 1998.

4 Murray's Law and the Vascular Architecture of Plants

Katherine A. McCulloh and John S. Sperry

CONTENTS

4.1 INTRODUCTION

In the move from the water to land, plants evolved traits that allowed them to cope with their new environment. Arguably, the most dramatic problem they faced was the relative dryness of air. Rapid water loss from photosynthetic tissues requires equally rapid water supply. Poor water transport capabilities limited the earliest nonvascular plants to small size [1]. The evolution of xylem vastly increased hydraulic conductance and contributed to the diversification of plant size evident today [2,3].

Xylem structure has changed and diversified considerably over time, presumably reflecting progressive adaptation [4]. "Measuring" this adaptation is a challenge because it is not always obvious what traits are being selected for and what constraints and trade-offs are limiting trait evolution. It seems likely, however, that a universally favorable trait for a water-conducting network is maximum hydraulic conductance per unit investment. Higher hydraulic conductance means greater volume flow rate of water per pressure drop across the network. The higher the hydraulic

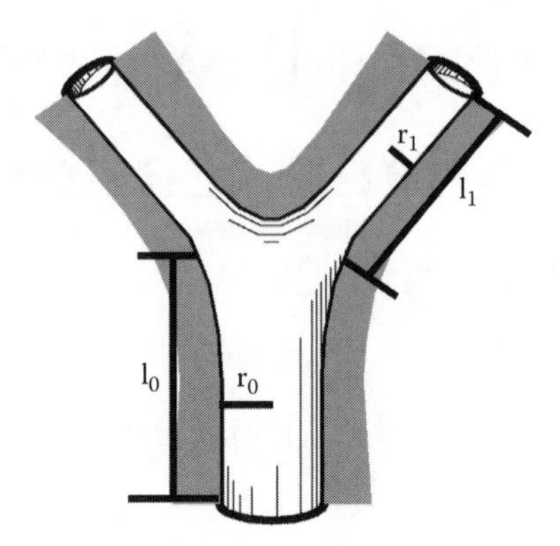

FIGURE 4.1 A bifurcating blood vessel for which Murray's law is derived in the text. Shown are the radii (r) and lengths (l) of mother (subscript 0) and daughter (subscript 1) branch ranks.

conductance, the more leaf area that can be supplied with water at a given water status and the greater potential for CO_2 uptake [5–10]. Minimizing vascular investment means that less of the assimilated carbon is required for the growth of vascular tissue, leaving more for reproduction and other functions. Evaluating the hydraulic conductance per investment criterion provides insight into the adaptive significance of diverse xylem anatomies throughout the plant kingdom.

In this chapter, we summarize the use of Murray's law for evaluating the conductance vs. investment trade-off across major xylem types. The results have appeared piecemeal elsewhere [11–14] but benefit from a unified summary. Of relevance is the high profile work of West and colleagues [15–17] who concluded that quarter-power scaling laws in biology (e.g., [18,19]) result from an energy-minimizing vascular structure that constrains metabolism. These models have been criticized as being mathematically flawed and based on inaccurate vascular anatomy [13,20,21]. Nevertheless, they have drawn attention to the "energy-minimizing" principle, which is equivalent to the maximizing of hydraulic conductance per vascular investment [22]. While West and colleagues assumed this principle holds empirically, we tested it using Murray's law.

4.2 MURRAY'S LAW

Murray's law was derived for animal vascular networks by Cecil Murray [23]. It predicts how the blood vessels should change in diameter across branch points to maximize hydraulic conductance for a given investment in vascular volume and a particular branching architecture. The derivation for a single bifurcating blood vessel (Figure 4.1) is useful for pointing out the underlying assumptions as highlighted in italics. An initial assumption is that the *volume flow rate is conserved* from mother to daughter branch ranks — no fluid is lost in transit. The total volume (V) *assuming cylindrical geometry* of the vessels is:

$$V = \pi\ (l_0 r_0^2 + F l_1 r_1^2) \tag{4.1}$$

where l is length, r is inner radius of vessel lumen, F is the number of daughter branches per mother ($F = 2$), and subscripts 0 and 1 designate mother and daughter ranks, respectively. We refer to this ratio as the "conduit furcation number." The total flow resistance (R, pressure drop per volume flow rate), *assuming laminar flow through cylindrical tubes,* is given by the Hagen–Poiseuille equation. Resistances are used rather than the reciprocal conductance because they are additive in series:

$$R = \frac{8\eta}{\pi}\left[\frac{l_0}{r_0^4} + \frac{l_1}{Fr_1^4}\right] \tag{4.2}$$

where η is the dynamic viscosity of the xylem sap.

Comparing Equations 4.1 and 4.2 demonstrates the basic conflict between minimizing flow resistance (and thus the energy consumed by the heart) while also minimizing vascular volume and the energy required to maintain the blood. Low flow resistance requires large radii ($R \propto 1/r^4$; Equation 4.2), but large radii make for larger and more expensive volume ($V \propto r^2 l$, Equation 4.1). If *the individual branch lengths and total network volume are held constant*, the relative change in conduit radii (r_0 and r_1) across branch ranks that minimizes the flow resistance per fixed volume can be determined.

Setting the volume equation to equal zero ($0 = V - C$, where C is a constant), Lagrange's theorem can be used to solve for the values of r_0 and r_1 that minimize the resistance:

$$\partial R/\partial r_0 = \lambda\ \partial V/\partial r_0 \tag{4.3a}$$

$$\partial R/\partial r_1 = \lambda\ \partial V/\partial r_1 \tag{4.3b}$$

where the Lagrange multiplier, λ, is a nonzero constant. From Equations 4.1 and 4.2, the partial derivatives are:

$$\partial R/\partial r_0 = -32\eta l_0/\pi r_0^5 \tag{4.4a}$$

$$\partial V/\partial r_0 = 2l_0\pi r_0 \tag{4.4b}$$

$$\partial R/\partial r_1 = -32\eta l_1/F\pi r_1^5 \tag{4.4c}$$

$$\partial V/\partial r_1 = 2Fl_1\pi r_1 \tag{4.4d}$$

Substituting Equations 4.4a and 4.4b into Equation 4.3a and solving for λ yields:

$$\lambda = -16\eta/\pi^2 r_0^6 \tag{4.5}$$

Inserting Equations 4.4c and 4.4d in Equation 4.3b and solving for λ yields:

$$\lambda = -16\eta/F^2\pi^2 r_1^6 \tag{4.6}$$

Substituting Equation 4.5 for λ into Equation 4.6, canceling terms, and rearranging, gives Murray's law:

$$r_0^3 = Fr_1^3 \tag{4.7a}$$

or more generally,

$$\Sigma r_0^3 = \Sigma r_1^3 \tag{4.7b}$$

which states that the hydraulic conductance of the fixed-volume network is maximized when the radius cubed of the mother vessel equals the sum of the radii cubed of the two daughter vessels. The shape of the conductivity optimum is shown for a network of three bifurcating branch ranks in Figure 4.2A.

The law does not require that all vessels of a branch rank be equal in radius as in our simple example; this can be seen by adjusting the equations accordingly and following the same derivation. The law also does not depend on F when conduit size is expressed as the conservation of Σr^3 across branch ranks (Equation 4.7b). Note that Murray's law is independent of the lengths of the branches, which cancel out in the derivation even if they are unequal within a branch level (e.g., at Equations 4.6 and 4.7). The law is equivalent to solving for the maximum hydraulic conductivity (volume flow rate per pressure *gradient*, and length-independent) for a fixed cumulative cross-sectional area (volume per unit length) summed across each branch rank. Measurements have largely supported Murray's law in animals, at least outside of the network of leaky capillaries and beyond the influence of pulsing pressures at the exit from the heart [24–26].

As the derivation shows, Murray's law does not solve for the optimal size or individual branch lengths — only the optimal tapering of the conduit diameter across a given branching topography. The law is less ambitious than the "West et al. model," which attempts to solve for both the optimum tapering and the relative lengths (l_0, l_1, etc.) of branches at each level. The limitations of the West et al. approach [20,21] recommend a more limited analysis based on Murray's law.

4.3 APPLYING MURRAY'S LAW TO XYLEM

Although Murray's law was derived for animal vascular systems, it recently has been shown to apply to plants given a few additional assumptions as italicized in the following [11]. The conduits in xylem, though roughly cylindrical, are not continuous ramifying tubes like cardiovascular vessels. Instead, tracheids and vessels in plants are unbranched and range in length from meters to less than a centimeter. They overlap longitudinally to form the branching network, and water has to flow between conduits through interconduit pits. The added flow resistance through these

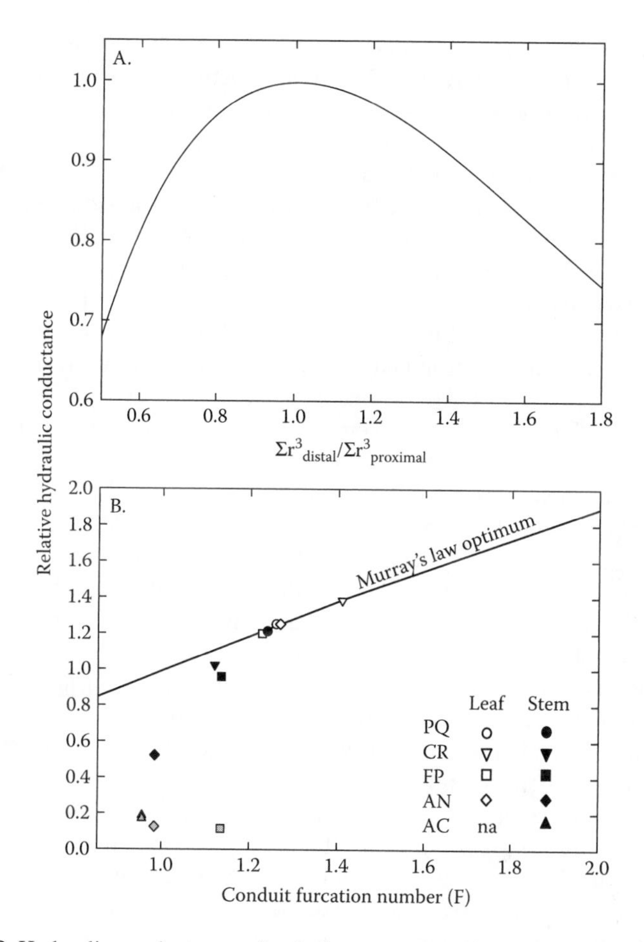

FIGURE 4.2 Hydraulic conductance of a "reference network" consisting of three ranks of bifurcating branches of fixed lengths vascularized by a conduit network of fixed volume and conduit furcation number (F). The distal-most branches contained a single conduit each — the minimum required to vascularize the network. For simplicity of calculation, the conduit radii were assumed constant within each branch level and to taper by a fixed percentage across branch levels. (A) Hydraulic conductance vs. the sum of the conduit radii to the third power across branch ranks (shown as distal Σr^3/proximal Σr^3). The Σr^3 ratio was altered by varying the percentage of conduit taper across branch ranks (F held constant). Conductance is shown relative to its maximum when the Σr^3 is conserved across branch ranks according to Murray's law. This result is independent of the choice of F. (B) The hydraulic conductance of the reference network as a function of conduit furcation number (F). Conductances are shown relative to the Murray optimum at $F = 1$. The solid "Murray's law optimum" line shows how conductance increases with F. Symbols defined in the legend are network conductances calculated for mean F and Σr^3 measurements from Figure 4.4. Species abbreviations: PQ = *Parthenocissus quinquefolia* (vine), CR = *Campsis radicans* (vine), FP = *Fraxinus pensylvanica* (ring-porous), AN = *Acer negundo* (diffuse-porous), and AC = *Abies concolor* (conifer). The gray symbols show the conductance of the three tree species when standardized by the entire wood cross-sectional area as opposed to just the conducting area. Figure modified from McCulloh, K.A. et al., *Nature*, 421, 939, 2003.

end walls means the hydraulic conductivity of xylem conduits is significantly below the value predicted by the Hagen–Poiseuille equation for open tubes [27]. The Murray's law derivation still holds, however, as long as the *actual conductivity is proportional to the Hagen–Poiseuille value* within a given network — this simply adds a proportionality constant to Equation 4.2. Recent work indicates that end walls contribute approximately half of the total resistance across a wide range of conduit sizes [28]. Flow in xylem under physiological pressures is laminar, so except for deviations from cylindrical geometry, the Hagen–Poiseuille equation accurately predicts conductivity [29].

In animals, the volume of the circulating blood is a major cost of transport. In plants, the xylem water is cheap and its volume is presumably irrelevant as a cost of the vascular system. Instead, the analogous cost in plants is the construction of the relatively thick conduit walls. Xylem sap is under significant negative pressures of 1 to 10 MPa, depending on the species and habitat [30]. Avoiding implosion of the conduits requires thick secondary walls stiffened with lignin [2]. Mechanics predict that the wall thickness be proportional to the lumen diameter to maintain a given safety margin against collapse at a given minimum negative pressure. Measurements confirm this prediction, showing that the wall thickness per lumen width scales with the cavitation pressure — the minimum negative pressure the xylem can withstand without breakage of water columns by vapor nucleation [31]. The Murray's law derivation is still valid as long as the *wall thickness is proportional to the inner radius of the conduit* as expected for a given cavitation pressure. This requires modification of Equation 4.1 but not the dependence of wall volume (hence cost) on r^2 on which the derivation depends.

Murray's Law may not necessarily hold *if conduit walls perform any additional function other than transporting water.* In some plants — conifers being the best example — conduit walls perform "double duty" by supporting the plant as well as transporting water. In these cases, wall volume cannot be viewed solely as a cost of transport and Equations 4.1 and 4.2 do not capture the intended conflict between conductivity and cost of investment.

4.4 IMPORTANCE OF THE CONDUIT FURCATION NUMBER (*F*)

A final distinction between animal and plant vasculature concerns the daughter to mother ratio in conduit number. In animals where the network consists of a single open tube that branches, $F \geq 2$. In plants, the network at every level from trunk to twig consists of numerous conduits running in parallel and $F > 0$. The increase in the number of tree branches moving from trunk to twig occurs independently of the change in the number of conduits in each branch. For any possible F, however, the optimal (maximum) conductivity occurs when the Σr^3 is conserved (Figure 4.2A, Ref. [11]).

The value of the conduit furcation number has adaptive significance for plants. The "Murray's law optimum" line in Figure 4.2B shows the conductance of a Murray's law network (Σr^3 conserved across three bifurcating branch ranks) that corresponds to different values of F when all else (volume and branching pattern) is held constant. The absolute value of the hydraulic conductance increases with F [11]. The increase in efficiency with F is greater with an increasing number of branch ranks in the network; Figure 4.2 is for a relatively small network of three branch ranks. Conductance increases because larger F means the trunk and major branches have fewer but wider and more conductive conduits for a given number of conduits in the twigs (held to the minimum of one per twig for the reference network in Figure 4.2). The plant becomes more like a branching aorta, which is the peak of efficiency. Thus, for the same investment in conduit volume, plants can increase their hydraulic conductance by having more conduits running in parallel in their twigs than in their trunk while also conforming to Murray's law.

The West et al. model assumed $F = 1$, citing the pipe model of tree form [32,33]. However, the "unit pipe" associated with each leaf in the pipe model does not refer to an actual fluid conducting pipe but rather to a strip of wood of constant cross-sectional area. The pipe model does not specify the anatomical composition of this strand of wood — how much of it is fibers, parenchyma, conduits, and so forth — because it is only concerned with biomass allocation and not hydraulics. There is no a priori value of F in plants except that it must be greater than zero.

4.5 DOES XYLEM FOLLOW MURRAY'S LAW?

Given that Murray's law is applicable to xylem, we hypothesized that plants should follow the law as long as the conduits were not providing structural support. We have tested this hypothesis in compound leaves (*Acer negundo*, *Fraxinus pensylvanica*, *Campsis radicans*, and *Parthenocissus quinquifolia*), vine stems (*C. radicans* and *P. quinquifolia*), and shoots (*Psilotum nudum*) [11,14]. In leaves and *Psilotum* shoots, structural support of the organ is primarily from nonvascular tissues. Vines, being structural parasites, require little in the way of self-supporting tissue. When the Σr^3 of the conduits in the petiolules and petioles of compound leaves were compared, they were statistically indistinguishable in four species as predicted by Murray's law (Figure 4.3). In the vine wood, one species (*P. quinquefolia*) complied with Murray's law and the other (*C. radicans*) deviated only slightly (Figure 4.4, compare y axis Σr^3 ratio with Murray value of 1).

Psilotum nudum is a stem photosynthesizer, so water is transpired continuously along the length of the flow path. This means that Q (the estimated xylem flow rate) declines from the single-stemmed base of the shoot to the tips of all the branch ends. Under these conditions, Murray's law predicts that the Σr^3 should diminish proportionally with Q [14]. The relative decline in Q from base to tip of *Psilotum* was estimated from shoot transpiration measurements, assuming steady-state conditions. Consistent with Murray's law, Q declined in direct proportion to the Σr^3 from base to tip. A log–log plot gave a slope indistinguishable from the Murray's law value of 1 (Figure 4.5).

The conduit furcation numbers of the compound leaves, vines, and *Ps. nudum* were all between 1.12 and 1.4, meaning an average increase in conduit number of between 12 and 40% from adjacent mother to daughter ranks. To compare the relative transport efficiency of these vascular networks, we calculated their position on

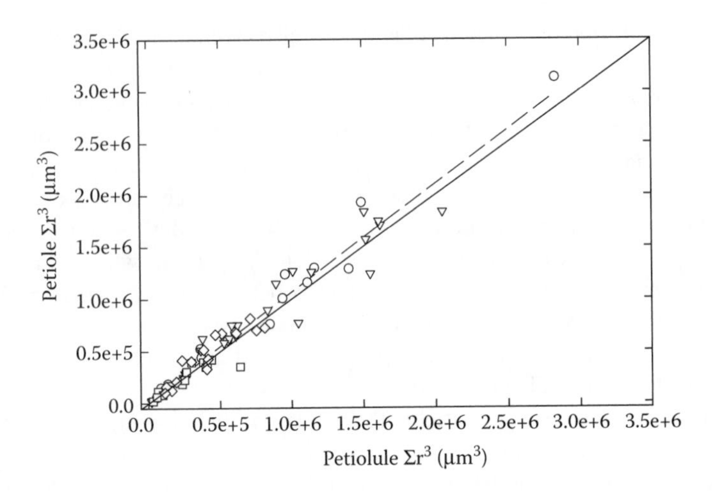

FIGURE 4.3 The sum of the conduit radii (Σr^3) in petioles of compound leaves vs. the Σr^3 of the petiolules they supply. Each symbol corresponds to a single leaf (*Acer negundo,* triangles; *Fraxinus pensylvanica,* circles; *Campsis radicans,* diamonds; and *Parthenocissus quinquifolia,* squares). The dashed line is a linear regression with a slope of 1.04. This is statistically indistinguishable from the Murray's law predicted slope of 1 (solid line). Data from McCulloh, K.A. et al., *Nature*, 421, 939, 2003.

Figure 4.2B — assuming the same "reference network" (constant volume and three bifurcating branch ranks) used to calculate the increase in the Murray optimum with *F*. Relative to the $F = 1$ value assumed by West et al., [15–17] the higher furcation numbers in leaf and vine networks show a shift toward a more efficient network (Figure 4.2B).

4.6 DOES TREE WOOD NOT FOLLOW MURRAY'S LAW?

As a further test of Murray's law, we predicted that the wood of freestanding trees should deviate from the law in proportion to how much of the wood was composed of transporting conduits. Wood holds up trees, and the more this wood is made of transporting conduits, the more these conduit walls are functioning in mechanical support — violating an assumption of Murray's law. Conifer wood is made of conduits, with over 90% of its volume in tracheids [12,34], and it should deviate the most from Murray's law by hypothesis. Ring-porous angiosperms, in contrast, have relatively few, large vessels produced in a narrow annual ring composed of nearly 10% of the wood area, the rest being fibers [12]. These trees should deviate the least. Diffuse-porous angiosperms are intermediate in wood structure and should be intermediate with respect to Murray's law. Except for the conifer species, we tested this hypothesis in wood of the same trees used for the compound leaf analysis.

The results supported the hypothesis (Figure 4.4). The conifer species, *Abies concolor*, had the highest fraction of wood devoted to conduit area (91%) and deviated the most from Murray's law (Figure 4.4, closed triangle). The deviation cannot be attributed to some inherent limitation of a tracheid-based conducting

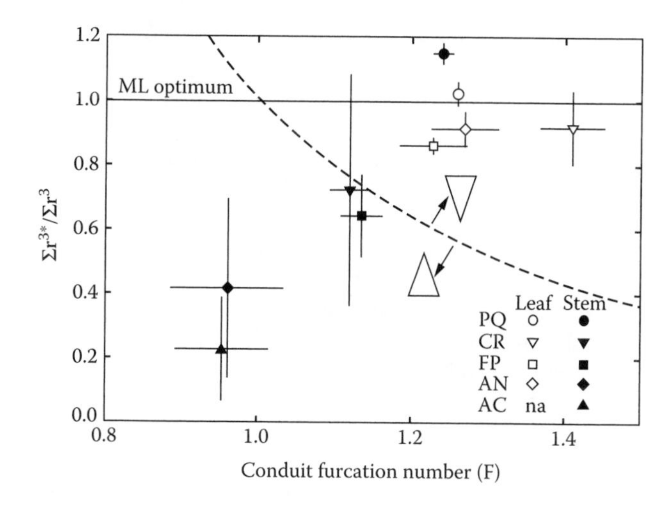

FIGURE 4.4 The ratio of the sum of the conduit radii cubed (Σr^3) versus the conduit furcation number (F) for branching ranks within leaves (open symbols — petiolule vs. petiole) and across the pooled means of three to four branching ranks within stems (closed symbols). The Σr^3 ratio is the most distal rank Σr^{3*}, which was the petiolule for the angiosperm species and the petiole for the conifer species, over the Σr^3 of progressively more proximal ranks. The conduit furcation number (F) was standardized to account for differences in branching architecture (McCulloh et al., *Nature*, 421, 939, 2003) and is always the mean for adjacent ranks. The horizontal "ML optimum" line at a Σr^3 ratio of 1 is for Murray law networks where Σr^3 is constant across ranks. The dashed curve is for networks where the Σr^2 is constant across ranks, i.e., a constant cross-sectional area of conduits. Above this line, conduit area increases from base to tip (inverted cone), while below the line, the conduit area diminishes in the same direction (upright cone). Symbols are grand means from three to four individuals. PQ = *Parthenocissus quinquefolia* (vine), CR = *Campsis radicans* (vine), FP = *Fraxinus pensylvanica* (ring-porous), AN = *Acer negundo* (diffuse-porous), and AC = *Abies concolor* (conifer). Data are from McCulloh, K.A. et al., *Funct. Ecol.*, 18: 931–938, 2004.

system because *Ps. nudum* possesses tracheids and followed Murray's law quite closely (Figure 4.5). The ring-porous species *F. pensylvanica* had the lowest conduit area fraction (12%) and had the lowest deviation from Murray's law (Figure 4.4, closed square). When statistics were performed on the ratio of the Σr^3 of the most distal to progressively more proximal ranks on a rank-by-rank basis (as opposed to overall means reported in Figures 4.2B and 4.4), the young wood of *F. pensylvanica* complied with Murray's law [11]. The diffuse-porous tree (*A. negundo*) had an intermediate conduit area fraction (24%) and was intermediate between the ring-porous and conifer tree with respect to Murray's law (Figure 4.4, closed circle).

Interestingly, the furcation numbers for wood of these self-supporting trees were generally smaller (0.98 to 1.1) than furcation numbers in vascular tissue that was not functioning in mechanical support (1.2 to 1.4) (Figure 4.4). Furthermore, when all the data are compared, a significant correlation exists between increasing conduit furcation number and increasing convergence on Murray's law (Figure 4.4, Ref. [12]). The implications for transport efficiency are evident in Figure 4.2B where the relative hydraulic conductance from the tree wood data is compared to the vine and

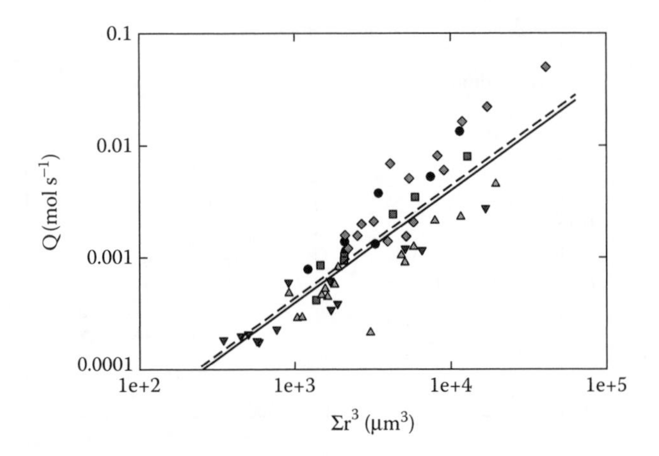

FIGURE 4.5 The sum of the conduit radii cubed (Σr^3) vs. the estimated xylem flow rate (Q) on a log–log plot. Murray's law predicts that $Q \propto \Sigma r^3$, indicating a slope of 1 (solid line). The pooled slope was determined using a linear mixed-effects model (slope = 1.01) and is statistically indistinguishable from this value. Symbols indicate measurements on all shoot segments from five *Ps. nudum* individuals. Figure modified from McCulloh, K.A. and Sperry, J.S., *Am J. Bot.*, 92: 985–989.

leaf data for the same reference network (Figure 4.2B, compare closed tree vs. vine and leaf data points). Not only is tree wood less efficient than the other xylem types because of its deviation from Murray's law, but it is also less efficient because of its lower furcation number. This result implies a conflict between the optimization of hydraulics vs. mechanics when the two functions are performed by vascular tissue.

4.7 NATURE OF THE MECHANICAL CONSTRAINT ON HYDRAULIC EFFICIENCY

Evidence for a support vs. transport conflict was most obvious in the conifer wood — where the two functions are performed by the tracheids that make up most of the wood volume. What prevents conifer wood from matching the efficiency of leaf xylem and vine wood? The answer is that if it did, conifer trees would likely fall over. The dashed line in Figure 4.4 represents "area-preserving" transport networks, i.e., vascular systems that have the same total cross-sectional area of functional xylem from base to tips. A conifer tree above the dashed line in company with the vines and leaves would be top heavy with more cross-sectional area and bulk in their upper branches than in its trunk.

This result exposes a basic support vs. transport trade-off. Achieving higher conductance for a given network volume requires "area-increasing" conduit networks (Figure 4.4, above dashed line). Achieving the tallest self-supporting structure for a given volume requires a tapered column with its area-decreasing pattern [35] (Figure 4.4, below dashed line). The two functions cannot be simultaneously optimized if the supporting cells are also transporting conduits. The conflict is not relevant for vine wood, which is less self-supporting, or for leaf and *Psilotum* shoot xylem, which is not involved in mechanical support. We interpret that these mechanically unconstrained networks are free to be area-increasing and more efficient. One of the vine species fell near the area-preserving line and also deviated more from Murray's law than the other (Figure 4.4, *C. radicans*). Vine stems can be transitional between self-supporting vs. tension structures, depending on species, stem position, and development [36,37].

The specialized wood of angiosperms can at least partially resolve the support vs. transport conflict by having separate support vs. transport cells. They can have an area-increasing conduit network while maintaining area-preserving or even area-diminishing branching of the wood as a whole for mechanical stability (Figure 4.4, *Fraxinus* datum). Ring-porous trees represent the extreme of this angiosperm strategy by localizing transport to a few large vessels embedded in a much greater area of fibers.

Sap velocity measurements reinforce our anatomical results. Relative sap velocity in trunk vs. twig is a proxy for relative cross-sectional area of functional conduits — assuming steady state flow conditions and conservation of mass flow. Conifers and diffuse-porous trees show accelerating sap velocity from trunk to twig [38–40], consistent with a reduction in conduit area and less efficient transport networks in the conductance per volume sense. Ring-porous trees, in contrast, tend to have decelerating velocity from trunk to twig [39], consistent with more efficient area-increasing conduit networks. Unfortunately, we know of no comparable data for vines or leaf xylem, but we would expect decelerating velocities based on the anatomy (Figure 4.4).

4.8 DA VINCI'S RULE

Our results bear on the venerable "rule" first postulated by Da Vinci that trees show area-preserving branching, meaning that the collective cross-sectional area of trunk, major branches, and minor branches is the same [27, 41]. Measurements generally confirm this rule, at least within the branches of the canopy [32,42]. Da Vinci based his rule on the explicit assumption that flow velocities were equal throughout the tree [13,41] and implicitly that branch cross-sectional area was proportional to the area of the water-conducting conduits. Neither of these assumptions must be true from first principles, nor are they true in fact. So why do trees tend to follow this rule? We suggest that trees follow Da Vinci's area-preserving branching pattern because it is the best compromise between opposing demands of hydraulics (area-increasing conduit networks) and mechanics (area-decreasing branching of wood in total).

4.9 DEVELOPMENTAL AND PHYSIOLOGICAL CONSTRAINTS ON TRANSPORT EFFICIENCY

In addition to the mechanical constraints on hydraulics discussed previously, there are other considerations that limit efficiency. To achieve Murray-type architecture in combination with higher conduit furcation numbers (i.e., moving up on the Murray optimum in Figure 4.2B), the xylem conduits in the major branches and main stem must be able to increase in diameter.

Developmental factors may limit conduit diameter growth. Intracellular transport and physiology can limit maximum cell size. This is probably part of the reason for the single-celled tracheid having a smaller maximum diameter [43] (about 80 μm) than vessels [27] (about 500 μm), which are composed of multiple single-celled vessel elements. Cambial development may also limit mature conduit width in secondary xylem. Diameter growth of a conduit beyond the width of its antecedent cambial initial may be limited by how much the development of flanking cambial derivatives can be modified.

Safety vs. efficiency trade-offs in transport function also limit effective conduit diameters. Large diameter vessels are known to be much more vulnerable to freezing-induced cavitation [44] and would seemingly be maladaptive where the growing season is frost prone. Larger vessels also tend to be more vulnerable to cavitation by water stress [45], in part because they tend to have greater area of interconnecting pits and hence greater chance of air-seeding cavitation [46]. These kinds of limitations probably combine to limit maximum conduit sizes and thus conduit furcation numbers and hydraulic efficiency.

4.10 COMPARATIVE EFFICIENCY OF CONIFER VS. ANGIOSPERM TREE WOOD

The data show that conifer wood can be less efficient hydraulically than angiosperm wood (Figure 4.2B, compare conifer vs. angiosperm closed symbols). Yet conifers remain competitive in certain habitats with angiosperm trees, suggesting that there are other considerations. The Murray's law definition of efficiency is a narrow one: maximum hydraulic conductance per unit volume of *conducting* cells. The difference between conifers and angiosperms could be much less if the definition of efficiency included the nonconducting volume required for mechanical support. By doing double duty, conifer tracheids may save on the sum of the tissue required for *both* hydraulic and mechanical functions when compared with the fiber vs. vessel cell types in angiosperm wood.

A theoretical analysis of this more comprehensive definition of efficiency is beyond the scope of this chapter. However, an approximation is to convert the Murray law conductances per fixed *conducting* volume to *total* volume (conducting plus nonconducting) by multiplying by the ratio of cross-sectional conducting area to total xylem area. This is equivalent to the analogous volume ratio for the same network dimensions, as long as the area ratio is the same for each branch rank. Doing this for the freestanding trees in Figure 4.2B (solid symbols) results in equivalent conductances per total volume across the wood types (Figure 4.2B, gray symbols). This preliminary calculation suggests that while angiosperm tree wood can pack more conductance into a given volume than conifer wood, when the additional volume required for fibers to support the tree is added, the two wood types can become more similar. Indeed, there is considerable overlap in xylem conductivity per total wood cross-sectional area between conifers and angiosperms

[47,48], although angiosperms as a group reach a higher maximum. Conifers are often competitive in stressful habitats [49] where avoidance of cavitation may limit vessel size, keeping conductivity per total wood area in angiosperms from significantly exceeding the conifer range.

4.11 CONCLUSIONS

In each case where conditions approached the Murray's law assumptions, the xylem conduit network was at or near the Murray law optimum. Even the tracheid-based conducting system of *Ps. nudum* conformed, suggesting the trait may be pervasive in the seedless vascular plants. Although *Psilotum* is allied to the ferns, its reduced morphology and central strand of primary tracheids strongly resembles fossils of the earliest of vascular plants [50]. If these ancestral vascular plants also obeyed Murray's law, the selective advantage of the trait must have been quite strong. An early origin of Murray architecture is consistent with the observation that ancestral vascular systems did not function significantly in mechanical support of the plant [51].

A major mechanical role of xylem probably evolved with the kind of secondary growth seen in extant conifers and angiosperms. Murray's law does not predict the optimal conduit structure in the wood of these self-supporting trees because it does not consider the mechanical role of the conduit network. Nor did the conifer and angiosperm woods measured match the Murray's law optimum. As predicted, however, the less the conduits contributed to wood volume and mechanical support (as in the ring-porous angiosperm wood), the less was the deviation from Murray's law. Although this means that the conduit portion of the wood can be inherently more efficient in angiosperms vs. conifers, the advantage is less evident when the entire wood volume required for both hydraulic and mechanical functions is considered (Figure 4.2B, gray vs. closed symbols). To better assess the trade-off between hydraulic vs. mechanical demands on wood structure, an analysis that moves beyond Murray's law is necessary.

ACKNOWLEDGMENTS

A Graduate Research Fellowship from the Graduate School at the University of Utah supported KAM. Some funding was obtained from National Science Foundation grant IBN-0112213 to JSS.

REFERENCES

1. Raven, J.A., Long-distance transport in non-vascular plants, *Plant Cell Environ.*, 26, 73, 2003.
2. Raven, J.A., The evolution of vascular land plants in relation to supracellular transport processes, *Adv. Bot. Res.*, 5, 153, 1987.
3. Sperry, J.S., Evolution of water transport and xylem structure, *Int. J. Plant Sci.*, 164, S115, 2003.

4. Niklas, K.J., The evolution of tracheid diameter in early vascular plants and its implications on the hydraulic conductance of the primary xylem strand, *Evolution*, 39, 1110, 1985.
5. Hubbard, R.M. et al., Stomatal conductance and photosynthesis vary linearly with plant hydraulic conductance in ponderosa pine, *Plant Cell Environ.*, 24, 113, 2001.
6. Meinzer, F.C. et al., Environmental and physiological regulation of transpiration in tropical forest gap species: The influence of boundary layer and hydraulic properties, *Oecologia,* 101, 514, 1995.
7. Meinzer, F.C. and Grantz, D.A., Stomatal and hydraulic conductance in growing sugarcane: Stomatal adjustment to water transport capacity, *Plant Cell Environ.*, 13, 383, 1990.
8. Saliendra, N.Z., Meinzer, F.C., and Grantz, D.A., Water potential in sugarcane measured from leaf segments in a pressure chamber, *Agron. J.*, 82, 359, 1990.
9. Sperry, J.S., Alder, N.N., and Eastlack, S.E., The effect of reduced hydraulic conductance on stomatal conductance and xylem cavitation, *J. Exp. Bot.*, 44, 1075, 1993.
10. Sperry, J.S. and Pockman, W.T., Limitation of transpiration by hydraulic conductance and xylem cavitation in *Betula occidentalis*, *Plant Cell Environ.*, 16, 279, 1993.
11. McCulloh, K.A., Sperry, J.S., and Adler, F.R., Water transport in plants obeys Murray's law, *Nature*, 421, 939, 2003.
12. McCulloh, K.A., Sperry, J.S., and Adler, F.R., Murray's law and the hydraulic versus mechanical functioning of wood, *Funct. Ecol.*, 18: 931–938, 2004.
13. McCulloh, K.A. and Sperry, J.S., Patterns in hydraulic architecture and their implications for transport efficiency, *Tree Physiol.*, 25: 257–267.
14. McCulloh, K.A. and Sperry, J.S., The evaluation of Murray's law in *Psilotum nudum* (Psilotaceae), an analogue of ancestral plants, *Am. J. Bot.*, in press.
15. Enquist, B.J., West, G.B., and Brown, J.H., Quarter-power allometric scaling in vascular plants: Functional basis and ecological consequences, in *Scaling in Biology*, Brown, J.H. and West, G.B., Eds., Oxford University Press, Oxford, 2000, p. 167.
16. West, G.B., Brown, J.H., and Enquist, B.J., A general model for the origin of allometric scaling laws in biology, *Science*, 276, 122, 1997.
17. West, G.B., Brown, J.H., and Enquist, B.J., A general model for the structure and allometry of plant vascular systems, *Nature*, 400, 664, 1999.
18. Kleiber, M., Body size and metabolism, *Hilgardia*, 6, 315, 1932.
19. Niklas, K.J. and Enquist, B.J., Invariant scaling relationships for interspecific plant biomass production rates and body size, *Proc. Natl. Acad. Sci. USA*, 98, 2922, 2001.
20. Dodds, P., Rothman, D., and Weitz, J., Re-examination of the "3/4-law" of metabolism, *J. Theor. Biol.*, 209, 9, 2001.
21. Kozlowski, J. and Konarzewski, M., Is West, Brown and Enquist's model of allometric scaling mathematically correct and biologically relevant? *Funct. Ecol.*, 18, 283, 2004.
22. West, G.B., Brown, J.H., and Enquist, B.J., The origin of universal scaling laws in biology, in *Scaling in Biology*, Brown, J.H. and West, G.B., Eds., Oxford University Press, Oxford, 2000, p. 87.
23. Murray, C.D., The physiological principle of minimum work. I. The vascular system and the cost of blood volume., *Proc. Natl. Acad. Sci. USA,* 12, 207, 1926.
24. LaBarbera, M., Principles of design of fluid transport systems in zoology, *Science*, 249, 992, 1990.
25. Sherman, T.F., On connecting large vessels to small: The meaning of Murray's law, *J. Gen. Physiol.*, 78, 431, 1981.
26. Vogel, S., *Life in Moving Fluids: The Physical Biology of Flow*, 2nd ed., Princeton University Press, Princeton, NJ, 1994.

27. Zimmermann, M.H., *Xylem Structure and the Ascent of Sap*, Springer-Verlag, Berlin, 1983.
28. Sperry, J.S., Hacke, U.G., and Wheeler, J.W., Comparative analysis of end wall resistance in xylem conduits, *Plant Cell Environ.*, 28: 456–465, 2005.
29. Zwieniecki, M.A., Melcher, P.J., and Holbrook, N.M., Hydraulic properties of individual xylem vessels of *Fraxinus americana*, *J. Exp. Bot.*, 52, 257, 2001.
30. Pockman, W.T. and Sperry, J.S., Vulnerability to xylem cavitation and the distribution of Sonoran Desert vegetation, *Am. J. Bot.*, 87, 1287, 2000.
31. Hacke, U.G. et al., Trends in wood density and structure are linked to prevention of xylem implosion by negative pressure, *Oecologia*, 126, 457, 2001.
32. Shinozaki, K., Yoda, K., Hozumi, K., and Kira, T., A quantitative analysis of plant form — the pipe model theory: II. Further evidence of the theory and its implications in forest ecology, *Jpn. J. Ecol.*, 14, 133, 1964.
33. Shinozaki, K., Yoda, K., Hozumi, K., and Kira, T., A quantitative analysis of plant form — the Pipe Model Theory: I. Basic analysis, *Jpn. J. Ecol.*, 14, 97, 1964.
34. Gartner, B.L., Patterns of xylem variation within a tree and their hydraulic and mechanical consequences, in *Plant Stems: Physiological and Functional Morphology*, Gartner, B. L., Ed., Academic Press, New York, 1995, p. 125.
35. Keller, J.B. and Niordson, F.I., The tallest column, *J. Math. Mech.*, 16, 433, 1966.
36. Gallenmuller, F. et al., The growth form of *Croton pullei* (Euphorbiaceae) — Functional morphology and biomechanics of a neotropical liana, *Plant Biol.*, 3, 50, 2001.
37. Speck, T. and Rowe, N.P., A quantitative approach for analytically defining size, growth form, and habit in living and fossil plants, in *The Evolution of Plant Architecture*, Kurmann, M.H. and Hemsley, A.R., Eds., Royal Botanic Garden, Kew, 1999, p. 447.
38. Andrade, J.L. et al., Regulation of water flux through trunks, branches, and leaves in trees of a lowland tropical forest, *Oecologia,* 115, 463, 1998.
39. Huber, B. and Schmidt, E., Weitere thermo-elektrische Untersuchungen uber den Transpirationsstrom der Baume., *Tharandt Forst Jb,* 87, 369, 1936.
40. McDonald, K.C., Zimmermann, R., and Kimball, J.S., Diurnal and spatial variation of xylem dielectric constant in Norway spruce (*Picea abies* [L.] Karst.) as related to microclimate, xylem sap flow, and xylem chemistry, *IEEE Trans. Geosci. Remote Sensing,* 40, 2063, 2002.
41. Richter, J.P., *The Notebooks of Leonardo da Vinci (1452-1519), Compiled and Edited from the Original Manuscripts*, Dover, New York, 1970.
42. Horn, H.S., Twigs, trees and the dynamics of carbon in the landscape, in *Scaling in Biology*, Brown, J.H. and West, G.B., Eds., Oxford University Press, Oxford, 2000, p. 199.
43. Lancashire, J.R. and Ennos, A.R., Modelling the hydrodynamic resistance of bordered pits, *J. Exp. Bot.*, 53, 1485, 2002.
44. Davis, S.D., Sperry, J.S., and Hacke, U.G., The relationship between xylem conduit diameter and cavitation caused by freeze-thaw events, *Am. J. Bot.*, 86, 1367, 1999.
45. Tyree, M., Davis, S., and Cochard, H., Biophysical perspectives of xylem evolution — Is there a tradeoff of hydraulic efficiency for vulnerability to dysfunction?, *IAWA J.*, 15, 335, 1994.
46. Wheeler, J.W. et al., Inter-vessel pitting and cavitation in woody Rosaceae and other vesseled plants: A basis for a safety vs. efficiency trade-off in xylem transport, *Plant Cell Environ.,* 28: 800–812, 2005.

47. Becker, P., Tyree, M.T., and Tsuda, M., Hydraulic conductances of angiosperm versus conifers: Similar transport sufficiency at the whole-plant level, *Tree Physiol.*, 19, 445, 1999.
48. Tyree, M.T. and Zimmermann, M.H., *Xylem Structure and the Ascent of Sap*, Springer, Berlin, 2002.
49. Bond, W.J., The tortoise and the hare: Ecology of angiosperm dominance and gymnosperm persistence, *Biol. J. Linn. Soc.*, 36, 227, 1989.
50. Kenrick, P. and Crane, P.R., The origin and early evolution of plants on land, *Nature*, 389, 33, 1997.
51. Vincent, J.F.V. and Jeronimidis, G., The mechanical design of fossil plants, in *Biomechanics and Evolution*, Rayner, J.M.V. and Wootten, R.J., Eds., Cambridge University Press, Cambridge, 1991, p. 21.

5 Plant–Animal Mechanics and Bite Procurement in Grazing Ruminants

Wendy M. Griffiths

CONTENTS

5.1 INTRODUCTION

Mammalian herbivores are major suppliers of the worlds' milk, meat, and fiber products to humans. Their existence on the various land types on Earth can be attributed to a behavioral mechanism geared toward maximizing fitness (i.e., proliferation of their genes via the production of progeny) and is commonly explained by the body of evolutionary theory known as Optimal Foraging Theory (OFT) [1]. The acquisition and assimilation of nutrients from food is of paramount importance to the ruminant because it is the fundamental process that enables survival, growth, and reproduction. Energy is the driving currency, but grazing ruminants face complex decisions in searching for and harvesting adequate forage to meet their energy requirements for survival, growth, and reproduction. Vegetation heterogeneity adds complexity to even the simplest of ecosystems and is itself a circular process shaped by the effects of herbivory on the environment. While food intake lies at the heart of the survival of animal species, the discrimination by animals between plant species and their morphological organs is central to the survival and regeneration of the plant population.

Despite the advances that have been made in understanding forage intake [2–4], mechanistic explanations for diet choice and observed behavior remain scarce. The application of materials science theory to understanding biological problems in herbivores has led to a revived interest in quantifying plant fracture mechanics, but parallel progress in understanding the mechanistic relationships between animal and plant mechanical properties and grazing strategies in ruminants, particularly of contrasting body size, has been much slower. The force that grazing animals exert in procuring a bite has received little attention despite the clear linkages with herbage intake. This probably reflects the difficulties associated with quantification of bite force. The objectives of this chapter are to: (1) discuss the ruminant species, their harvesting apparatus, and the process that these herbivores use to harvest food, (2) clarify the terminology used to describe fracture mechanics as they apply to ruminants, (3) demonstrate how bite force can be quantified and discuss the problems and opportunities facing the researcher, and (4) introduce the concept of biting effort.

5.2 RUMINANT SPECIES

Ruminant species constitute a suborder of Artiodactyla, the hoofed mammals, with the most significant anatomical difference between ruminants and other mammals being the four-chambered (rumen, reticulum, omasum, and abomasum) digestive system that allows ruminants to derive 60% of their energy requirements from the microbial fermentation in the rumen–reticulum of the constituents of plant cell walls. Furthermore, the presence of the rumen–reticulum permits the distinguishable "cud chewing" cycle known as rumination. The literature has been dominated by the scheme of ruminants being grouped into three ecophysiological types [5] according to the predominant food type they consume: grazers (e.g., cattle, *Bos taurus*), browsers (e.g., moose, *Alces alces*), and intermediate feeders (e.g., red deer, *Cervus elaphus*). Furthermore, ruminant species also exhibit strong variations in the ability to digest fiber, a feature that has been attributed to ecophysiological differences between species [6], although more recently it has been proposed [7] that the morphophysiological contrasts in digestive capability merely reflect the contrasts in body size and not feeding type as historically presented.

5.3 HARVESTING APPARATUS

Physically, the harvesting apparatus is housed within an elongated and bluntly pointed skull and is an important structure within the ruminants' body. The jaws are the housing to which the teeth and muscles are attached. The upper jawbone, often called the "maxilla," is fused to the skull, and the lower jawbone, termed the "mandible," is hinged at each side to the bones of the temple by ligaments. Common

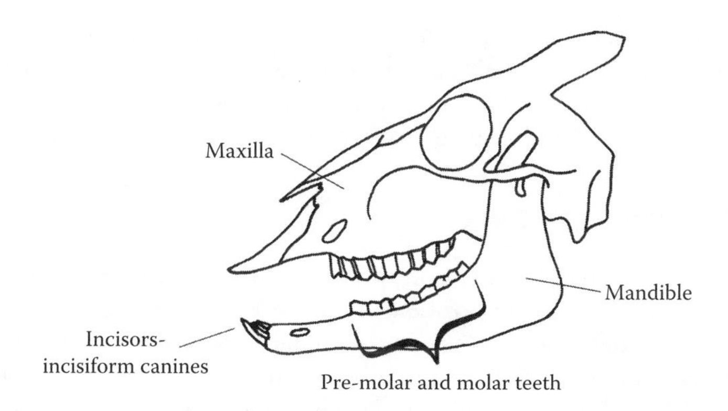

FIGURE 5.1 Harvesting apparatus of a sheep, illustrating the maxilla and mandible.

to all ruminants are the four-paired anterior teeth consisting of true incisors and incisiform canines set on the lower jaw, believed to have evolved for harvesting of plant material. On the upper jawbone, above the incisors, a thick pad of connective tissue (the dental pad) is present. Unlike odd-toed ungulates (the equids, e.g., horses, asses, zebras, and rhinos), ruminants do not possess incisors on the upper jawbone. Toward the back of the mouth, ruminants have sets of molars and premolars (Figure 5.1) that are flat and lined with sharp ridges of enamel. Because tooth shape governs functionality, these posterior teeth generally do not make contact with the bulk of the grasped forage during prehension, and their pivotal role lies in chewing the severed bite contents.

Jarman [8] recognized the functional interrelationships between the size and dispersion of food items in the environment and animal species' body size. Smaller animals have a smaller harvesting apparatus in absolute terms and can remove smaller bites, but relative to body mass, smaller animals require a diet of higher nutritional quality compared with larger animals to meet the higher metabolic requirements per unit of body weight ($W^{0.75}$). It is believed that small animals have, therefore, evolved a jaw configuration that is narrower relative to larger-bodied animals and additionally supported by prehensile and mobile lips that permit the selection of leaves, which in the extreme scenario may come from thorny browse species. By contrast, larger-bodied animals have a wide jaw configuration, and irrespective of whether these species can perceptually discriminate between leaf and stem, they are constrained by the inability to selectively remove leaf from stem because of the constraints of the wide muzzle. A long prehensile tongue that primarily serves to sweep forage toward the center of the bite, increasing the effective bite area, aids larger-bodied species.

The harvesting apparatus and body mass of animals accounts for much of the variation that exists in selection strategies between species. There is, however, extensive overlap in body mass between species within the three feeding types, as documented by Gordon and Illius [9] who presented an excellent examination of the jaw configuration of 34 species of grazers, 27 species of intermediate feeders, and 19 species of browsers, varying in body mass from 3 to 1200 kg. Their results provided compelling evidence that large-bodied grazers have a broad and flat incisor arcade, where incisor arcade is defined as the distance between the outer edges of the incisiform canines on the right and left ramus, compared to large-bodied browsers of a similar body mass who have a more narrow and pointed arcade. Similar patterns are evident between small-bodied grazers and browsers of a similar body mass.

5.4 BITE PROCUREMENT

Interest in the application of engineering principles to the understanding of biological systems such as foraging behavior stems from the knowledge that these principles are embedded in the everyday behavior of animals. Ruminants forage in diverse environments with available forage offering relatively low levels of nutrients per ingested bite. They face considerable challenges in procuring a large number of bites during a 9- to 10-hr day of grazing activity (approximately 30,000 bites for cattle); consequently, the "bite" is considered the building block of daily herbage intake [10,11].

Alternating periods of grazing, rumination, and rest constitute the diurnal activity of a ruminant. During grazing, the location of potential bites while the animal's forelegs are stationary is known as a "feeding station" [12] and is defined as the semicircular area in front of and to each side of the animal. The establishment of a feeding station implies that one or both of the peripheral senses — sight and smell — have been activated, while the senses of taste and touch influence subsequent behavior following the procurement of initial bites [13]. Procurement of a bite is initiated when the animal lowers its head in search of food. A bite is then removed when a series of manipulative jaw movements (with or without protruding tongue sweeps) gathers herbage, which is gripped by the incisors biting against the dental pad, with forage material effectively running across the incisal edge, allowing for severance to result from the animal jerking its head in a characteristic and timely fashion. On dense foliage or swards of strong phenological contrast, foliage may be lost during the jerking of the head since stiff stems increase the probability that the foliage will spring back, evading the clamping action of the jaws. Furthermore, the bite may necessitate several swinging or jerking motions of the head (i.e., one or more tugs) to sever the bite. The principle of any foraging strategy is dependent upon how the ruminant animal decides where to select bites from, across the habitat as well as from within the sward canopy, and this entails a series of complex mechanisms that have yet to be unraveled.

5.5 PLANT FORM AND FRACTURE MECHANICS AT THE PLANT LEVEL

Plants are the staple source of the mammalian herbivore diet. The leaves are generally flat and engineered to capture sunlight for photosynthesis, the primary process that leads to the production of energy, a source that animal subsistence and production is dependent upon. Plants themselves are complex but can be divided into three main morphological organs — roots, stems, and leaves — which are each exposed to environmental forms of mechanical strain. Leaves of monocots are constructed from vascular tissue that forms in parallel strands (veins) extending along the long axis of the leaf. This vascular tissue is supported by mesophyll tissue (i.e., sclerenchyma, storage parenchyma, and clorenchyma cells) and is covered by waxy epidermal tissue that reduces water loss from evaporation. Sclerenchyma is of great interest in materials science because these cells have thick, rigid, nonstretchable secondary walls that confer strength to the plant. There is wide variation in the interveinal distance between plant species, but a small interveinal distance does not necessarily imply that a leaf will be less digestible to livestock [14]. It is the organization of the sclerenchyma bundles that determines fracture properties, and this fact formed the basis of the comment from Wright and Illius [15] that the properties determining digestion were essentially those influencing fracture mechanics. Although fibers may only constitute a very small proportion (5%) of the leaf cross-sectional area, that seemingly small proportion accounts for 90 to 95% of longitudinal stiffness [16].

Animal scientists interested in digestive function have long been interested in the fracture of cellular material as an indicator of its susceptibility to crushing and shearing forces during rumination. As a result, strong working relationships have been forged between animal scientists and plant breeders, with much of the work instigated by plant breeders aimed at selecting for plant traits that increased feeding value (FV), and this has been reflected in the strong focus on screening for low shear strength [17]. Plant fracture properties have also been assessed and related to forage avoidance and/or preference and forage intake [18–21], trampling resistance [22], and plant uprooting ("pulling") [23]. Additionally, the impact of environmental constraints on plant fracture properties have been evaluated [16,24] alongside the relationships with bite force [25,26] and bite dimensions [27,28]. Studies predominantly investigate the fracture mechanics of leaves since leaves are innately the preferred morphological organ. Nevertheless, there have been important contributions from examination of the stem properties of monocotyledons [15] and dicotyledons [18,29,30].

There is a broad range of terminology within the subject of fracture mechanics in plants. In an agricultural context, one of the pioneering studies examining plant fracture properties was the work by Evans [31], but like many other studies, this research has attracted widespread criticism for the inconsistency in adhering to the fundamental engineering principles underlying fracture mechanics. Several published studies and review articles have addressed the confusion in the use of descriptive terms and the units of expression for defining the fracture mechanics in plants. This has generally led to better application of terminology, but incorrect parameters and units still surface in the literature [22,23], largely due to the subjective nature of the experimental objectives.

Briefly, fracture in a test specimen involves both the initiation and propagation of a crack. Cracks can be propagated by three contrasting modes: mode I is by tension (crack opening), mode II is by shear (edge-sliding or in-plane shearing movement), and mode III is caused by tearing (out-of-plane shearing movement) [32]. Where ruminants are concerned, mode I fracture tests best describe the harvesting of forage in a predominantly vertical dimension while mode III fracture represents the mechanisms of fracture that take place when forage is crushed and ground against the molars during chewing.

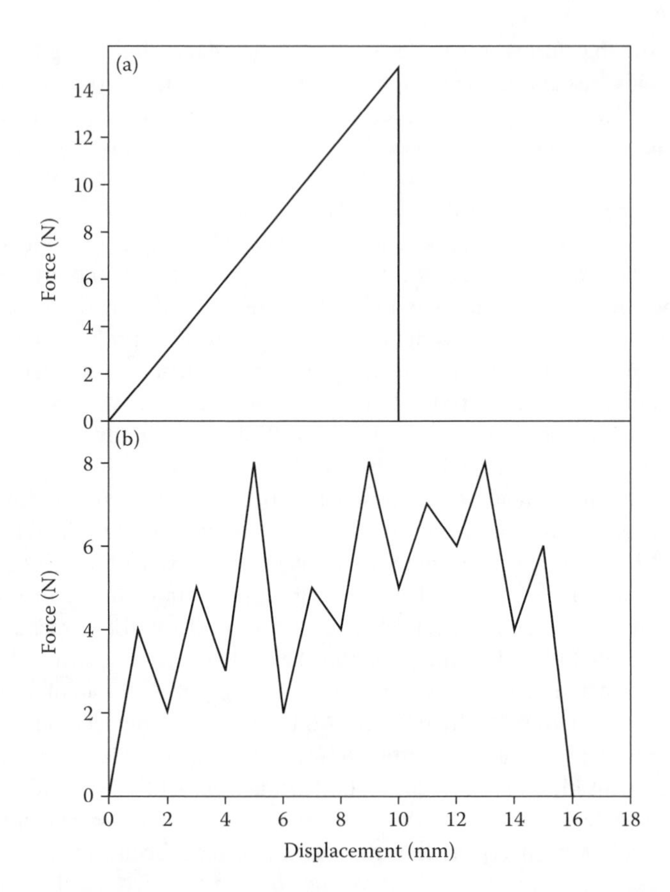

FIGURE 5.2 Simulated force-displacement curves for lamina from (a) tensile (mode I) and (b) out-of-plane shear (mode III) fracture tests.

It can also be helpful to be familiar with how materials perform under load. Figure 5.2a and 5.2b illustrate simplified representations of plant lamina when tested under tension (mode I) and out-of-plane shear (mode III) modes, respectively. The triangular-shaped force-displacement curve in Figure 5.2a illustrates the dynamics of a leaf under tension. The curve represents a steady linear increase in force, the slope being an indicator of the leaf's stiffness, until the leaf specimen fractures, at which point the material ceases to be elastic, resulting in a sudden decrease in force to zero immediately after fracture. By contrast, the spiky force–displacement curve in Figure 5.2b illustrates the dynamic relationship of the shearing of a leaf specimen, where the force is constantly changing in a controlled manner as a crack is propagated across the specimen.

The objective of this chapter is to review and discuss the role of plant–animal mechanics in understanding bite procurement. The focus of the remainder of this chapter, therefore, concerns itself with only mode I fracture where relevant. Three plant-based terms of interest in understanding the procurement constraints facing grazing ruminants were summarized by Griffiths and Gordon [33]:

- *Fracture force* is a measure of the force required to fracture a plant organ under tension and can be assessed from the maximum force recorded on the force–displacement curve that produces fracture.
- *Tensile strength* is the fracture force under tension per unit of cross-sectional area of the plant specimen.
- *Resistance* is a plant-based term that has no underlying engineering concept, but it carries importance in the application of plant fracture mechanics to predicting foraging strategies in ruminants. It can be defined as the accumulated force required by the animal to sever all the plant organs encompassed within the bite. It can be argued that tensile strength is a measure of resistance, but in relation to ruminants, we are interested in the resistance of the bite contents to rupture under load, and hence the accumulation of plant material.

5.6 INSTRUMENTATION FOR MEASURING PLANT FRACTURE MECHANICS UNDER TENSION

Tensile (mode I fracture) tests have arguably received less attention than out-of-plane shear (mode III) tests in the literature. While this reflects the greater attention that has been given to the importance of chewing, it also, in part, reflects the fact that tensile tests are more awkward to perform successfully. Nevertheless, the increased reporting of tensile tests with reference to ruminants is recognition that forages place different food procurement constraints on ruminants, which feed by grasping and tearing herbage with their muscle mass, as opposed to invertebrates that chew between the fibers of leaves.

Tensile tests are commonly conducted by securing the test piece between two clamps and breaking the specimen by longitudinal pull. Examination of the literature, however, shows profound deficiencies in the reporting of instrumentation and procedures for assessment of tensile strength in grassland studies [15,21,22,26,29,31,34–36]. The instrumentation of Sun and Liddle [22] was a modification of that used by Evans [31]. The apparatus consisted of a pivoting beam, with a clamp setup on one side and a bucket hung on the other side into which sand was poured until fracture of the specimen occurred. Spring-tensioned instruments used by Diaz et al. [21] and Adler et al. [37], modeled on that described by Hendry and Grime [38], are of similar design to a manually operated fiber-testing machine. Plant material is clamped between screw-type clamps, and tension is applied to the plant material by winding up a spring-operated crank until fracture results. These two forms of instrumentation provide a subjective measure of fracture force and can fulfill the objectives of an experiment designed to compare tensile strength across a range of plant species or genera under a prescribed set of environmental conditions. Translation to understanding grazing mechanics is, however, limited. The Instron testing instruments reported by Henry et al. [39] and Wright and Illius [15] offer tighter control over acceleration and greater precision in recording fracture force. Additionally, the machines are compatible with computers and/or plotters that plot the force–displacement curve for each test specimen, which provides visual reinforcement of the timing selection of fracture.

Following the choice of apparatus, a clamp that minimizes slippage while simultaneously minimizing damage from the compression force applied to the test specimen at the site of the clamp has been a serious obstacle in acquiring reliable and repeatable estimates of tensile strength. Samples that fracture at the vicinity of the clamp should be discarded, with good reason, because their inclusion will lead to erroneous data. Not all studies detail the clamp type used, but square clamps are often surfaced with rubber and/or emery paper [31]. Griffiths [40] used one clamp surfaced with emery paper, and a second clamp with one side surfaced with rubber that closed against a solid square cross bar, displaced 10 mm from the top of the clamp, to simulate the incisor grip. Henry et al. [39] devised cylindrical clamps and argued that the clamp method eliminated stress concentration by allowing a gradual increase in the transmitted force to the specimen around the periphery of the cylinder, avoiding fracture at the clamp. However, cylindrical clamps necessitate long specimen lengths and restrict the opportunities for assessment of the fracture mechanics of short vegetative forage material. Vincent [41] recommended that specimens be glued to tabs of aluminum, which could then be held by clamps, and a more recent study [34] described a "glue and screw" technique where the test specimen was glued into the slotted heads of screws.

Notching has been used to control the site of fracture, involving the creation of a small notch at the edge of the test piece using a needle or razor blade. Many monocotyledons with their parallel venation do not transmit shear and are considered notch-insensitive [16], although there are exceptions, and notch insensitivity should not be assumed to apply to all genera. Notch insensitivity implies that a single fiber can be broken without affecting the strength of the test specimen since the stress is distributed evenly among the remaining fibers. Given the ease with which notching can be carried out and the advantages that it offers in minimizing the number of samples fracturing in the region of the clamp, it is perhaps rather surprising the procedure has not been more widely utilized. Wright and Illius [15] assessed the fracture properties of leaf and pseudostem in the same manner, although the pseudostem samples could not be notched. By contrast, to assess the tensile strength of culms, Hongo and Akimoto [25] used the chuck of an electric drill as the clamp rather than jaw clamps (specifications not given) that had been used for leaves. Culms were wrapped with emery paper and enclosed within a thin rubber tube with one end inserted into the chuck. A further concern over the use of clamps is the need to standardize clamp compression force between specimen tests. Screw-type clamps [21,42] lead to inconsistency in the compression force between tests whereas pneumatic clamps, often found on floor-positioned or bench-top Instron testing machines [15], eliminate this problem.

Implicit in materials science is the principle that when any load is applied to an object, strain energy will be stored in that object. Atkins and Mai [43] suggested that it was critical that the test specimen be unloaded prior to specimen failure, and the most appropriate method to ensure that that this has occurred is to conduct mechanical tests at a slow and constant speed. Wide ranges of extension rates have been reported from as low as 5 mm/min [15] to 10 to 15 mm/min [25,26,39] through to 50 mm/min [34,44]. However, it must be noted that the removal of elastic strain energy from the test specimen is only a prerequisite where the stress–strain relation is to be assessed, and, therefore, the work to fracture is calculated using the work–area method. For estimates of fracture force, the use of faster rates of extension would more closely mimic the fast rates of head acceleration used by ruminants during bite severance [45].

The majority of studies utilize the youngest fully expanded leaf, which is usually the first leaf to make contact with the animal's mouth. However, some studies have involved measurements on older leaves, which require greater force to fracture [15,39], so caution must be exercised when comparing studies. Although an intact, whole leaf is usually tested, there have been reports of tests performed on an excised strip of leaf, running parallel with the midrib [34], reinforcing the point that it is critical to assess what the tensile strength estimate is related to. Moreover, the site of fracture can hamper comparisons across studies. Leaves are not homogeneous along their length, and thus the position of measurement can influence the estimate of tensile strength. This was the reason why Evans [31] assessed tensile strength as being equal to the breaking load divided by the dry weight of a 5 cm length, despite having correctly defined tensile strength in the introduction to the study. MacAdam and Mayland [35] showed that the position of maximum leaf width does not equate with the midpoint of the leaf and that there is a region, approximately 50 to 80 mm long, of constant maximal width in fully expanded tall fescue (*Festuca arundincea*) leaves. This approach contrasts with that of Zhang et al. [26] who used 20 cm lengths of the central portion of orchard grass (*Dactylis glomerata*) leaves and Sun and Liddle [22] who clamped leaves one-third of the distance from each end. It is interesting to note that Wright and Illius [15] did not assess tensile strength; rather, they quantified the energy required to fracture the specimen standardized for cross section, avoiding any confounding variation due to contrasting sites of fracture between plant species.

The instrumentation described has all involved plant material being cut from the field or from pots in chamber-grown complexes. A portable instrument consisting of modified pliers with a strain gauge to assess tensile strength of plant specimens growing *in situ* was developed by Westfall et al. [46]. However, it was not clear how clamp compression and acceleration of the longitudinal pull were controlled, factors that have been discussed previously, and the apparatus probably offers little advantage over the other forms of instrumentation other than the fact that plants are naturally anchored.

5.7 APPLICATION OF PLANT FRACTURE MECHANICS TO FORAGING STRATEGIES

Why are plant ecologists interested in the application of materials science theory in understanding bite mechanics? It is energetically profitable for animals to penetrate deep into the sward canopy [27,47], and yet empirical evidence has shown that ruminants forage using a stratum-orientated depletion style at the patch scale, where a stratum is defined as a depth of sward canopy confined between two distinct lines. Such a strategy implies that bites from one stratum are removed before penetration into a second stratum [48,49], with the depth of the stratum determined by the magnitude of structural complexity [50,51]. Understanding the conceptual basis of bite depth has been the subject of ongoing research over the past decade with a strong focus on the linkage with bite force. The hypotheses of *Summit Force, Balancing Reward* against the cost of bite procurement and *Marginal Revenue* have all been tested, but evidence to support any of the hypotheses is weak [33]. The original Summit Force theory implied that once a maximum force was attained, the bite dimensions, primarily bite area, would be moderated to maintain a constant bite force. Evidence does suggest that animals moderate the bite area when faced with increased strength of plant components or increased tiller density — and thus bite resistance — but the adjustment is much smaller than the magnitude of the increase in bite resistance [2,47,52,53]. Further, several studies have shown no constancy in the force per bite relative to the reward attained [27,52,54]. A study by Illius et al. [27] found that goats offered a group of broad-leaved grasses grazed to different residual sward heights — and so contrasting bite depths — and exerted variation in bite force to sever the bites, with bite depth being equated to common marginal revenue. However, in comparison with a group of fine-leaved grasses, the value for common Marginal Revenue differed, leaving insufficient evidence for the acceptance of the Marginal Revenue hypothesis.

In understanding the mechanical interactions and grazing strategies of ruminants, much of the interest lies in defining the force that animals exert in severing a bite. Griffiths and Gordon [33] contended that there are two important animal-based terms when assessing the magnitude of force animals exert in procuring forage:

- Bite force
- Biting effort

Peak bite force represents the maximum force that an animal exerts in three-dimensional space to sever a bite of herbage and has been referred to in studies with other vertebrates as the "maximal bite capacity" [55]. Care is required to differentiate between bite force and the force generated by the masseter muscle during the mechanical action of clamping forage between incisors and dental pad. Likewise, bite force should be differentiated from the force applied during food comminution, where substantial forces are required during occlusal motions of crushing and grinding the food bolus against the molars. The peak force exerted in severance of forage material is thought of as a response to muscle moving against a fixed anchor of body mass, and to generate the cyclic patterns depicted on force–time curves from biomechanical force plates, animals must move the mass of the head in rhythm. Biting effort, as defined by Griffiths and Gordon [33], is primarily determined by bite force but is regulated by the components of plant resistance and animal resistance, e.g., head resistance. It is conceivable that other animal anatomical characteristics will shape biting effort. These authors argued that biting effort represents a holistic approach to understanding the dynamic nature of the animal's response to severance constraints arising from forage complexity.

How can biting effort be measured in real terms? The area under a force–time curve, as output from a biomechanical force plate, can be used to represent the work done. Since the forces that a grazing animal exerts in severing a bite of herbage are related to time, an accurate assessment of energy is somewhat difficult to derive in the absence of a measure of displacement. A conservative assumption on displacement from force–displacement curves could be used [47], but in the absence of more evidence, the validity of the approach could be questioned. However, we would expect that bites of a longer duration would utilize more muscular effort than those bites that result in rapid fracture. The integral of force over time can be used as a measure of bite effort. This appears valid on the assumption that muscles use energy for isometric contractions and that energy is approximately proportional to force [56], so the term "bite effort" carries a physical meaning in a mechanical sense, even if it is somewhat subjective.

5.8 INSTRUMENTATION FOR MEASURING BITE FORCE AT THE ANIMAL LEVEL

Two approaches in quantitatively measuring the force that ruminants exert in the removal of a bite of forage have been reported:

- Indirectly from plant fracture properties of leaf and stem
- Directly from biomechanical force instruments

5.8.1 Prediction of Bite Force from Assessment of Plant Fracture Properties

The translation of plant fracture properties into the bite force exerted appears attractive, but few researchers have attempted to relate fracture properties to bite force. A classic case of integration across the science and engineering disciplines is given in the two studies by Wright and Illius [15] and Illius et al. [27] on the application of the fracture mechanics in a group of broad- and narrow-leaved grasses and measured bite dimensions in goats. These authors determined the predicted bite force by substituting the residual heights from boxed grazed swards into polynomial regression equations summarizing force–canopy structure relationships based on the measured fracture properties. With a similar focus to that of Illius and colleagues, Tharmaraj et al. [47] estimated the resistance of material *in situ* within a fixed bite area of 100 cm^2 (an average value for cattle) progressively down the sward canopy profile using a tensile apparatus. Their polynomial curves showed marked changes in bite force around 0.6 to 0.7 of sward height, which is in agreement with other published literature on pseudostem height in grazed swards [57]. The residual heights from paddock-scale grazing sessions (greater than 1 h) were used to predict bite force after adjustment for the measured bite area.

Despite these inspiring studies, there are limitations with the indirect approach. The most obvious is the assumption involved in modeling bite force as the summation of the strength of leaf and stem, and the number of organs severed in the bite [27,47] although Illius et al. [27] had noted that canopy structure has a greater bearing on bite force than individual plant strength. Nevertheless, the scaling of bite force with number of leaves harvested as determined by the area of sward encompassed within a bite, using a constant, might be ambiguous. A constant scaling factor would assume

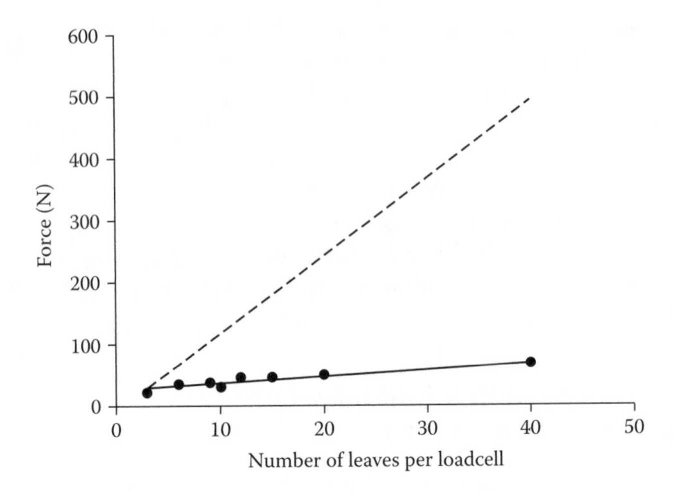

FIGURE 5.3 Relationship between the number of leaves grazed and exerted force per bite (solid circles and fitted linear relationship) by sheep and the predicted bite force using a 1:1 relationship (dashed line) based on the strength of a single leaf from tensile tests.

that all leaves fracture simultaneously under load. From aural observations we know that this characteristic does not hold true; output generated from biomechanical force instruments (see the next section) consistently show that the force–time curve can consist of multiple peaks. Only very recently has there been quantification of the variation between predictions of bite force from fracture properties and the actual forces exerted by the animal. Data from Zhang et al. [26] and unpublished data by W.M. Griffiths and also by E.A. Laca indicate that there is a large discrepancy between the indirect and direct estimates of bite force. Figure 5.3, drawn from the data of Zhang et al. [26], clearly show that the force exerted to sever a bite of a known number of leaves was not proportional to the sum of the strength of individual leaves. However, it is conceivable that the timing of leaf fracture is only one contributor to the force per leaf response curve, and we are left wondering about mechanisms of fracture at the animal level.

A stiff tensile testing machine capable of recording in one dimension is restricted to modeling the action of severance by vertical pull, excluding causes of friction between forage material and the incisors. The procurement action in three-dimensional space has been clearly illustrated in the work of Griffiths and Gordon [33] and Zhang et al. [26]. A plant tiller can essentially be viewed as a moment arm, and creation of an angle perpendicular to the tiller from the rooted position allows for greater friction between the incisors and dental pad, avoiding an action where the incisors would merely scrape against the fibrous veins in the lamina. This leads us to think about the dynamics of vein fracture. According to materials science theory, many grass leaves do not become weaker with the introduction of a notch [41], and, therefore, the biting action of ruminants will not function as a concentrator of stress, making the grass easier to break. Bending flexibility of stems has been assessed as an index for trampling resistance [22], with marked variation between plant species in the angle at which stems break. Zhang et al. [26] reported that plant material was generally severed when the angle was around 60° perpendicular to the vertical plane. While there has been no quantitative evidence for or against the argument of a reduction in strength of plant organs after damage inflicted from being held at an angle over the incisors and exposed to compression stress during the clamping action against the dental pad, it is plausible that the clamping action induces damage to the outer fibers of the lamina from buckling. This would cause the fibers to weaken in much the same way as compression creasing in wood. The implication of such an action is that tensile strength may not be the only plant mechanical parameter determining the ease of bite procurement [39].

Furthermore, compared with the tensile testing instrumentation, animals are soft machines constructed from muscles containing fibers that are extremely elastic and, therefore, have a large capability to store elastic-strain energy [58]. There does not appear to have been any examination of the storage of strain energy in the muscles of ruminant species, but it is conceivable that ruminants may exert sufficient force to create cracks in some of the grasped forage, and the energy stored within muscles is then used to propagate a crack [59]. However, such a mechanism is likely to only be valid if the forage being harvested is considered notch-sensitive (e.g., *Stipa gigantea*) [60]. Rate-dependent effects have been given as a more likely explanation for notch-insensitive species [61]. King and Vincent [61] suggested that the stretching of grasped leaves allows for the storage of elastic-strain energy in the respective leaves, and the quick jerking motion of the head then facilitates the generation of a shock wave through the leaves. This subsequently influences the fracture of the remaining fibers, most probably as a result of insufficient time for the stress to be redistributed among the remaining fibers, giving rise to an overload in strain energy [15]. This would go some way to explaining the faster rates of head acceleration in small-bodied ruminants, with smaller muscle mass than larger species like cattle [45], and suggests very clearly that acceleration of the head plays a major contributing role in the effort that animals exert in severing a bite.

The fibrous nature of forage material ingested and frequent contamination with soil particles is primarily responsible for tooth wear in ruminants, lending to variation in the incisor–pad configuration [62]. However, there is no consistent evidence of the relation between incisor wear and reduced levels of animal performance [63], although a relationship between dental erosion of the molars and animal growth has been documented [64]. Quantification of the role of dentition on procurement capability and mechanism is largely unexplored, but Hongo et al. [65] showed that sheep that had shed their two central temporary incisors found it difficult to increase the number of leaves procured with increasing sward density, a result that was reversed when the sheep grazed identical swards after the eruption and growth of the two permanent central incisors [65]. Although the study was conducted with a small number of animals ($n = 3$) that had little opportunity to adapt to foraging without the temporary central incisors, the results clearly suggest that incisors do inflict damage weakening plant tissue and/or forming a point of friction. This result is supported by the finding that in the absence of the two temporary central incisors, the force exerted per grasped leaf was significantly higher than when the animals grazed with complete dentition.

In summary, although it might be argued that the predicted bite forces in the study of Illius et al. [27] were within the range measured directly by Hughes et al. [66] for sheep grazing broad-leaved grasses, the discussion above does not lend support for the indirect approach (sum of individual plant strength and the number of leaves severed) for quantifying bite force. There are clear indications that tensile strength can provide an indication of plant resistance to fracture, but this information cannot be translated using bite dimensions to predict the force exerted by ruminants. Plant fracture properties will, however, continue to provide valuable information on leaf and stem characteristics that can assist with screening for plant traits that lead to improved animal performance.

5.8.2 Biomechanical Force Instruments

Instruments that measure force have had widespread use, having been extensively used to provide insights into the mechanics of locomotion, for example, in understanding the jumping behavior of animals [67] and in gait and foot conformation studies [68]. Furthermore, force transducers have been used in assessing maximal bite force in lizards [69], turtles [70], and bats [55], and also for determining the magnitude of resistance to fracture of individual prey [69]. In ruminants, bite force has been assessed for goats, sheep, deer, and cattle [25,26,44,52,66,71]. Portable force plates used in studies with ruminants have had load transducers mounted underneath a flat surface, and the forces and moments acting on the top surface of the plate using strain gauge technology were recorded. The plates avoid many of the constraints associated with indirect assessment of bite force. Further, they exhibit a high capability to record frequent and swift movements with accuracy, and the forces and moments can be broken down into three vectors acting along the three coordinate system axes: x, y, and z. The plates can be positioned in a pit [66], or alternatively animals graze from an elevated platform [44] with swards secured to the plate such that the animals' feet are level with the base of the sward, simulating a natural grazing angle.

The threading of individual leaves through holes arranged in wooden modules created the sward board pioneered by Black and Kenney [72]. In an interesting development, Hongo and co-workers [25,26] have transformed the sward board into a force plate where the leaves threaded through each hole are attached to a transducer and the force exerted per load-cell can be recorded. The sward board offers control over determination of the number of leaves removed per bite and the force per leaf compared with studies that present animals with boxed swards of herbage secured to a force plate. Boxed swards, however, present a natural sward that is not confounded by structured spaces between clumps of leaves, the number of potential bites is greater, and there are not the time limitations on board (module) preparation that are common to hand-constructed sward boards.

The greatest limitation to using biomechanical plates to measure force lies with the maximum available herbage that can be grazed at any single time, limiting grazing sessions to shorter than 2 to 3 min. Daily intake has been shown to be of some magnitude lower than that predicted from scaling up the instantaneous rate of intake by daily grazing time, and the inability to predict whether animals maintain instantaneous bite forces across temporal scales or lower the exerted force to maintain a biting momentum is a limitation with such instrumentation. Nevertheless, force boards and plates avoid the confounding issues associated with one-dimensional laboratory-based instrumentation and allow for an objective measure of the force exerted by the animal. However, with the recent attention given to quantification of bite force in ruminants, it has become apparent that there are issues surrounding output interpretation that need resolving. Output generated from force plates has illustrated very clearly the incidence of multiple peaks, with at least two and four peaks recorded for sheep and cattle, respectively [25,26]. The incidence of multiple peaks would be expected to correlate with capture rate since a faster capture rate allows for accurate representation of the dynamics of the force being applied to the forage material, and likewise a slow capture rate will see the force accumulated into fewer data points. This raises concerns as to how bite force should be determined. A more representative assessment might be to describe force in terms of the number of peaks rather than force per bite, but in many instances, this may only be relevant for bites removed from dense and/or mature swards. Calculating a reward to cost ratio using the peak bite force recording can lead to erroneous predictions, which can only leave us wondering whether peak bite force is the parameter we seek.

On vegetative swards where the upper stratum offers little impediment to the prehension of leaf, the vertical vector is dominant, but with increasing sward structural complexity, the relative magnitude of the longitudinal and lateral vectors increases [44]. Unpublished data [44] from deer, sheep, and cattle suggests that the relative change in the magnitude of the force vectors, in response to the spatial arrangement of morphological organs, appears to be less significant for smaller bodied animals. There is no comparable data since the individual vectors appear to be independent in the data sets generated by Hongo and co-workers [25,26]. Nevertheless, clarification of the direction of head movement and dominant force vector across animal body size could provide insights into understanding bite procurement. Descriptive observations in the literature suggest that cattle and sheep sever plant material using a backward jerk of the head [73,74], and deer tear herbage with an upward and downward jerking action [75], although these observations may be confounded by the physical structure of the vegetation that animals were grazing. Hongo and Akimoto [25] suggested that the observation that cattle moved their heads in a backward direction when grazing mixtures of perennial ryegrass and reed canary grass culms was a strategy to minimize the ingestion of stems because the jerking action in a backward plane permitted stiff stems to spring back from the clamped material, and these authors described the foraging action as a "comb-out" strategy.

5.9 BITING EFFORT

The quantification of biting effort is an under-researched area deserving further investigation, particularly in view of the mounting evidence on the concern over the validity of peak bite force from recent studies in ruminants. Biting effort is an index of efficiency that takes into account the activity budget for gathering as well as severance of forage.

One of the exciting challenges ahead lies in understanding the mechanistic basis for the foraging strategy contrasts between animals of different body mass. As the bite composition of a large-bodied ruminant generally contains more fiber relative to that of a small-bodied ruminant, larger-bodied ruminants must rely on their strength to harvest forage material to meet their requirements and in accordance with forage supply. Animal mechanics permit this since it is believed that bite force scales allometrically with body mass on the premise that force generated by muscle is proportional to the muscular cross-sectional area ($W^{0.67}$). Consequently, the costs of severance for a given force are lower for animal species of greater body mass. This begs the question of how small-bodied ruminants cope when faced with the same constraints as large-bodied ruminants. When presented with identical swards, sheep and deer [28] have been observed to exhibit a very similar grazing strategy despite a threefold difference in body mass between the two species. While it appears that there is a fixed force exerted on a bite of forage directly from the momentum of the head, as supported by the observation that force does not increase linearly with increasing number of leaves procured at very low leaf density [26], there has been little quantification of the effort and costs incurred in upholding the tension in the muscles of the neck to maintain head momentum, although Illius et al. [27] commented that it must be considerable.

Allometric relationships between elements of feeding behavior in ruminants and body mass have been studied in detail [76–78], and it would be reasonable to think that an allometric relationship between bite force and body mass has evolved. Figure 5.4 depicts the allometric relationship between body mass and peak bite force for a range of animal species (goat, sheep, deer, cattle, and horse) from the limited data available in the literature [25,26,44,52,65]. The calculated exponent (0.77 ± 0.18) indicates that peak bite force scales allometrically with body mass. However, the high standard error reflects the greater variation among the smaller-bodied animals, and the reasons for such responses should be investigated. Further, there have been inconsistent reports on the inter- and intraspecific allometric relationships between body mass and bite depth [78,79]. In a horse–cattle comparison, Hongo and Akimoto [25] hypothesized that horses, which have anterior incisors on their upper jaw, would be more efficient grazers on complex swards consisting of perennial ryegrass leaves and reed canary grass culms, but instead they found that horses exhibited little competitive advantage over cattle.

A number of studies have shown that incisor breadth scales with body mass [9,78], and total masseter weight, which reflects masseter tissue size, is correlated with body mass [80]. However, unlike in some carnivores [81], there appears to be no compensation between mandibular length and masseter muscle mass in ruminants despite the fact that mandibular length scales with body mass. Given that the mandible acts as a moment, the absence of any correlation between mandibular length and masseter muscle mass implies a weaker gripping force at the incisors. Consequently, animals with a high mandibular length to body mass ratio (e.g., goats and moose) are, theoretically, mechanically disadvantaged when procuring material that is tall and/or of high tensile strength. However, little is known about the mechanical consequences of changes in mandibular length on the torque around the temporomandibular joint in ruminants. Given that torque is the force over distance,

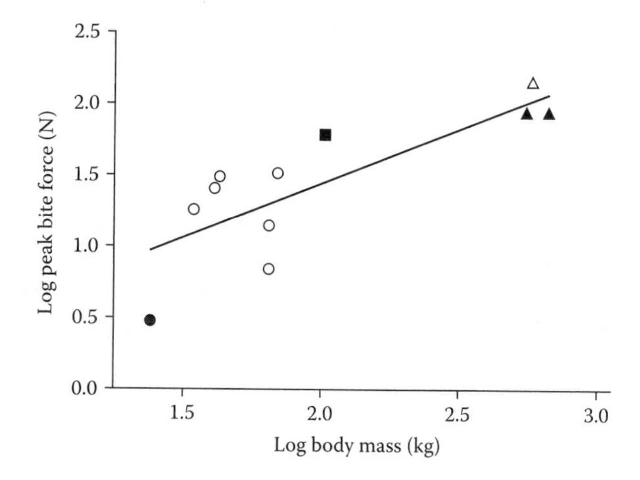

FIGURE 5.4 Relationship between peak bite force (N) and body mass (kg) for a range of animal species. Solid circle: goat; open circle: sheep; solid square: deer; solid triangle: cattle; open triangle: horse. Solid line represents the fitted linear relationship ($y = 0.08 + 0.77x \pm 0.18$, $p = 0.002$, $R^2 = 0.818$).

it could well be that the balance between incisor breadth and the magnitude of protrusion that has evolved in ruminants not only caters for selectivity but also minimizes the competitive advantage of grazers over browsers of a similar body mass. Compensation mechanisms for body mass are regularly observed in other species, e.g., in flying insects [82], so it is more than likely that forms of mechanical design, aside from rates of acceleration, contribute to the competitiveness of smaller-bodied ruminants, and the challenge is to think of approaches to investigate those features.

Quantification of biting effort would be a step forward, but success will be determined by the accuracy of defining the endpoints of bites on force–time curves. Since electric equipment generates noise, the potential for the location of endpoints to become clouded can be significant, particularly when the ratio between exerted forces and noise is weak. There have only been a few studies that have examined biting effort [44,47,52], limiting any generalization across plant species and animal body mass. Hughes [52] found that the effort expended by sheep grazing prairie grass (*Bromus willdenowii*) and tall fescue (*Festuca arundincea*) swards was similar but lower than that on perennial ryegrass (*Lolium perenne*) swards. Unpublished preliminary data of Griffiths [44] showed that the across-bite contrast in biting effort from cattle grazing swards of Italian ryegrass (*L. multiflorum*) exceeded the contrast in peak bite force. Nevertheless, the data showed evidence of similarity in bite effort for individual bites where there were strong contrasts in peak bite force. Further research on biting effort would help to clarify the consistency of biting effort across animal species and sward complexity. The challenge is to investigate whether animal mechanics can assist with understanding the mechanisms for diet selection and the bite dimensions. It will only be after we have addressed questions concerning biting effort that we will be able to draw conclusions regarding the mechanistic basis for the contrast or similarity in grazing strategies across species of contrasting body mass and understand functionality of cospecies grazing in ecosystems.

5.10 CONCLUSION

In summary, although there has been renewed interest in quantifying plant fracture mechanics, particularly plant tensile strength, indications are that the procurement of forage material is far more complex than can be defined indirectly from assessment of plant mechanical parameters. Looking to the future, there is a call for a more holistic approach that includes the quantification of biting effort from assessment of mechanical properties directly. There is tremendous scope for applying engineering principles in understanding the foraging strategies used by animals of contrasting body mass, and while much of this research in the short term will be conducted at the small scale, it is foreseeable that in the longer term, the work will need to span temporal scales to achieve wider application across ecosystems.

ACKNOWLEDGMENTS

I would like to thank J. Hodgson and I.J. Gordon for their constructive comments and J. Perez-Barberia, who kindly drew Figure 5.1.

REFERENCES

1. Charnov, E.L., Optimal foraging: The marginal value theorem, *Theor. Popul. Biol.*, 9, 129, 1976.
2. Laca, E.A. et al., Effects of sward height and bulk-density on bite dimensions of cattle grazing homogeneous swards, *Grass Forage Sci.*, 47, 91, 1992.
3. Ginnett, T.F. et al., Patch depression in grazers: The roles of biomass distribution and residual stems, *Funct. Ecol.*, 13, 37, 1999.
4. Baumont, R. et al., A mechanistic model of intake and grazing behaviour in sheep integrating sward architecture and animal decisions, *Anim. Feed Sci. Tech.*, 112, 5, 2004.
5. Hofmann, R.R. and Stewart, D.R.M., Grazers and browsers: A classification based on the stomach structure and feeding habits of East African ruminants, *Mammalia*, 36, 226, 1972.
6. Iason, G.R. and Van Wieren, S.E., Digestive and ingestive adaptations of mammalian herbivores to low quality forage, in *Herbivores: Between Plants and Predators*, Olf, H., Brown, V.K. and Brent, R., Eds., Blackwells Scientific Publications, Oxford, 1999, p. 337.
7. Perez-Barberia, F.J. and Gordon, I.J., Relationships between oral morphology and feeding style in the ungulata: A phylogenetically controlled evaluation, *Proc. R. Soc. Lond. B-Biol. Sci.*, 268, 1023, 2001.
8. Jarman, P.J., The social organisation of antelope in relation to their ecology, *Behaviour,* 48, 215, 1974.
9. Gordon, I.J. and Illius, A.W., Incisor arcade structure and diet selection in ruminants, *Funct. Ecol.*, 2, 15, 1988.

10. Ungar, E.D., Ingestive behaviour, in *The Ecology and Management of Grazing Systems*, Hodgson, J. and Illius, A.W., Eds., CAB International, Wallingford, 1996, p. 185.
11. Demment, M.W. and Laca, E.A., The grazing ruminant: Models and experimental techniques to relate sward structure and intake, *Proc. 7th World Conf. Anim. Prod.*, 1993, p. 439.
12. Ruyle, G.B. and Dwyer, D.D., Feeding stations of sheep as an indicator of diminished forage supply, *J. Anim. Sci.*, 61, 349, 1985.
13. Illius, A.W. and Gordon, I.J., Constraints on diet selection and foraging behaviour in mammalian herbivores, in *Behavioural Mechanisms of Food Selection*, Hughes, R.N., Ed. Springer-Verlag, Berlin, 1990,p. 369.
14. Wilman, D., Mtengeti, E.J., and Moseley, G., Physical structure of twelve forage species in relation to rate of intake by sheep, *J. Agric. Sci. (Cambridge),* 126, 277, 1996.
15. Wright, W. and Illius, A.W., A comparative study of the fracture properties of 5 grasses, *Funct. Ecol.*, 9, 269, 1995.
16. Vincent, J.F.V., The influence of water content on the stiffness and fracture properties of grass leaves, *Grass Forage Sci.,* 38, 107, 1983.
17. Inoue, T. et al., Effects of leaf shear breaking load on the feeding value of perennial ryegrass (*Lolium perenne*) for sheep. 1. Effects on leaf anatomy and morphology, *J. Agric. Sci.,* 123, 129, 1994.
18. Hendricksen, R. and Minson, D.J., The feed intake and grazing behaviour of cattle grazing a crop of *Lablab purpureus* cv. Rongai, *J. Agric. Sci. (Cambridge),* 95, 547, 1980.
19. Theron, E.P. and Booysen, P.D.V., Palatability in grasses, *Proc. Grass. Soc. South Afr.,* 11, 111, 1966.
20. Mayland, H.F. et al., Nonstructural carbohydrates in tall fescue cultivars: Relationship to animal preference, *Agron. J.,* 92, 1203, 2000.
21. Diaz, S., Noy-Meir, I., and Cabido, M., Can grazing response of herbaceous plants be predicted from simple vegetative traits? *J. Appl. Ecol.,* 38, 497, 2001.
22. Sun, D. and Liddle, M.J., Trampling resistance, stem flexibility and leaf strength in nine Australian grasses and herbs, *Biol. Conser.,* 65, 35, 1993.
23. Thom, E.R. et al., Relationship of tillering and morphological characteristics of two perennial ryegrass lines to "pulling" when grazed by dairy cows, *N.Z. J. Agric. Res.,* 46, 15, 2003.
24. Henry, D.A., Simpson, R.J., and Macmillan, R.H., Seasonal changes and the effect of temperature and leaf moisture content on intrinsic shear strength of leaves of pasture grasses, *Aust. J. Agric. Res.,* 51, 823, 2000.
25. Hongo, A. and Akimoto, M., The role of incisors in selective grazing by cattle and horses, *J. Agric. Sci.,* 140, 469, 2003.
26. Zhang, J., Akimoto, M., and Hongo, A., Effects of applied nitrogen and leaf density of orchardgrass (*Dactylis glomerata* l.) on the grazing behavior of sheep, *Grass. Sci.,* 49, 563, 2004.
27. Illius, A.W. et al., Costs and benefits of foraging on grasses varying in canopy structure and resistance to defoliation, *Funct. Ecol.,* 9, 894, 1995.
28. Mitchell, R.J., Hodgson, J., and Clark, D.A., The effect of varying leafy sward height and bulk density on the ingestive behaviour of young deer and sheep, *Proc. N.Z. Soc. Anim. Prod.,* 51, 159, 1991.
29. Halyk, R.M. and Hurlbut, L.W., Tensile and shear strength characteristics of alfalfa stems, *Trans. ASAE,* 11, 256, 1968.

30. Iwaasa, A.D. et al., A shearing technique measuring resistance properties of plant stems, *Anim. Feed Sci. Tech.*, 57, 225, 1996.
31. Evans, P.S., A study of leaf strength in four ryegrass varieties, *N.Z. J. Agric. Res.*, 7, 508, 1964.
32. Vincent, J.F.V., Fracture, in *Biomechanics – Material: A Practical Approach*, Vincent, J.F.V., Ed., Oxford University Press, New York, 1992, p. 192.
33. Griffiths, W.M. and Gordon, I.J., Sward structural resistance and biting effort in grazing ruminants, *Anim. Res.*, 52, 145, 2003.
34. Aranwela, N., Sanson, G., and Read, J., Methods of assessing leaf-fracture properties, *New Phytol.*, 144, 369, 1999.
35. MacAdam, J.W. and Mayland, H.F., The relationship of leaf strength to cattle preference in tall fescue cultivars, *Agron. J.*, 95, 414, 2003.
36. O'Reagain, P.J., Haller, M., and Zacharias, P.J.K., Relationship between sward structure and dietary quality and intake in cattle grazing humid sour grassveld in South Africa, *Proc. XVII Inter. Grass. Cong.*, 730, 1993.
37. Adler, P.B. et al., Functional traits of graminoids in semi-arid steppes: A test of grazing histories, *J. Appl. Ecol.*, 41, 653, 2004.
38. Hendry, G.A.F. and Grime, J.P., *Methods in Comparative Plant Ecology: A Laboratory Manual*, 1st ed., Chapman & Hall, London, 1993.
39. Henry, D.A., Macmillan, R.H., and Simpson, R.J., Measurement of the shear and tensile fracture properties of leaves of pasture grasses, *Aust. J. Agric. Res.*, 47, 587, 1996.
40. Griffiths, W.M., Sward structural characteristics and selective foraging behaviour in dairy cows, Ph.D. thesis, Massey University, New Zealand, 1999.
41. Vincent, J.F.V., The mechanical design of grass, *J. Mat. Sci.*, 17, 856, 1982.
42. Perez-Harguindeguy, N. et al., Leaf traits and herbivore selection in the field and in cafeteria experiments, *Austral. Ecol.*, 28, 642, 2003.
43. Atkins, A.H. and Mai, Y.W., *Elastic and Plastic Fracture*, 1st ed., Ellis Horwood Ltd, Chicester, U.K., 1985.
44. Griffiths, W.M., personal communication.
45. Chambers, A.R.M., Hodgson, J., and Milne, J.A., The development and use of equipment for the automatic recording of ingestive behavior in sheep and cattle, *Grass Forage Sci.*, 36, 97, 1981.
46. Westfall, R.H. et al., The development of a leaf tensilmeter for *in situ* measurement of leaf tensile strength, *J. Grass. Soc. South. Afr.*, 9, 50, 1992.
47. Tharmaraj, J. et al., Defoliation pattern, foraging behaviour and diet selection by lactating dairy cows in response to sward height and herbage allowance of a ryegrass-dominated pasture, *Grass Forage Sci.*, 58, 225, 2003.
48. Ungar, E.D. and Noymeir, I., Herbage intake in relation to availability and sward structure – grazing processes and optimal foraging, *J. Appl. Ecol.*, 25, 1045, 1988.
49. Ungar, E.D. and Ravid, N., Bite horizons and dimensions for cattle grazing herbage to high levels of depletion, *Grass Forage Sci.*, 54, 357, 1999.
50. Bergman, C.M., Fryxell, J.M., and Gates, C.G., The effect of tissue complexity and sward height on the functional response of wood bison, *Funct. Ecol.*, 14, 61, 2000.
51. Griffiths, W.M., Hodgson, J., and Arnold, G.C., The influence of sward canopy structure on foraging decisions by grazing cattle. II. Regulation of bite depth, *Grass Forage Sci.*, 58, 125, 2003.
52. Hughes, T.P., Sward structure and intake of ruminants, Ph.D. thesis, Lincoln University, New Zealand, 1990.

53. Mitchell, R.J., The effects of sward height, bulk density and tiller structure on the ingestive behaviour of Red deer and Romney sheep, Ph.D. thesis, Massey University, New Zealand, 1995.
54. Tharmaraj, J., Chapman, D.F., and Egan, A.R., Bite breaking properties of some pasture species, *Anim. Prod. Aust.,* 22, 360, 1998.
55. Aguirre, L.F. et al., Ecomorphological analysis of trophic niche partitioning in a tropical savannah bat community, *Proc. R. Soc. Lond. B-Biol. Sci.,* 269, 1271, 2002.
56. Peters, R.H., *The Ecological Implications of Body Size*, 1st ed., Cambridge University Press, Cambridge, 1983.
57. Parsons, A.J. et al., Diet preference of sheep — effects of recent diet, physiological-state and species abundance, *J. Anim. Ecol.,* 63, 465, 1994.
58. Alexander, R.M. and Bennet-Clark, H.C., Storage of elastic strain energy in muscle and other tissues, *Nature,* 265, 114, 1977.
59. Wright, W. and Vincent, J.F.V., Herbivory and the mechanics of fracture in plants, *Biol. Rev. Cambridge Phil. Soc.,* 71, 401, 1996.
60. Vincent, J.F.V., Strength and fracture of grasses, *J. Mat. Sci.,* 26, 1947, 1991.
61. King, M.J. and Vincent, J.F.V., Static and dynamic fracture properties of the leaf of New Zealand flax *Phormium tenax* (Phormiaceae: Monocotyledones), *Proc. R. Soc. Lond. B-Biol. Sci.,* 263, 521, 1996.
62. Northey, R.D. and Hawley, J.G., The tooth to pad relationship in sheep — some mechanical considerations, *N.Z. J. Agric. Res.,* 18, 133, 1975.
63. Dove, H. and Milne, J.A., An evaluation of the effects of incisor dentition and of age on the performance of lactating ewes and their lambs, *Anim. Prod.,* 53, 183, 1991.
64. Rogers, G.M. et al., Dental wear and growth performance in steers fed sweet potato cannery waste, *J. Am. Vet. Med. Assoc.,* 214, 681, 1999.
65. Hongo, A. et al., Changes in biting force with incisor dentition of sheep, *Grass Forage Sci.,* 59, 293, 2004.
66. Hughes, T.P. et al., The influence of sward structure on peak bite force and bite weight in sheep, *Proc. N.Z. Soc. Anim. Prod.,* 51, 153, 1991.
67. Toro, E. et al., A biomechanical analysis of intra- and interspecific scaling of jumping and morphology in Caribbean *Anolis* lizards, *J. Expt. Biol.,* 206, 2641, 2003.
68. Webb, N.G. and Clark, M., Livestock foot-floor interactions measured by force and pressure plate, *Farm Build. Prog.*, 23, 1981.
69. Herrel, A. et al., The implications of bite performance for diet in two species of lacertid lizards, *Can. J. Zool.,* 79, 662, 2001.
70. Herrel, A., O'Reilly, J.C., and Richmond, A.M., Evolution of bite performance in turtles, *J. Evol. Biol.,* 15, 1083, 2002.
71. Laca, E.A. et al., An integrated methodology for studying short-term grazing behavior of cattle, *Grass Forage Sci.,* 47, 81, 1992.
72. Black, J.L. and Kenney, P.A., Factors affecting diet selection by sheep 2. Height and density of pasture, *Aust. J. Agric. Res.,* 35, 565, 1984.
73. Ellis, J.E. and Travis, M., Comparative aspects of foraging behaviour of pronghorn antelope and cattle, *J. Appl. Ecol.,* 12, 411, 1975.
74. Arnold, G.W. and Dudzinski, M.L., *Ethology of Free-Ranging Domestic Animals*, 1st ed., Elsevier, New York, 1978.
75. Willms, W., Forage strategy of ruminants, *Rangemans J.,* 5, 72, 1978.
76. Shipley, L.A. et al., The scaling of intake rate in mammalian herbivores, *Am. Nat.,* 143, 1055, 1994.
77. Shipley, L.A. et al., The dynamics and scaling of foraging velocity and encounter rate in mammalian herbivores, *Funct. Ecol.,* 10, 234, 1996.

78. Gordon, I.J., Illius, A.W., and Milne, J.D., Sources of variation in the foraging efficiency of grazing ruminants, *Funct. Ecol.,* 10, 219, 1996.
79. Cangiano, C.A. et al., Effect of liveweight and pasture height on cattle bite dimensions during progressive defoliation, *Aust. J. Agric. Res.,* 53, 541, 2002.
80. Axmacher, H. and Hofmann, R.R., Morphological characteristics of the masseter muscle of 22 ruminant species, *J. Zool. (London),* 215, 463, 1988.
81. Jaslow, C.R., Morphology and digestive efficiency of red foxes (*Vulpes-vulpes*) and gray foxes (*Urocyon-cinereoargenteus*) in relation to diet, *Can. J. Zool.,* 65, 72, 1987.
82. Lehmann, F.O., The constraints of body size on aerodynamics and energetics in flying fruit flies: An integrative view, *Zoology,* 105, 287, 2002.

6 Biomechanics of *Salvia* Flowers: The Role of Lever and Flower Tube in Specialization on Pollinators

Martin Reith, Regine Claßen-Bockhoff, and Thomas Speck

CONTENTS

6.1 INTRODUCTION

6.1.1 BIOMECHANICS AND BEE POLLINATION

Biomechanical interactions between bees and flowers have been known to be involved in pollen transfer in plant species for more than two centuries [e.g., 1–6]. Bees are very important pollinators for many plant species. However, because bees feed their offspring nearly exclusively with pollen, plants have evolved several mechanisms to avoid overexploitation by pollen-collecting bees and to ensure pollen transfer [7–11]. Many of these mechanisms involve specific mechanical features. Four of the most important are buzz pollination and the piston, brush, and lever mechanisms.

Buzz pollination is well-known in many economically important members of the nightshade family (Solanaceae) [12]. Examples include tomato, *Solanum lycopersicum*; pepper, *Capsicum annuum*; and eggplant, *S. melongena*. However, buzz pollination is much more widespread and known to occur in at least 65 plant families [13]. The thecae of many of these plants dehisce only partially, and they only open small pores through which pollen can be released (poricidal anthers). To collect pollen from these plants, bees place their body near the small openings and vibrate the stamen by rapid contraction of their indirect flight muscles. Buchmann and Hurley [13] model the process of pollen release. The bee's buzzing and the consequential vibration of the anther wall transmit energy to the pollen grains inside the anther. Thus, the energy content of the anther increases while pollen grains accumulate more and more kinetic energy by repeated interactions with the walls and with each other. Release of pollen on the other hand diminishes the energy content of the anther. If buzzing continues, the number of pollen grains in the anther and the number of pollen grains that escape the anther diminish, but their average kinetic energy increases. Although this model is an extreme simplification of flower morphology, it is a cornerstone in the understanding of buzz pollination. In the case of buzz pollination, overexploitation of the pollen is prevented because only a limited amount of pollen can be "buzzed out" during a single visit. It would be interesting to test if the size of buzzing pollinators and the energy they can produce are related to morphological characters of the poricidal anthers and the energy needed to buzz the flowers of a given plant species, thereby limiting "mechanically" the pollinator spectrum.

The piston mechanism [14] often found in the pea family (Fabaceae) acts, from a functional point of view, as a pollen pump. The stamens that are hidden in the carina have coalesced filaments and form a tube that encloses the style. Flower visitors have to deform the flower actively to reach the nectar. The forces exerted by visiting insects and the concomitant deformations of the flower cause a relative forward movement of style and stigma in the staminal tube that squeezes a portion of the pollen mass out of the staminal tube. By this mechanism, pollen is transferred to the ventral part of the pollinator's body. During a single visit, only a small amount of pollen is extruded by the pollen pump, ensuring pollen dispensing.

The brush mechanism, which also occurs in the pea family, works in a similar manner [15]. In *Lathyrus latifolius*, a stylar brush underneath the stigma takes the pollen up during flower development. Later when the ripe flower is deformed by forces exerted by a visiting insect, the carina is lowered and the stigma and the stylar brush touch the insect consecutively, thus avoiding self-pollination. In flowers of *Lathyrus latifolius*, 100 millinewtons (mN) are needed to trigger this mechanism [16] (measured with a spring balance). Pollen dispensing is realized by the brush, which deposits only a dosed amount of pollen on each visitor.

Staminal levers in the broadest sense, i.e., stamina that can be tilted with or without a hinge, occur not only in the genus *Salvia* (see Section 6.1.2) but are also found in the Lamiaceae–Prostantheroideae [17,18] and in other families such as the Zingiberaceae. An example from the latter family is *Roscoea purpurea* [19]. As the author points out, the structure of its stamen differs significantly from the *Salvia* lever. Flowers of *R. purpurea* have only one fertile stamen; the two thecae are extended; and their basal parts are sterile and block the flower entrance. The upper part of the thecae is fertile and produces pollen. *Salvia*, in contrast, has two fertile stamens and a different stamen morphology, as described in more detail below. In both taxa, a flower visitor releases the mechanism by pushing against the lever arm, which extends into the flower tube. Thereby, it is loaded with pollen on its head, neck, or back.

6.1.2 A Case Study: The Staminal Lever Mechanism in *Salvia*

The previous examples show that there exist many bee-pollinated plant species in which sophisticated mechanisms are involved in pollen presentation, pollen dispensing, and/or pollen transfer. In the present paper, we focus on the staminal lever mechanism found in sages (genus *Salvia*). Form, function, and ecology of flowers of the genus *Salvia* have been recently reviewed [20,21]. We therefore only summarize briefly the basic morphological and biomechanical parameters of the *Salvia* flower and its staminal lever mechanism, which are important for our studies. Flowers are sympetalous and mainly bilabiate with an upper and a lower lip (Figure 6.1). Four stamens are usually formed, the upper (adaxial) pair being reduced to small to minute staminodes and the lower (abaxial) pair forming the lever mechanism. In each stamen, the lever arms are formed by the extended connective that is fixed to the filament by a miniature joint-like ligament. The lower lever arm is sterile in many species, while the upper one is always fertile, producing one theca. The two stamens are often partially fused, thus forming a functional unit. Pollinators in a "typical" *Salvia* flower push against the base of the sterile lever arm (BC in Figure 6.1), causing a swing of the lever around the jointlike ligament (JL) so that the visitor is touched by the pollen sacs located at the upper, fertile lever arm (UC) and becomes loaded with pollen on its back or head. In later flower development, stigmas are generally orientated to occupy the same position as the previously functional pollen sacs. A pollinator carrying pollen delivered from the staminal levers is therefore likely to transfer the pollen to the stigma of a flower of the same species.

Mechanical barriers that limit the access to floral food sources are one of the most important mechanisms that promote specialization in pollination systems [22,23]. Since *Salvia* flowers provide nectar as a main food source to their pollinators, a crucial question is if and how *Salvia* flowers restrict access to the nectar to a limited range of flower visitors. Having the typical structure of a *Salvia* sympetalous flower with the staminal lever mechanism and nectar presentation at the base of a

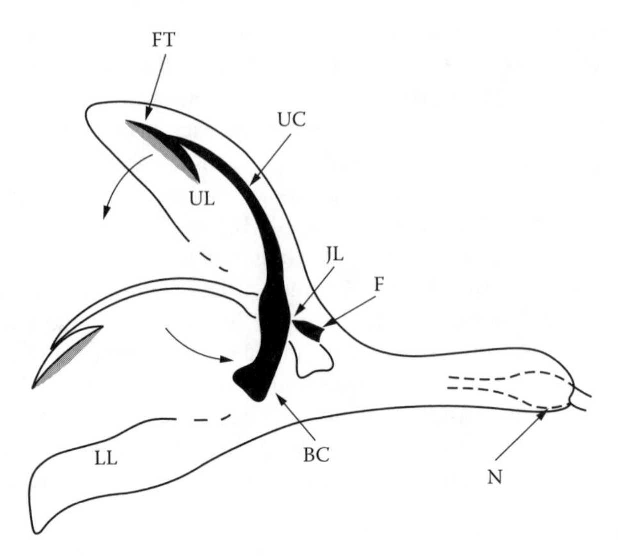

FIGURE 6.1 Schematic longitudinal section of a "typical" *Salvia* flower showing the functioning of the staminal lever mechanism. For clarity, the style and second stamen are omitted. Black: staminal lever arm in unreleased position, white: in released position. UL: upper lip of the flower, LL: lower lip of the flower, F: filament, JL: jointlike ligament, UC: upper part of the elongated connective with fertile theca (FT), BC: basal, sterile part of the connective, N: nectar glands. (Modified from Speck, T. et al., in *Deep Morphology: Toward a Renaissance of Morphology in Plant Systematics*, T. Stuessey, T., F. Hörandl, F., and Mayer, V., Eds., Koeltz, Königstein, 2003, p. 241.)

corolla tube in mind, there are two different mechanical exclusion mechanisms possible: (1) the length and diameter of the flower tube might be an important factor in restricting access to the nectar [24,25, Claßen-Bockhoff and Kuschwitz (in preparation)], and (2) flower visitors that are not able to move the staminal lever may be mechanically excluded from the nectar by this rigid barrier [21,26]. The methodological approaches we have developed allow for a quantitative measurement of the forces insects are able to exert while gaining access to an artificial nectar source. The methods we have developed allow us to measure (1) the forces that insects are able to exert while gaining access to a custom-made artificial nectar source (Figure 6.3), and (2) the forces necessary for insects to trigger the staminal lever and to force themselves along the corolla tube towards the nectar source (Figure 6.4). This enables us to test the different barrier hypotheses and to reconsider the functional importance of different flower structures for flower-pollinator specialization.

6.2 MATERIALS AND METHODS

6.2.1 MATERIALS

We studied honeybees (*Apis mellifera*) and bumblebees (*Bombus terrestris*) as exemplary species. *Apis mellifera* workers have a body length of 12.1 ± 0.3 mm (n = 10, mounted specimens), and a mean proboscis length between 6.05 and 6.40 mm in the three most often cultured subspecies [27]. The European bumblebee *B. terrestris* was used as an example of a large bee species. Bumblebees vary greatly in size because workers can be much smaller than queens. Body length of the tested workers was 12.0 ± 0.8 mm (*n* = 10, mounted specimens), but queens can reach a body length of 19 mm or more. Compared with other European bumblebee species, *B. terrestris* has a rather short proboscis of 8 to 9 mm length [28]. The bumblebee colonies were bought from a commercial supplier (Re-natur GmbH, D-24601 Ruhwinkel, Germany). Forces of bumblebees were measured in the laboratory, and forces of *A. mellifera* were measured in a private bee yard near Freiburg, Germany.

Flowers of the following *Salvia* species were studied: *S. amplexicaulis* Lam., *S. forskahlii* L., *S. glutinosa* L., *S.* cf. *microphylla* Kunth, *S. nilotica* Juss. ex Jacq., *S. nubicola* Wall. ex Sweet, *S. phlomoides* Asso, *S. pratensis* L., *S. sclarea* L., *S. transsylvanica* (Schur ex Griseb.) Schur, *S. uliginosa* Benth., *S. verbenaca* L., and *S. viridis* L. We took the flowers from plants that were cultivated in the Freiburg Botanical Garden (Southwestern Germany). Before flowers were analyzed mechanically, internal structure and morphology of each species was studied in a number of longitudinal sections and cross sections. The species we were especially interested in are:

Jupiter's distaff, *S. glutinosa* L., is a yellow-flowered species that occurs in European and Asian mountain forests. *S. glutinosa* has a simple lever morphology in the sense of Himmelbaur and Stibal [29]. The connective is strongly curved, and the flower entrance is not completely blocked by the lever (Figure 6.2A and Figure 6.7A). Flower-visiting insects on *S. glutinosa* were studied during the flowering periods in 2002 to 2004 in the Freiburg Botanical Garden and in 2003 to 2004 on a further site near Eichstetten (southwestern Germany) where this species grows naturally. Flowers are usually visited by bumblebees (see Table 6.1) and not by honeybees or other midsized bees except for some pollen-thieving or nectar-robbing visits.

Clary sage, *S. sclarea* L., is a well-known, light blue or pink to white flowering Mediterranean species that is widely cultivated for its essential oils and as an ornamental. This species is a representative of the genus with a derived staminal lever in the sense of Himmelbaur and Stibal [29]. The lever is spoon shaped and blocks the flower entrance completely (Figure 6.2B and Figure 6.7B). In 2003 and 2004, we observed flower-visiting insects on *S. sclarea* in the Freiburg Botanical Garden and in private gardens in Boetzingen (southwestern Germany) and Schwanau (southwestern Germany) where this plant was cultivated. Clary sage is visited mainly by carpenter bees *Xylocopa violacea* and, as in *S. glutinosa*, not by honeybees or other midsized bees (Table 6.1).

6.2.2 Forces of Flower-Visiting Bees

Forces that insects are able to exert to gain access to a food source were measured during visits to a custom-made "artificial flower" for which the insects can be trained (Figure 6.3). The artificial flower consists of a base plate connected to a highly

(A)

(B)

FIGURE 6.2 (A) *S. glutinosa* flowers with *B. hortorum* leaving a flower after a visit; original habitat near Eichstetten. (B) *S. sclarea* visited by *X. violacea* (private garden in Schwanau).

sensitive force transducer (Burster Präzisionsmeßtechnik GmbH, D-76593 Gernsbach, Germany) and a retractable tube behind the posterior wall of the construction. The retractable tube contains a sugar solution (apiinvert, Südzucker AG, D-97195 Ochsenfurt, diluted to 50%) as a food source. The posterior wall is covered by a foam pad with a central hole giving access to the tube with the sugar solution. As the bees are feeding, the thin tube containing the sugar solution is manually retracted. By trying further to reach the food source, the insects start pressing their heads against the foam pad. The induced reactive force is transmitted via the bees' legs and the platform to the force transducer that measures this force with an accuracy of ± 50 μN. The force transducer converts the forces into a voltage output, which is stored online by a laptop computer. The tested bumblebees were continuously fed with air-dried pollen, and the supply of the colony with sugar solution was interrupted 24 hr before the experiments were started.

TABLE 6.1
Flower Visitors of the Specialized *S. sclarea* and *S. glutinosa*, and the Less Specialized *S. pratensis*

literature	*Salvia glutinosa*										*Salvia sclarea*				*Salvia pratensis*						
	21	51	52	33	55	41/56	X	32	34	57	51	33	X	57	21	50	51	53	54	41/56	57
Bees (Apoidea)																	0				
Andrena carbonaria																					0
Andrena dorsata															+						0
Andrena hattorfiana															+						
Andrena labiata																					0
Andrena nigripes																					0
Anthidium manicatum										0			0	0						+	
Anthophora ssp.											[+]										
Anthophora fulvitarsis																					0
Anthopora quadrimaculate																					0
Apis mellifua						N		p	p						+	0		0	n	[0 n]	
Bombus ssp.			+						0		+						0				
Bombus sp.		[0]				+N															
Bombus argillaceus		0																			
Bombus hortorum	+	0			+		0	+							+	0		0	+	0	
Bombus humilis																				[0]	
Bombus hypnorum															+						
Bombus lapidarius															+	0		0		0	
Bombus muscorum																				0	
Bombus pasquorum	+	[0]					0	+							+	0		0	+		
Bombus pratorum															+			0	n		
Bombus ruderarius															+					[0]	
Bombus sylvarum															+			0	N	+	
Bombus terrestris															+			0			
Bombus wurfleini		[N]				[N]		[N]												[N]	
Eucera ssp.										+											

TABLE 6.1 (CONTINUED)
Flower Visitors of the Specialized *S. sclarea* and *S. glutinosa*, and the Less Specialized *S. pratensis*

	Salvia glutinosa										*Salvia sclarea*				*Salvia pratensis*						
literature	**21**	**51**	**52**	**33**	**55**	**41/56**	**X**	**32**	**34**	**57**	**51**	**33**	**X**	**57**	**21**	**50**	**51**	**53**	**54**	**41/56**	**57**
Eucera nigrescens															+						
Halictus leucaheneus																					0
"*Halictus nitidus*"																				n0	
Hylaeus sp.							pn														
Hylaeus communis																				[n0]	
Lasioglossum sp.							p									0					
Lasioglossum convexiusculum																					0
Lasioglossum majus																					0
Lasioglossum malachurum																					0
Lasioglossum minutissimum																					0
Lasioglossum morio																				[n0]	
Lasioglossum nitidiusculum																				[n0]	
Lasioglossum pauxillum																					0
Lasioglossum pygmaeum																					0
Lasioglossum sexstrigatum																				[n0]	
Lasioglossum xanthopus																					0
Megachile sp.							0						0								
Megachile circumcincta								pN													
"*Megachile pyrina* (fasciata)"																				+	
Megachile parietina																					0
Osmia aurulenta																					0
Osmia bicolor																					0
Osmia bicornis																				[+]	[0]
Osmia caerulescens															+						0
Osmia pilicornis																					0
Osmia uncinata																					0

Xylocopa itis													0
Xylocopa violacea	0			+	+	0	0						
Hoverflies								0	0			0	
Butterflies					n				0				
Vanessa cardui												0	
Euphydryas aurinia										0			
Mellicta parthenoides										0			
Lasiommata petropolitana												0	
Whites and Yellows (Pieridae)											n		
Anthocharis cardamines												0	
Aporia crataegi												0	
Colias hyale												0	
Pieris brassicae												0	
Pieris rapae												0	
Moths									0				
Siona lineata										0			
Plusia gamma												0n	
Hemaris sp.												[0]	
Macroglossum stellatarum		n	0		n							0	
Hyles euphorbiae												[0]	

Source: Data was obtained either from the literature (21,32–34,41,50–57) or from this paper's authors, represented by X. Activity by the flower visitors was evaluated by the respective authors or was based on descriptions from the references: + pollinator; 0 visitor (or no details given); N nectar robber; n nectar thief; p pollen thief; [ ] the original author used a synonym.

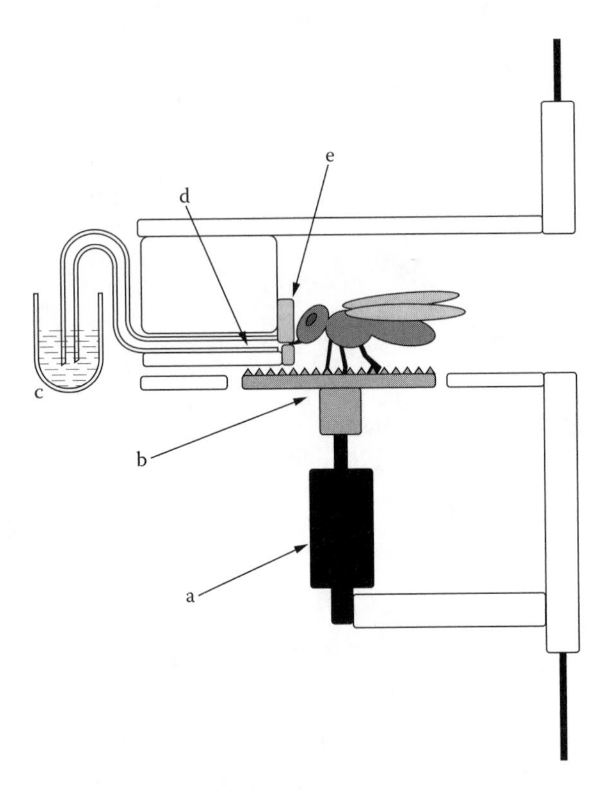

FIGURE 6.3 Schematic drawing of the artificial flower, a custom-made device for measuring forces exerted by insects to gain access to a food source (for description, see the text). (a) force transducer, (b) base plate, (c) sugar solution, (d) retractable food source, and (e) foam pad.

6.2.3 Force Measurements on *Salvia* Flowers and Staminal Levers

The forces a bee would have to exert during a visit on a *Salvia* flower were measured with a custom-built force-measuring device (Figure 6.4) that mechanically mimics the visit of a pollinator [30]. The very sensitive force transducer (Burster Präzisionsmeßtechnik) and a displacement transducer (Burster Präzisionsmeßtechnik) are both mounted on a micromanipulator that is driven by a low vibration DC-micromotor and a gearing system (Märzhäuser GmbH, D-35579 Wetzlar, Germany). The computerized instrument is powered by rechargeable batteries. The measuring device is mounted on a portable tripod with a coordinate system setting that allows a straight positioning of the force transducer and the force sensor to the flower. The movements of the sensor are controlled by a laptop computer that also stores the force–displacement data. This allows for a constant, predefined movement of the sensor into the flower and a measurement of forces even in fragile objects like *Salvia* flowers with an accuracy of ± 50 μm and ± 50 μN, respectively.

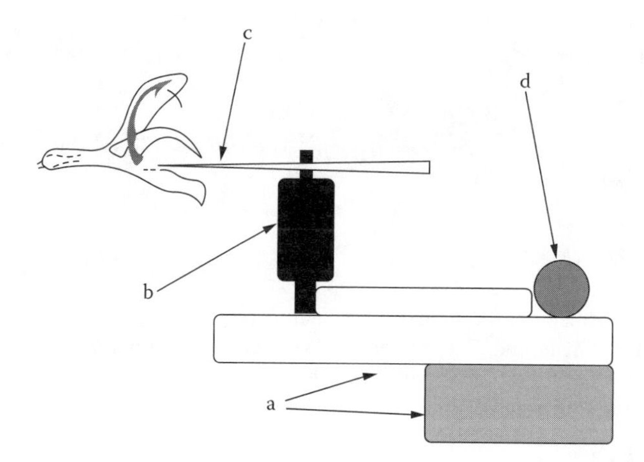

FIGURE 6.4 Schematic drawing of the "pollination simulator," a custom-made device for measuring forces necessary to release the staminal levers and to get access to the nectar gland of *Salvia* flowers (for description, see the text). (a) Micromanipulator with distance transducer, (b) force transducer, (c) sensor, and (d) low vibration motor.

For the initial tests on different *Salvia* species, the sensor was a thin metal rod with a diameter of 0.60 mm. This should mimic a bee's proboscis (proboscis width in *A. mellifera* = 0.61 ± 0.05 mm, n = 10). In a first insertion cycle, we controlled for correct positioning of sensor and flower. Furthermore, we determined the internal, biologically sensitive flower-tube length, as indicated by a strong increase of the measured forces when reaching the flower bottom. When further measurements on the same flower were conducted, the cycling distance was chosen according to the internal flower-tube length. For these measurements of forces necessary to release the staminal lever, the flowers were fixed in their original orientation on their pedicel.

To test whether honeybees might be excluded from the nectar in *S. sclarea* by the dimensions of the corolla tube, we measured the forces that bees with a defined proboscis length have to exert to reach the nectar. For this purpose, we used a modified sensor (Figure 6.5). A honeybee's head was mounted on the thin metal rod. We used a sliding caliper to adjust the distance between the sensor's tip and the bee's head (edge of clypeus) to match the proboscis length that should be simulated. Before each measurement, the bee's head and the sensor were wetted with silicone oil to reduce friction and sticking due to the lack of head motility. During the first insertion cycle, a sufficient proboscis length was chosen for the sensor tip so as to avoid contact of the bee's head with the flower tube; this allowed us to determine the internal flower-tube length. In subsequent measurements on the same flower, this internal flower-tube length was used as a maximum cycling distance. For this type of measurements, the flowers were fixed on their lower lip, which reflects the natural situation in which bees typically use the lateral lobes of the lower lip as grips when collecting pollen and nectar. This type of fixation allows for repeated measurements as the flowers show only very small spatial movements even under the repeated exertion of (comparatively) high forces.

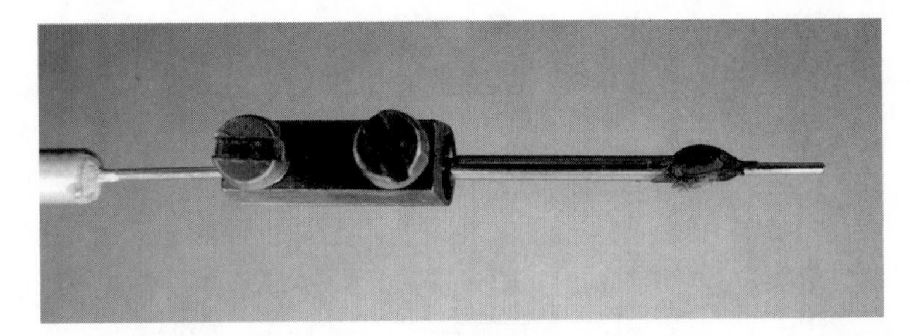

FIGURE 6.5 Modified sensor used in the "pollination simulator," showing a bee head mounted on a thin metal rod. The distance of the insect's head from the tip of the sensor is adjustable to account for different proboscis lengths.

6.3 RESULTS

6.3.1 FORCES EXERTED BY *B. TERRESTRIS* AND *A. MELLIFERA*

For workers and queens of two *B. terrestris* colonies, our data show that the actual force an individual bumblebee exerts may greatly differ in sequential attempts. The mean values were 24.6 ± 14.8 mN (n = 103, 44 individuals) with a maximum of 59 mN for bumblebee workers (Figure 6.6A) and 46.8 ± 25.5 mN (n = 8) with a maximum of 90 mN for bumblebee queens (Figure 6.6B). In honeybee workers (*A. mellifera*), measured forces had a mean value of 14.0 ± 7.4 mN (n = 8) and a maximum value of 29 mN. A detailed analysis of the forces exerted by insects to gain access to the food source in the artificial flower testing device will be the subject of a separate paper [31].

6.3.2 FORCES AND FLOWER VISITORS OF *SALVIA*

In 13 *Salvia* species, the forces necessary to release the staminal lever have been measured, and for 10 of these species, quantitative data are presented for the first time. The data range between 0.5 mN for *S. nilotica* and 10 mN for *S.* cf. *microphylla* (Table 6.2).

A schematic drawing of a *S. glutinosa* flower is shown in Figure 6.7A. The mean force necessary to trigger the lever is 1.6 ± 0.8 mN (n = 10). In this species the corolla tube is narrowed by a broad ring of hairs and the force necessary to pass the hairy ring with the sensor is 2.5 ± 0.7 mN (n = 8) (Figure 6.7C). In *S. glutinosa*, it was not possible to measure the release of the staminal lever and the forces necessary to pass the hairy ring within one continuous measurement because the lever and the flower tube were not arranged linearly in the flowers. As a result, the measurements cannot be depicted in the continuous force–distance diagram. We observed *B. hortorum* visiting *S. glutinosa* flowers at both study sites and *B. pasquorum* and one species of leaf cutter bee *Megachile* sp. visiting at the Freiburg Botanical Garden. Several pollen and/or nectar thieves were observed that did not trigger the lever, including a small sweat bee and a yellow-faced bee (*Lasioglossum* sp. and *Hylaeus* sp.).

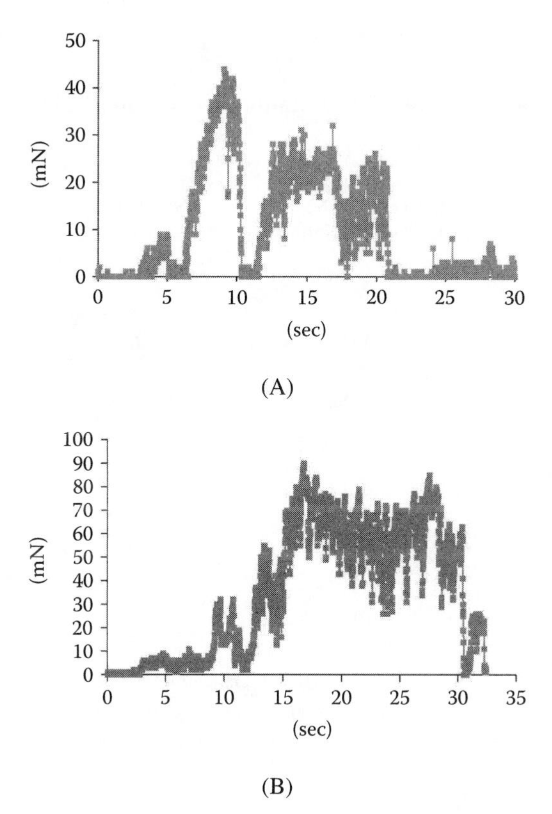

FIGURE 6.6 Forces exerted by individual bumblebees (*B. terrestris*) to gain access to the food source in the artificial flower testing device. (A) Worker, and (B) queen.

A schematic drawing of a *S. sclarea* flower is shown in Figure 6.7B. The force necessary to release the staminal lever of *S. sclarea* is 5.2 ± 3.0 mN (n = 43) (Figure 6.7D). The triggering of the lever often occurred during a steep increase in force. The peak value of this force increase is relatively variable and shows a mean value of 14.0 ± 9.9 mN (n = 33). An internal barrier narrows the entrance to the basal region of the flower tube. The force necessary to pass this additional internal barrier is 5.3 ± 1.7 mN (n = 17). At all three localities, the carpenter bee *X. violacea* was observed visiting *S. sclarea* flowers. In Boetzingen, we also observed a leaf cutter bee *Megachile* sp. and in Schwanau a wool carder bee *Anthidium manicatum* visiting this sage.

To quantify the forces an insect has to produce to reach the nectar in *S. sclarea*, which is located at the base of the corolla tubes, measurements with the modified sensor were carried out (cf., Figure 6.5). On the metal sensor, the head of a honeybee (*A. mellifera*) was fixed at various distances from the sensor tip to mimic the influence of different proboscis lengths. Proboscis lengths of 5, 6, 7, 8, and 11 mm have been simulated for the measurements (Figure 6.8). To account for different flower sizes found in *S. sclarea*, we used a large (internal flower-tube length, 9.7 mm) and a small flower (internal flower-tube length, 9.4 mm) for our measurements.

TABLE 6.2
Forces Necessary to Release the Staminal Lever in 13 *Salvia* Species

Salvia Species	Average Force Necessary to Trigger the Lever (mN) ± Standard Deviation	Number of Measurements	Number of Flowers
S. amplexicaulis	1.2 ± 0.9	17	10
S. forskahlii	2.9 ± 0.9	65	10
S. glutinosa	1.6 ± 0.8	10	3
S. cf. *microphylla*	10.0 ± 5.3	22	3
S. nilotica	0.5 ± 0.2	25	2
S. nubicola	1.0 ± 0.7	26	4
S. phlomoides	1.1 ± 0.2	25	2
S. pratensis	2.7 ± 0.3	28	5
S. sclarea	5.2 ± 3.0	43	7
S. transsylvanica	3.0 ± 0.7	20	3
S. uliginosa	0.9 ± 0.3	6	1
S. verbenaca	2.7 ± 1.1	49	5
S. viridis	1.5 ± 0.5	15	3

Note: Measurements were performed with a metal sensor of 0.6-mm diameter, mimicking a bee's proboscis. In plants where the lever is not triggered by the bees proboscis but by the bee's head (*S. forskahlii*, *S. glutinosa*, *S. nilotica*, and *S. nubicola*), a wider sensor (1 to 3 mm) was used to trigger the lever.

A maximum force of 407 mN was recorded with a simulated proboscis length of 6 mm and a maximum force of 167 mN for a simulated proboscis length of 8 mm in big *S. sclarea* flowers with 9.7 mm flower-tube length. The maximum force recorded for small flowers (flower-tube length 9.4 mm) was 260 mN and 45 mN for a proboscis length of 6 mm and 8 mm, respectively (Figure 6.8). Using even shorter proboscis lengths was not possible because the flowers were destroyed by the head mounted on the sensor if the proboscis length was shorter than 6 mm. The shorter the length of the simulated proboscis, the earlier the insect's head has to be squeezed into the tight corolla tube, resulting in an earlier onset of force increase and in a higher maximum value of force needed to reach the nectar at the base of the corolla tube (Figure 6.8).

6.4 DISCUSSION

6.4.1 Insect Forces

The maximum forces measured were 29 mN for honeybees (*A. mellifera*), 90 mN for bumblebee queens, and 59 mN for bumblebee workers (*Bombus terrestris*); thus, our data indicate that the maximum forces exerted by midsized bees and bumblebees during a flower visit range between 20 and 100 mN. According to the current state

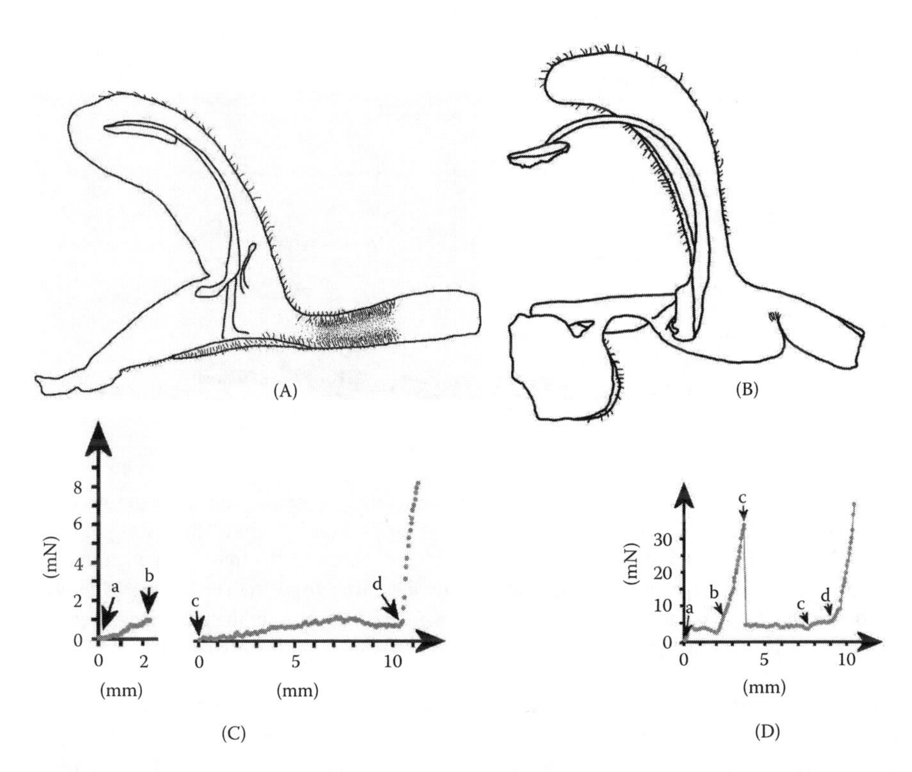

FIGURE 6.7 A and B show schematic drawings of a longitudinal section of a flower, showing one stamen (style and ovary not drawn). (A) *S. glutinosa*, and (B) *S. sclarea*. C and D show force–distance diagrams of the movement of a metal sensor (sensor diameter 0.6 mm) in a flower: (a) sensor touches lever, (b) lever released, (c) sensor touches internal barrier, and (d) sensor touches bottom of the corolla tube. (C) *S. glutinosa* flower; left: releasing and passing the staminal lever, right: passing the internal ring of hairs. (D) *S. sclarea* flower.

of knowledge, the bee flowers that are the hardest to trigger are those of *L. latifolius*, which require forces of 100 mN to be triggered [16]. Therefore, we hypothesize that even large bees do not exert forces that significantly exceed 100 mN during flower visits. The forces bees and bumblebees can exert on flowers are much larger than the weight forces that correspond to their body mass [16]. For example, in the tested bumblebee workers that can exert forces up to 59 mN (mean force measured: 24.6 ± 14.8 mN), a mean body mass of 199 ± 42 mg (n = 56) is found. This means bumblebees can produce a maximum force that is about 30 times their weight force (a mass of 199 mg induces a weight force of 1.95 mN).

6.4.2 Observed Flower Visitors

During our observation periods, *S. glutinosa* and *S. sclarea* were visited only by a small number of bee species. Most of these (exceptions, e.g., the pollen and nectar thieves, *Lasioglossum* sp. and *Hylaeus* sp.) are already known to be flower visitors of *S. glutinosa* or *S. sclarea* (Table 6.1). Though there are numerous publications

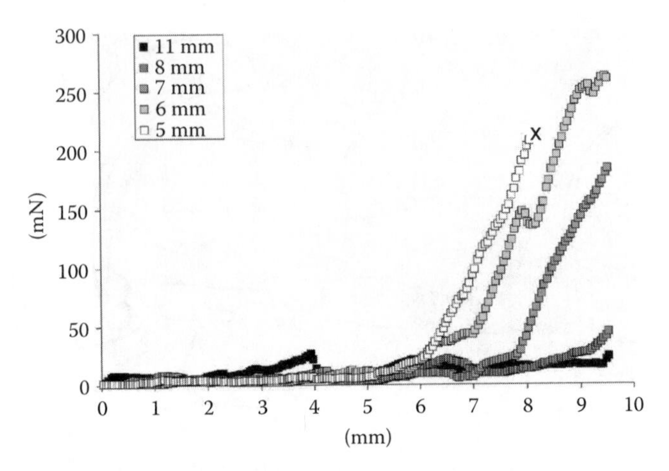

FIGURE 6.8 Force–distance diagrams showing the forces necessary to reach the nectar at the bottom of the corolla tube depending on different proboscis lengths. Series of measurements in the same flower of *S. sclarea* (internal flower-tube length 9.4 mm) using the modified sensor with an insect's head mounted at various distances from the sensor tip. Mimicked proboscis length (black to white, see insert): 11, 8, 7, 6, and 5 mm (in the latter measurement, the flower tube was destroyed at the end point of the force–distance curve marked with an x).

on the pollination biology of these two *Salvia* species, there are no reports that honeybees have ever been seen triggering the staminal lever while foraging for nectar in flowers of these species (see Refs. [32–34] and our unpublished observations). The observation that only a small number of bee species visits these two sages supports the assumption that *S. glutinosa* and *S. sclarea* are specialized (in terms of species number of flower-visiting insects). Honeybees seem to be excluded from successfully (with regard to nectar feeding) visiting these flowers by either the staminal lever, the additional internal barriers, or the length of the flower tube.

6.4.3 Forces Measured in *Salvia* Flowers

6.4.3.1 Critical Discussion of the Applied Methods

Our methodological approach allows quantitative analysis of the forces involved in flower visits to different sage species and compares for the first time these results with the forces that different flower-visiting bee species are able to produce to gain access to a food source. In addition, the force measurements in *Salvia* flowers can be correlated with the functional morphology of the corolla tube and the structure of the staminal lever.

The first method used was to imitate the proboscis of a bee as it is inserted into a flower. This test procedure cannot be considered a perfect imitation of the actual process taking place during the visit of a bee because the functional mode of our testing device forces us to use a stiff metal rod with the same diameter as the bee's proboscis. A bee's proboscis is more flexible than a stiff rod, and a bee should be able to avoid adhering and becoming lodged within flower structures (at least part of the time) and piercing floral tissue during a normal flower visit. Because of the complex morphology of the flowers of some *Salvia* species, we were unable to avoid piercing the floral tissue during some of our measurements. But sticking and piercing events were easy to recognize after the force measurements when careful dissection revealed the resulting damage to the plant tissue. Because the exact position of the sensor in the flower is measured in the force–distance diagrams, we were able to trace the corresponding force peaks exactly in our data set and to exclude them from further biological interpretation. The high peak that followed the triggering of the staminal lever of *S. sclarea* is the consequence of such a sticking event. In addition to our morphological investigations, this interpretation is supported by the very high variability of this peak. In this species, the sterile base of the staminal lever arm is strongly sculptured (Figure 6.7B), and the sensor often sticks to these structures. In *S. glutinosa*, the lever arm is much less sculptured, and so the sensor typically passes smoothly.

The force–distance diagrams that resulted from the measurements are rich in information about internal flower morphology and the functional importance of the various flower parts during the activities of a visiting insect. Even very small structures that are found in a direct line from the flower entrance to the flower base where the nectar is present often caused a typical, detectable peak in force measurements when the sensor touched them. Peaks and structures are easy to correlate if one compares the position of the peak on the diagram with the structure's position in the flower.

6.4.3.2 Comparison of Levers and Internal Barriers in Flowers

Comparing the lever and other internal barriers in terms of forces needed to release or to overcome them points to the potential importance of the staminal lever as a mechanical barrier. In *S. glutinosa*, the force needed to trigger the staminal lever is lower by about 40% than the force necessary to pass the artificial proboscis through the additional internal barrier, built in this species by a hairy ring in the narrow part of the flower tube. This indicates that in terms of forces, the lever does not represent the main device to exclude insects from gaining access to the nectar, and, more likely, the narrow corolla tube with its hairy ring has this function in *S. glutinosa*.

The situation in *S. sclarea* is different. In this species (even after exclusion of the peak due to sticking events), the staminal lever is the main barrier in terms of the force required to protrude the artificial proboscis to the nectar at the flower base.

Claßen-Bockhoff et al. [21] measured the forces that are necessary to trigger the levers of eight *Salvia* species; three of these species were also included in this analysis. The previously published values are identical to those found in our present studies [in brackets] for two species: *S. pratensis* 2.98 ± 2.43 mN (n = 1140) [2.7 ± 0.3 mN (n = 28)] and *S. glutinosa* 1.47 ± 1.05 mN (n = 780) [1.6 ± 0.8 mN (n = 10)]. The only exception is *S. sclarea* for which a tentative value of 31.8 mN based on two measurements was given previously by Claßen-Bockhoff et al. [21]. This is much higher than the value of 5.2 ± 3.0 mN (n = 43) for the more detailed measurements presented in this paper. This difference might be due to the possibility that the actual release of the staminal lever is very difficult to distinguish from the sticking event at the sculptured sterile lever base that caused the large force peak in the present measurements.

6.4.4 Comparing Insect Forces to the Barriers in Flowers

The force needed to insert a proboscis and trigger the lever in all measured *Salvia* flowers did not exceed 5.2 ± 3.0 mN (*S. sclarea*) in bee-pollinated species. This is clearly below the maximum value of 29 mN that even the weakest measured bee species (honeybee) can exert. This means that honeybees as well as bumblebees are strong enough to insert their proboscis in the respective flower, and that these bees are not excluded by the force that is needed to trigger the staminal lever of any measured sage flowers. This even holds true for the additional internal barriers of sage flowers under investigation. Neither the protrusion narrowing the corolla tube in Clary sage (*S. sclarea*) nor the hairy ring in *S. glutinosa* flowers forms a barrier that mechanically blocks the access to the nectar for honeybees or bumblebees.

The hypothesis that honeybees and bumblebees are not excluded by force from visits of a given *Salvia* flower is not new. Meadow sage (*S. pratensis*) is less specialized than other sages because it is pollinated by many species of insects including bumblebees and honeybees (Table 6.1). Therefore, the lever of meadow sage evidently does not exclude these insects by force [35]. Our results indicate that even sages that are pollinated by big bees (Table 6.1) do not exclude smaller bee species by the magnitude of the force necessary to release the staminal lever. This implies that the forces needed to trigger the staminal lever are of minor significance in the specialization of these *Salvia* species for specific flower visitors.

6.4.5 Proboscis Length, Flower-Tube Length, and Forces Exerted by Visiting Bees

The results on proboscis length and flower-tube length indicate that the amount of force that a bee has to exert to reach the nectar at the flower base of *S. sclarea* largely depends on its proboscis length. Bees with a longer proboscis have to exert a much smaller force than bees with a short proboscis, because the latter have to press their heads deeper in the corolla tube to reach the nectar. The forces that a bee with a short proboscis (6 mm) needs to reach the nectar of *S. sclarea* (260 mN to 407 mN, for a short or a big flower, respectively) are much higher than the forces honeybees can exert (maximum measured force 29 mN), and even if a honeybee could exert forces comparable to the ones measured in the strongest bumblebee queen (90 mN), they would not be able to reach the nectar. This means that the length of the proboscis of a bee has to fit the flower-tube length of a visited *Salvia* species. Honeybees with their short proboscis (6.05 to 6.40 mm) are not able to reach the nectar in *S. sclarea* flowers and thus are excluded from the nectar by the length and narrowness of the flower tube. Therefore, flower-tube length seems to be an important factor in specialization of sages on a certain set of flower visitors because the forces required to reach the nectar drastically increase with decreasing proboscis length.

The important role of flower-tube length or spur length in specialization on flower visitors is known for many plants, e.g., ecotypes of *Satyrium hallackii*, the *Disa draconis* complex, and many other species [36–39]. Several authors have mentioned the possible role of flower-tube length for specialization in *Salvia*. Dieringer et al. [40] discuss that *S. mexicana* might be specialized on a single pollinator, the bee *Deltophila elephas* because the very long corolla tube of this sage species corresponds to the long tongue of this bee species [40]. Westerkamp [34] writes that *S. glutinosa* is a bumblebee flower, and that honeybees are not able to reach the nectar of this sage because of their short proboscis. Claßen-Bockhoff et al. [20] emphasize that for a definition of pollination niches in *Salvia* species, a variety of flower characters including length and width of the corolla tube are essential (see also [24,41]). This was recently confirmed by Kuschewitz [25] who distinguished 11 flower types in 39 European sages that differ in floral proportions and thus in the spatial patterns of pollen transfer.

Our results, based on the first quantitative measurements of the forces involved in flower visits by bees, support these hypotheses, which were based on morphological characteristics, and allow a more detailed discussion of the function of various flower parts in the genus *Salvia*.

6.5 CONCLUSION

To test the evolutionary significance and the function of the staminal lever in *Salvia* species, we quantitatively analyzed the "force–barrier-hypothesis" [21,26] using a biomechanical approach. This hypothesis states that the interspecific variation in the forces needed to release the lever in *Salvia* are the main reason for interspecific differences in the pollinator spectra. A *Salvia* population that undergoes an evolutionary change in these forces should therefore either specialize on a smaller number of strong flower visitors or generalize on a wider range of visitors that include weak ones. Small changes in joint anatomy could therefore lead to great differences in pollinator spectra and in populations of a *Salvia* species, which differ largely in specialization on pollinators. Because they have different pollinators, these populations might even become genetically isolated and be a starting point of speciation.

This interesting idea is not supported by the data presented in this paper. Despite the expected interspecific variability among *Salvia* species in the forces necessary to trigger the lever, these forces are generally low compared with the forces bees can exert. In two *Salvia* species specialized on bumblebees or carpenter bees and not visited by the honeybee, honeybees were, in contrast to the predictions of this hypothesis, strong enough to trigger the levers. Instead of the force required to trigger the lever, it is the flower-tube length that seems to exclude honeybees from the nectar of these two species. Nonetheless, the relative importance of the flower-tube length or the lever forces for specialization and speciation in *Salvia* cannot be finally evaluated because until now only 2% of all *Salvia* species have been investigated. In a large number of sages, the lever does not close the flower entrance and can therefore be circumvented easily (e.g., *S. glutinosa*). Furthermore, in species with a more derived flower morphology in which the lever closes the flower entrance (e.g., *S. sclarea*), the forces that have to be exerted while triggering the lever are low.

Therefore, we propose that these forces are of minor importance for pollinator shifts and specialization in the genus *Salvia*. We propose that the length of the flower tube in *Salvia* has the same role as in many other plant groups with tubular flowers or nectar spurs. Flower-tube length restricts access to nectar and is an important factor in recruiting a certain set of flower visitors.

With respect to the evolution of the staminal levers found in *Salvia*, we favor two possible functions involved in pollen transfer that are not mutually exclusive.

1. *Pollen dispensing*. Portioning of pollen by dispensing multiple small portions of pollen on successive visitors is found in many other plant species as an important mechanism for maximizing male fitness [42–44]. The hypothesis that staminal levers in *Salvia* may be involved in pollen dispensing is supported by the finding that the staminal levers can be released many times (more than 20 times in all tested sage species), resulting in (nearly) exactly the same force–distance curve with each release [21].
2. *Exact pollen placement on the pollinator body*. Exact pollen placement may be important for the reproductive success of plants for three reasons.
   a. In subsequent visits of flowers of the same *Salvia* species, pollen placement at exactly the site where the style touches the insect enhances reproductive success by increasing the probability of pollen transfer [10,21].
   b. In subsequent visits of an insect to different *Salvia* species, exact pollen placement on the pollinator's body as achieved by the staminal lever could avoid interspecific pollen transfer and thereby hybridization. This design would allow sympatric *Salvia* species to share a large number of pollinating species [20,45].
   c. Exact, nototribic pollen placement as performed by the staminal lever in most *Salvia* species is thought to reduce pollen losses through grooming because bees seem to groom less effectively on their dorsal rather than on their lateral sides [46]. Nototribic pollen placement, in adddition to the fact that the thecae (and hence the pollen) in most bee-pollinated *Salvia* species are hidden in the upper lip of the flower during a bee's approach, should additionally reduce the influence of a flower visitor on its pollen uptake [47].

The previously mentioned possible functions of the staminal lever, i.e., optimizing pollen dispension and exact pollen placement, could allow *Salvia* species to specialize on certain bees as pollinators, which are known as "high-removal low-deposition" pollinators in the sense of Thomson et al. [48]. Therefore, the staminal lever of *Salvia* can be included in a list of floral features that optimize male reproductive success by controlling pollen release and pollen placement in bee-pollinated plants and that include buzz pollination, and the piston and brush mechanisms [9,10,14,49]. The immense species richness of the genus *Salvia* and the high morphological and functional diversity of its levers can then be interpreted as a by-product of selection on male reproductive success in a variable pollinator environment.

ACKNOWLEDGMENTS

We thank the technical workshop of the University of Freiburg (Biological Institutes II/III) for constructing the technical devices, the team of the Botanical Garden of Freiburg for providing plant material, Rainer Oberle, and Hans and Johanna Bumbel for granting access to plants in private gardens, and Randy Cassada for improving the readability of this text. The Deutsche Forschungsgemeinschaft (SPP 1127) is gratefully acknowledged for financial support (Cl 81/9–1, Sp 534/5–1 and 5–2).

REFERENCES

1. Sprengel, C.K., *Das entdeckte Geheimnis der Natur im Bau und in der Befruchtung der Pflanzen*, Vieweg, Berlin, 1793.
2. Darwin, C., *The Various Continuances by Which Orchids Are Fertilised by Insects*, John Murray, London, 1862.
3. Hildebrand, F., Ueber die Befruchtung der Salviaarten mit Hilfe von Insekten, Jahrb. *Wiss. Bot.*, 4, 451, 1865.
4. Müller, H., *The Mechanism of Flowers*, Macmillan, London, 1883.
5. Correns, C., Zur Biologie und Anatomie der Salvienblüthe, *Jahrb. Wiss. Bot.*, 22, 190, 1891.
6. Troll, W., Über Staubblatt- und Griffelbewegungen und ihre teleologische Deutung, *Flora*, 115, 191, 1922.
7. Dukas, R. and Dafni, A., Buzz-pollination in three nectariferous boraginaceae and possible evolution of buzz-pollinated flowers, *Plant Syst. Evol.*, 169, 65, 1990.
8. Harder, L.D. and Barclay, R.M.R., The functional significance of poricidal anther and buzz pollination: controlled pollen removal from Dodecatheon, *Funct. Ecol.*, 8, 509, 1994.
9. Harder, L.D. and Thomson, J.D., Evolutionary options for maximizing pollen dispersal of animal-pollinated plants, *Am. Nat.*, 133, 323, 1989.
10. Yeo, P.F., *Secondary Pollen Presentation: Form Function and Evolution*, Springer, New York, 1992.
11. Müller, A., Morphological specializations in Central European bees for the uptake of pollen from flowers with anthers hidden in narrow corolla tubes (Hymenoptera: Apoidea), *Entomol. Gener.*, 20, 43, 1995.
12. Asada, S. and Ono, M., Crop pollination by Japanese bumblebees, *Bombus* spp. (Hymenoptera: Apidae): tomato foraging behavior and pollination efficiency, *Appl. Entomol. Zool.*, 31, 581 1996.
13. Buchmann, S.L. and Hurley, J.P., A biophysical model for buzz pollination in angiosperms, *J. Theor. Biol.*, 72, 639, 1978.
14. Westerkamp, C., Keel blossoms: bee flowers with adaptions against bees. *Flora*, 192, 125, 1997.
15. Faegri, K. and Van der Pijl, L., *The Principles of Pollination Ecology*, Pergamon Press, Oxford, 1979.
16. Westerkamp, C., The co-operation between the asymmetric flower of *Lathyrus latifolius* (Fabaceae-Vicieae) and its visitors, *Phyton (Horn)*, 33, 121, 1993.
17. Huck, R.B., Overview of pollination biology in the Lamiaceae, in *Advances in Labiatae Science*, Harley R.M. and Reynolds T., Eds., Royal Botanic Gardens, Kew, 1992, p. 167.

18. Tweraser, E. and Claßen-Bockhoff, R., Hemigenia eutaxioides C.R.P.Andr. — the "invers *Salvia* flower" from Australia, *Abstractband Botanikertagung 2002*, Selbstverlag Botan. Institut, Universität Freiberg, Freiburg i.Br.. 2002, p. 437.
19. Troll, W., *Roscoea purpurea* SM., eine Zingiberacee mit Hebelmechanismus in den Blüten. Mit Bemerkungen über die Entfaltungsbewegungen der fertilen Staubblätter von Salvia, *Planta*, 7, 1, 1929.
20. Claßen-Bockhoff, R., Wester, P., and Tweraser, E., The staminal lever mechanism in *Salvia* L. (Lamiaceae): a review, *Plant Biol.*, 5, 33, 2003.
21. Claßen-Bockhoff, R. et al., The staminal lever mechanism in *Salvia* L. (Lamiaceae): a key innovation for adaptive radiation? *Org. Divers. Evol.*, 4, 189, 2004.
22. Manning, J. C. and Goldblatt, P., The *Moegistorhynchus longirostris* (Diptera: Nemestrinidae) pollination guild: long-tubed flowers and a specialized long-proboscid fly pollination system in southern Africa, *Plant Syst. Evol.*, 206, 51, 1997.
23. Lunau, K., Adaptive radiation and coevolution — pollination biology case studies, *Org. Divers. Evol.*, 4, 207, 2004.
24. Dafni, A., Advertisement, flower longevity, reward and nectar protection in Labiatae, *Acta Hortic.*, 288, 340, 1991.
25. Kuschewitz, M., Diversität und mechanische Isolation europäischer Salvia-Arten (Lamiaceae) — eine datenbankgestützte Evaluation, Staatsexamensarbeit Universität Mainz (unpublished), 2004.
26. Claßen-Bockhoff, R. and Speck, T., Diversity and evolution in *Salvia* — presentation of a new research project, *Vitex*, 1, 3, 2000.
27. Ruttner, F., Geographic variability and classification, in *Bee Genetics and Breeding*, Rinderer, T.E., Ed., Academic Press, Orlando, 1986, p. 23.
28. Knuth, P., *Handbuch der Bluetenbiologie*, W. Engelmann, Leipzig, 1898.
29. Himmelbaur, W. and Stibal, E., Entwicklungsrichtungen in der Blütenregion der Gattung Salvia L. II., *Biologia Generalis*, 9, 129, 1934.
30. Speck, T. and Claßen-Bockhoff, R., Biomechanik und Funktionsmorphoplogie der Staubblätter von insektenblütigen und vogelblütigen Salbei-Arten, *BIONA-Report*, 14, 153, 1999.
31. Reith, M., Speck, T. and Claßen-Bockhoff, R., Forces produced by bumble bees (*Bombus terrestris*) during nectar feeding, (unpublished).
32. Schremmer, F., Blütenbiologische Beobachtungen an Labiaten: nektar und Pollendiebstahl, *Öesterr. Bot. Z.*, 100, 8, 1953.
33. Kugler, H., Zur Bestäubung von *Salvia sclarea* L. durch Holzbienen (*Xylocopa violacea* L.), *Oesterr. Bot. Z.*, 120, 77, 1972.
34. Westerkamp, C., Honeybees are poor pollinators — why?, *Plant Syst. Evol.*, 177, 71, 1991.
35. Thimm, S. et al.., Pollinators of *Salvia pratensis* are not excluded by physical force — evidence from biomechanical measurements and field investigations, Conference Proceedings, Abstracts 4th International Plant Biomechanics Conference, W.J. Beal Botanical Garden, Michigan State University (East Lansing), Conference Proceedings, 2003, p. 21.
36. Johnson, S.D., Pollination ecotypes of *Satyrium hallackii* (Orchidaceae) in South Africa, *Bot. J. Linn. Soc.*, 123, 225, 1997.
37. Johnson, S.D. and Steiner, K.E., Long-tongued fly pollination and evolution of floral spur length in the *Disa draconis* complex (Orchidaceae), *Evolution*, 51, 45, 1997.
38. Hodges, S.A. and Arnold, M.L., Spurring plant diversification: are floral nectar spurs a key innovation? *Proc. R. Soc. Lond. B.*, 262, 343, 1995.

39. Dohzono, I. and Suzuki, K., Bumblebee-pollination and temporal change of the calyx tube length in *Clematis stans* (Ranunculaceae), *J. Plant. Res.*, 115, 355, 2002.
40. Dieringer, G., Ramamoorthy, T.P., and Tenorio Lezama, P., Floral visitors and their behavior to sympatric *Salvia* species (Lamiaceae) in Mexico, *Acta Botànica Mexicana*, 13, 75, 1991.
41. Müller, H., *Die Befruchtung der Blumen durch Insekten und die gegenseitigen Anpassungen beider*, W. Engelmann, Leipzig, 1873.
42. Percival, M.S., The presentation of pollen in certain angiosperms and its collection by *Apis mellifera*, *New Phytol.*, 54, 353, 1955.
43. Erbar, C. and Leins, P., Portioned pollen release and the syndromes of secondary pollen presentation in the Campanulales-Asterales-complex. *Flora*, 190, 323, 1995.
44. Harder, L.D., Pollen removal by bumblebees and its implications for pollen dispersal, *Ecology*, 71, 1110, 1990.
45. Ramamoorthy, T.P. and Elliot, M., Mexican Labiatae: diversity, distribution, evolution and endemism, in *Biological Diversity of Mexico: Origins and Distribution*, Ramamoorthy, T.P. et al., Eds., Oxford University Press, New York, 1993, p. 513.
46. Müller, A., Convergent evolution of morphological specializations in Central European bee and honey wasp species as an adaptation to the uptake of pollen from nototribic flowers (Hymenoptera, Apoidea and Masaridae), *Biol. J. Linn. Soc.*, 57, 235, 1996.
47. Westerkamp, C., Pollen in bee-flower relations: some considerations on melittophily, *Bot. Acta*, 109, 325, 1996.
48. Thomson, J.D. et al., Pollen presentation and pollination syndromes, with special reference to Penstemon, *Plant Species Biol.*, 15, 11, 2000.
49. Leins, P., Flower pollination by animals: remarks on external and self-pollination and on pollen portioning, *Heidelb. Jahrb.*, 35, 133, 1991.
50. Corbet, S.A. et al., Native or exotic? Double or single? Evaluating plants for pollinator-friendly gardens, *Ann. Bot.*, 87, 219, 2001.
51. Gams, H. (1927). Labiatae, in *Illustrierte Flora von Mittel-Europa*, Hegi, G., Ed., Bd. V. Teil 4, Carl Hauser, München, 1927, p. 2255.
52. Kerner von Maurilaun, A., Pflanzenleben. 2. Band: Geschichte der Pflanzen, Bibliographisches Institut, Leipzig, Wien, 1891.
53. Kratochwil, A., Pflanzengesellschaften und Blütenbesucher-Gemeinschaften: biozönologische Untersuchungen in einem nicht mehr bewirtschafteten Halbtrockenrasen (Mesobrometum) im Kaiserstuhl (Südwestdeutschland), *Phytocoenologia*, 11, 455, 1984.
54. Kwak, M.M., Bumblebees as flower visitors: pollinators and profiteers, *Entomol. Ber. (Amsterdam)*, 62, 73, 2002.
55. Loew, E., Beiträge zur Kenntniss der Bestäubungseinrichtungen einiger Labiaten, *Ber. d. dtsch. bot. Ges.*, 4, 128, 1886.
56. Müller, H., *Alpenblumen, ihre Befruchtung durch Insekten und ihre Anpassungen an dieselben*, W. Engelmann, Leipzig, 1881.
57. Westrich, P., *Die Wildbienen Baden-Württembergs*, 1st ed., Ulmer, Stuttgart, 1989.
58. Speck, T. et al., The potential of plant biomechanics in functional biology and systematics, in *Deep Morphology: Toward a Renaissance of Morphology in Plant Systematics*, Stuessey, T., Hörandl, F., and Mayer, V., Eds., Koeltz, Königstein, 2005, p. 241.

7 Do Plant Waxes Make Insect Attachment Structures Dirty? Experimental Evidence for the Contamination Hypothesis

Elena Gorb and Stanislav Gorb

CONTENTS

7.1 INTRODUCTION

In many higher plants, primary surfaces of aerial organs are covered with crystalline epicuticular waxes. An extremely thick wax layer usually appears as pruinescence on the stems, leaves, and fruits in various plants [1–3]. Generally, waxes were reported to be important for interactions between plant surfaces and the environment

[3,4]. It has been previously suggested that wax coverage on plant surfaces decreases locomotory activity of insects, and this prevents insects from robbing nectar and other resources [5–8], protects plants against insects and herbivores [7,9–19], and precludes the escape of insects from traps in carnivorous plants [7,13,20–24].

A few studies have experimentally shown that the wax crystalline coverage affects insect attachment. It was found for several insect species that they can attach well to smooth substrates, without wax or with wax bloom removed, whereas they slip from the waxy surfaces [8,11–13,16,19,22,23,25–27]. The experiments with various plant substrates (99 surfaces of 83 plant species), among them smooth, hairy, feltlike, pruinose, and glandular ones, and the beetle *Chrysolina fastuosa* showed that the insects were not able to attach to any of 17 pruinose surfaces, whereas they performed well on all other surface types [28]. After having contact with pruinose substrates, insects were observed to clean their attachment devices, and within a short time, they regained full attachment ability.

To explain the reduction of insect attachment on pruinose substrates, Gorb and Gorb [28] had previously proposed four hypotheses. (1) Wax crystals cause the microroughness, which considerably decreases the real contact area between the substrate and setal tips of insect adhesive pads. (2) Wax crystals are easily detachable structures that contaminate the pads. (3) Insect pad secretion may dissolve wax crystals; this will result in the appearance of a thick layer of fluid, making the substrate slippery. (4) Structured wax coverage may absorb the fluid from the setal surface. In previous studies, some data confirming the hypotheses, e.g., the roughness hypothesis [29], the contamination hypothesis [11,12,20,21,23,30,31], and the wax-dissolving hypothesis [23], were obtained. These data usually dealt with just one or a few substrates and did not analyze wax features, which might be responsible for a particular effect. So far there is no direct evidence to support the fourth hypothesis.

This study was undertaken to show the influence of the plant surfaces structured with crystalline waxes on insect adhesive pads of the hairy type. The aim was to prove experimentally the contamination hypothesis using a model insect species and compare a number of plant surfaces. The beetle *Chrysolina fastuosa* Scop. (Coleoptera, Chrysomelidae) was tested on pruinose plant surfaces that differed in the structure of their wax coverage. Young stems, adaxial surfaces of leaves, and fruits of 12 plant species were used as substrates. The same insect and plant species were previously used in the experimental study on the insect attachment ability on various plant substrates [28]. The following questions were asked: (1) Do all plant waxes contaminate insect pads? (2) Do the degree and the nature of contamination vary depending on the plant species? (3) How do the shape and the dimensions of wax crystals affect pad contamination?

To obtain detailed information about plant substrates, we employed scanning electron microscopy (SEM). For SEM, parts of plant organs with pruinose (covered with wax bloom) surfaces were air dried, mounted on holders, sputter coated with gold-palladium (8 nm), and examined in a Hitachi S-800 scanning electron microscope at 2 to 20 kV. The type of wax projections in each plant species was identified according to the recent classification of plant epicuticular waxes [32] and published earlier [28]. To calculate aspect ratios of wax crystals, the following variables were quantified from digital images: the largest linear dimension (the length in

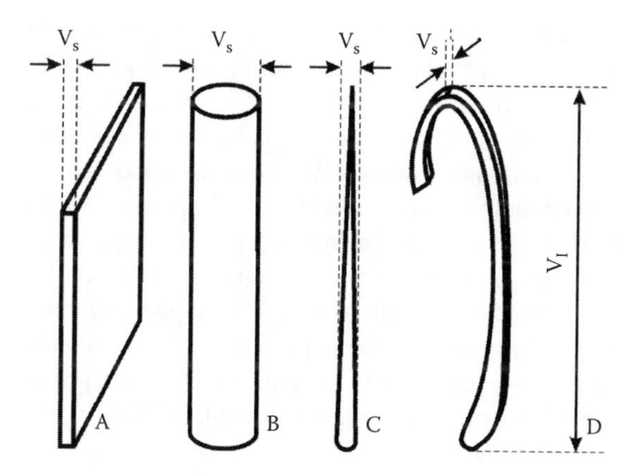

FIGURE 7.1 Diagram of the variables used to calculate aspect ratios of wax crystals having various shapes. (A) Platelets (platelike shape). (B) Tubules (tubelike shape). (C) Terete rodlets (filamentous shape). (D) Polygonal rodlets (ribbonlike shape). V_l, the largest linear variable; V_s, the smallest linear variable.

filamentous, ribbonlike, tubelike, and platelike structures) and the smallest linear dimension (the diameter in filamentous, tubelike structures; the thickness in platelike and ribbonlike structures) (Figure 7.1). For measurements, SigmaScan 5.0 software (SPSS) was employed.

7.2 MATERIAL AND METHODS

7.2.1 Plant Surfaces and Other Substrates

Twelve common wild and domestic plants species were used in the experiments: *Acer negundo* L., *Aloe vera* (L.) Webb. & Berth., *Aquilegia vulgaris* L., *Berberis vulgaris* L., *Brassica oleracea* L., *Chelidonium majus* L., *Chenopodium album* L., *Iris germanica* L., *Lactuca serriola* Torner, *Prunus domestica* L., *Trifolium montanum* L., and *Vitis vinifera* L. These include two trees, one scrub, one liana, and eight herbs. Young stems, leaves, or fruits of these plants were collected in the deciduous forest, meadows, gardens, or by roadsides near Jagotin (Kiev District, Ukraine). In addition, clean glass surfaces and the glass surfaces covered with dust particles were used as control substrates for comparison.

7.2.2 Model Insect Species and Experiments

The beetle *Chrysolina fastuosa* was used as a model insect species because (1) it occurred in great numbers at the study site and (2) our previous experiments, proving insect attachment ability on various plant surfaces, have been carried out with this insect species [28]. An insect was allowed to walk on the tested surface for 1 min. The beetle was transferred on a clean glass plate, and its legs (foreleg, midleg, and hindleg of one side) were immediately cut off with a sharp razor blade and mounted on SEM holders using conductive carbon double-sided adhesive tabs. On each plant surface, at least five insects were tested. Insects demonstrating cleaning behavior of the tarsi were excluded from the further microscopic procedures. Preparations were air dried, sputter coated with gold-palladium (10 nm), and examined in a Hitachi S-800 scanning electron microscope at 20 kV. Only males were used in the experiments because their pads bear mushroomlike hairs having round or oval-shaped flat tips, and this facilitated quantification of the contamination.

A portion of the contaminated area of the setal tips was quantified from digital images. At least 10 setae having flat, round, or oval-shaped tips were taken for measurements. For this purpose, the most contaminated part on the ventral surface of the third tarsomere (the most distal attachment pad) was selected. Since the tarsomere was not regularly contaminated, portion of setae covered with wax crystals and dust particles were evaluated from three to five SEM images of the central part of the pad (120 μm × 100 μm each) taken at low magnification. For measurements, SPSS was used.

7.3 RESULTS

7.3.1 Pruinose Plant Surfaces

Surfaces of plant species studied are either very densely (most species) or rather sparsely (*L. serriola*, *T. montanum*, and *V. vinifera*) covered with wax crystals (Figure 7.2). In most species, the wax coverage is relatively uniform. However in *L. serriola*, crystals are arranged in starlike clusters (rosettes) (Figure 7.2L). In some species (*B. oleracea* and *A. negundo*), crystals form compact networks, where single crystals are hardly discerned (Figures 7.2D and 7.2H). Depending on the plant species, crystals vary in shape, size, and aspect ratio (Table 7.1).

7.3.2 Adhesive Pads of the Beetle *Chrysolina Fastuosa*

The beetle *C. fastuosa* possesses hairy adhesive pads [33]. The ventral side of the three first proximal tarsomeres (1 to 3) is covered by tenent setae [28,34]. There are several types of setae. The first type includes setae with curved sharp tips (Figures 7.3A to 7.3C). The second one is characterized by the platelike thickening on the tip so that the seta has a mushroomlike shape (Figures 7.3A to 7.3D). The third one includes setae with spatula-like tips (Figure 7.3C). There are also setae with transitional shapes. Setae of the first type are located on the second and at margins of the first and third tarsomeres. Setae of the second type are situated in the middle of the first and third tarsomeres. Setae of the third type are located in the distal region of the third tarsomeres. Setae with transitional shapes cover areas between setae of different types.

7.3.3 Pad Contamination

Intact pads of *C. fastuosa* beetles were not absolutely clean but bore some contaminating particles of unknown origin on the surface of their setal tips (Figures 7.3A to 7.3D). After walking on the glass plate covered with dust, pads were rather strongly contaminated by large dust particles covering large areas of pads (Figures 7.3E and 7.3F). The particles usually got stuck between setae and contaminated setal tips only

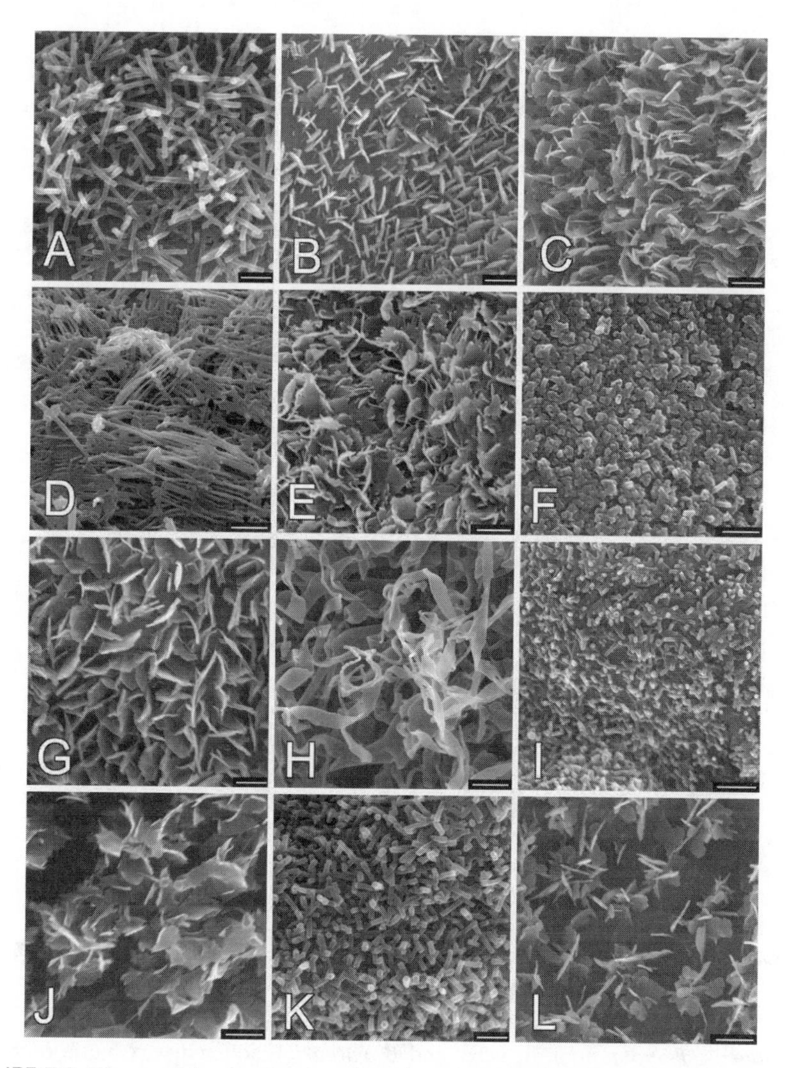

FIGURE 7.2 Wax crystals of pruinose surfaces in various plant species. (A) Leaf of *Aquilegia vulgaris* L. (B) Leaf of *Trifolium montanum* L. (C) Leaf of *Iris germanica* L. (D) Leaf of *Brassica oleracea* L. (E) Leaf of *Aloe vera* (L.) Webb. & Berth. (F) Ripe fruit of *Prunus domestica* L. (G) Leaf of *Chenopodium album* L. (H) Young stem of *Acer negundo* L. (I) Leaf of *Berberis vulgaris* L. (J) Ripe fruit of *Vitis vinifera* L. (K) Leaf of *Chelidonium majus* L. (L) Leaf of *Lactuca serriola* Torner. Scale bars: 1 μm for (A), (B), (G), and (K); 2 μm for (C) to (F), (H) to (J), and (L).

at some places. In the case of plant wax contamination, wax material adhered only to the setal tips' surface (Figures 7.3G to 7.3L, Figure 7.4, and Figure 7.5). Dust and wax contaminations may be distinguished easily because of the particles' dimensions and the nature of the contamination.

Pad contamination by wax crystals differed greatly in both nature and degree, depending on the plant species (Figures 7.3G to 7.3L, 7.4, and 7.5). In some plants

TABLE 7.1
Morphometrical Variables and Aspect Ratio of Wax Crystals Constituting Pruinose Surfaces in 12 Plant Species

Plant Species	Organ	Wax Projection Type	Crystal Shape	Variable, μm		Aspect Ratio
				Largest	Smallest	
Acer negundo	Stem	Polygonal rodlets	Ribbon	~20	0.34	~ 100
Aloe vera	Leaf	Membraneous platelets	Plate	1.11	0.09	12.33
Aquilegia vulgaris	Leaf	Tubules	Tube	0.58	0.17	3.41
Berberis vulgaris	Leaf	Tubules	Tube	0.73	0.16	4.56
Brassica oleracea	Leaf	Terete rodlets	Filament	2.44	0.06	40.67
Chelidonium majus	Leaf	Tubules	Tube	0.83	0.18	4.61
Chenopodium album	Leaf	Irregular platelets	Plate	0.89	0.07	12.71
Iris germanica	Leaf	Irregular platelets	Plate	1.74	0.08	21.75
Lactuca serriola	Leaf	Irregular platelets	Plate	0.55	0.06	9.17
Prunus domestica	Fruit	Tubules	Tube	0.58	0.26	2.23
Trifolium montanum	Leaf	Irregular platelets	Plate	0.58	0.06	9.67
Vitis vinifera	Fruit	Membraneous platelets	Plate	3.17	0.14	22.64

(*A. vulgaris*, *T. montanum*, *A. vera*, and *B. vulgaris*), adhered wax had a relatively homogeneous texture with almost unstructured or weakly structured surface, and wax crystals were not recognizable anymore (Figures 7.3H, 7.3J, 7.4D, and 7.4L). The waxes of most plants (*B. oleracea*, *C. album*, *A. negundo*, *C. majus*, and *L. serriola*) caused clearly structured contaminating coverages or homogeneous coverages with well-structured surfaces, where a few crystals or their contours may be identified (Figures 7.4B, 7.4H, 7.4J, 7.5D, and 7.5F). Waxes of *I. germanica* and *V. vinifera* appeared on the setal tips as relatively homogeneous layers with numerous crystals visible on surfaces (Figures 7.3L and 7.5B).

In the case of *A. vulgaris* and *B. vulgaris*, the setae remained almost clean, rarely bearing a few contaminated points or single tiny spots of wax on their tips (Figures 7.3G, 7.3H, 7.4K, and 7.4L). Crystals of *C. majus* displayed only weak contamination appearing as several spots on the setal surface (low degree of contamination) (Figures 7.5C and 7.5D). In *B. oleracea*, *P. domestica*, and *L. serriola*, contaminated

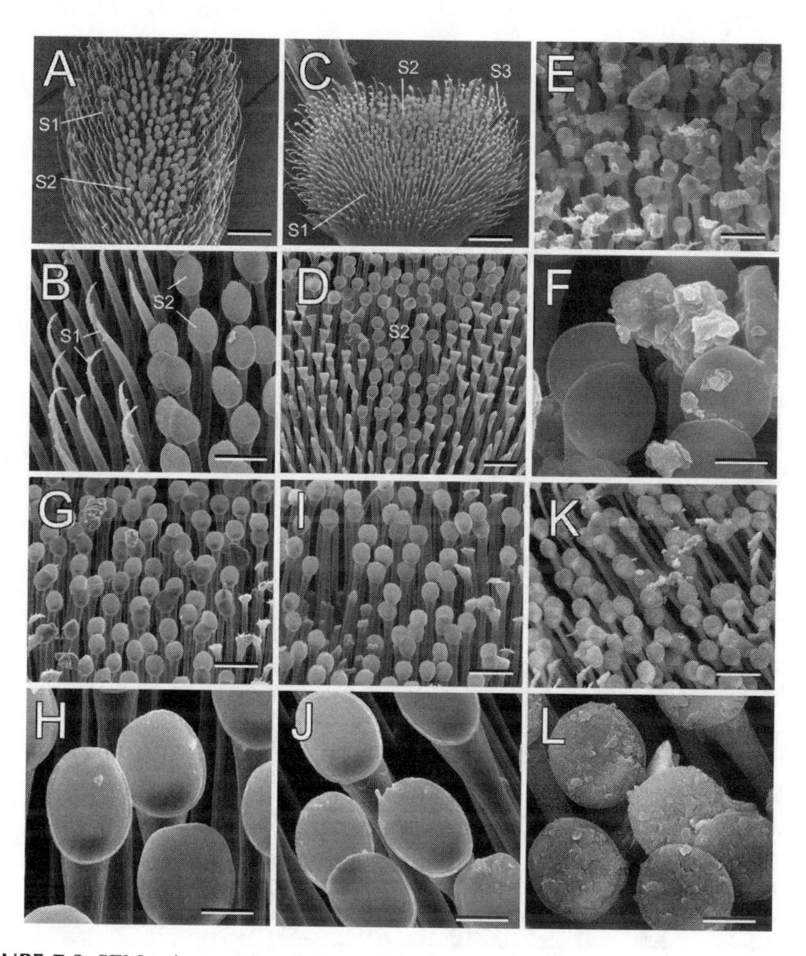

FIGURE 7.3 SEM micrographs of the pad surface in the first (A and B) and the third tarsomeres (C to L) of the beetle *Chrysolina fastuosa* after walking on various surfaces. (A) to (D) Clean glass plate. (E) and (F) Dust particles. (G) and (H) Leaf of *Aquilegia vulgaris*. (I) and (J) Leaf of *Trifolium montanum*. (K) and (L) Leaf of *Iris germanica*. S1, S2, and S3 correspond to setae of the first, second, and third type, respectively (see the text). Scale bars: 50 μm (A); 10 μm (B); 100 μm (C); 20 μm (D), (E), (G), (I), and (K); 5 μm (F), (H), (J), and (L).

spots might cover up to a half of the setal tip area (medium degree of contamination) (Figures 7.4A, 7.4B, 7.4E, 7.4F, 7.5E, and 7.5F). Waxes of *T. montanum*, *I. germanica*, *A. vera*, and *C. album* might cover the whole setal tips (strong contamination) (Figures 7.3I to 7.3L, 7.4C, 7.4D, 7. 4G, and 7.4H). In cases of *A. negundo* and *V. vinifera*, wax often spread even over the setal tips' margins (very strong contamination) (Figures 7.4I, 7.4J, 7.5A, and 7.5B).

To quantify the influence of wax structure on the degree of contamination, we used the following two parameters: (1) the portion of setal tip surface covered with contaminating particles and (2) the portion of setae covered with the particles. Plant waxes studied differed greatly according to both parameters (for the portion of setal tip surface: degrees of freedom (d.f.) = 11, $F = 8.963$, $P < 0.001$, one-way ANOVA)

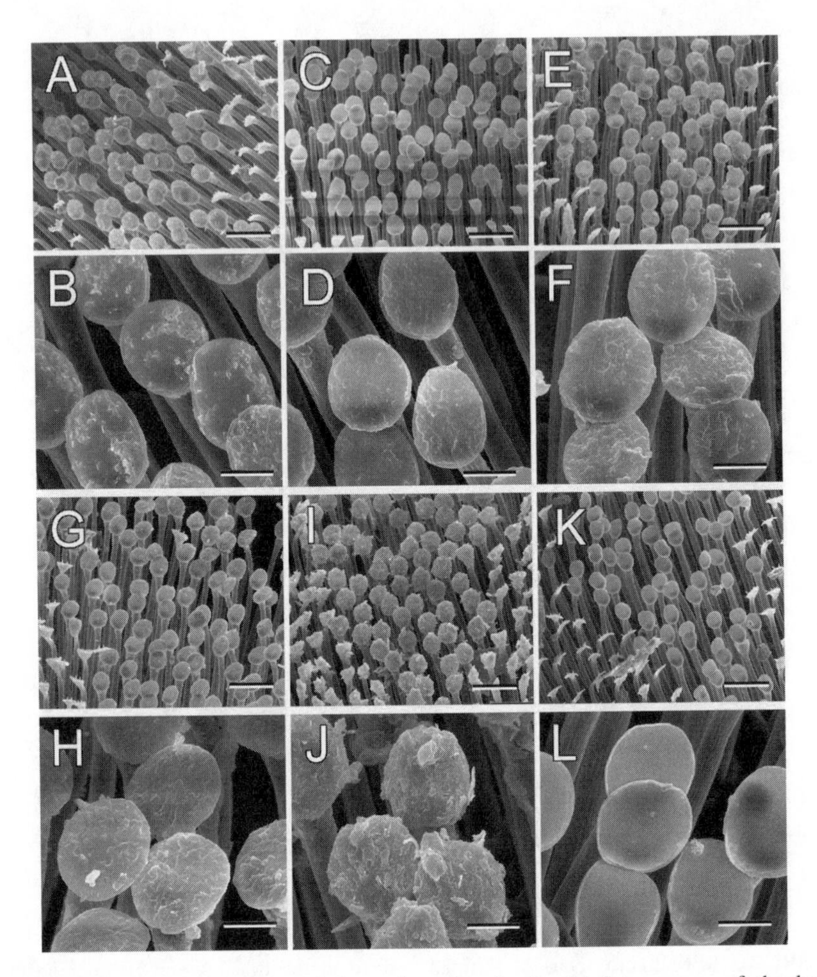

FIGURE 7.4 SEM micrographs of the pad surface in the third tarsomere of the beetle *Chrysolina fastuosa* after walking on various plant surfaces. (A) and (B) Leaf of *Brassica oleracea*. (C) and (D) Leaf of *Aloe vera*. (E) and (F) Ripe fruit of *Prunus domestica*. (G) and (H) Leaf of *Chenopodium album*. (I) and (J) Young stem of *Acer negundo*. (K) and (L) Leaf of *Berberis vulgaris*. Scale bars: 20 μm (A), (C), (E), (G), (I), and (K); 5 μm (B), (D), (F), (H), (J), and (L).

(Figures 7.6A and 7.6B). A positive correlation (NS) between the portion of the setal tip surface covered with particles and the portion of setae contaminated was found (linear regression: $y = 10.332 + 0.978x$, $R^2 = 0.574$; d.f. = 1, $F = 13.502$, $P = 0.004$, one-way ANOVA) (Figure 7.6C). The positive correlation between the two parameters was observed not only in the whole set of plant species having various types of wax crystals, but also in plants with the same crystal type, e.g., platelets and tubes. Therefore, only one of these parameters, the portion of setal tip surface contaminated, was used in our further evaluations.

The portion of the setal tip surface contaminated with plant waxes depended on size of wax crystals. The aspect ratio was found to be statistically significant with

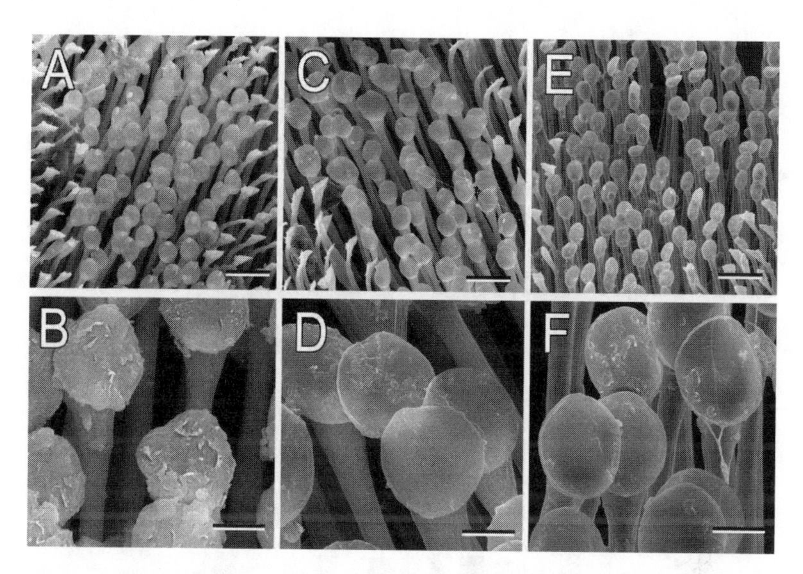

FIGURE 7.5 SEM micrographs of pad surface in the third tarsomere of the beetle *Chrysolina fastuosa* after walking on various plant surfaces. (A) and (B) Ripe fruit of *Vitis vinifera.* (C) and (D) Leaf of *Chelidonium majus.* (E) and (F) Leaf of *Lactuca serriola.* Scale bars: 20 μm (A), (C), and (E); 5 μm (B), (D), and (F).

an increase in the largest dimension of crystals (linear regression: $y = 6.112+3.955x$, $R^2 = 0.766$; d.f. = 1, $F = 32.774$, $P < 0.001$, one-way ANOVA) (Figure 7.7A). The portion of the setal tip surface contaminated correlated positively (NS) with both the largest dimension (linear regression: $y = 44.013+2.509x$, $R^2 = 0.294$; d.f. = 1, $F = 4.173$, $P = 0.068$, one-way ANOVA) and aspect ratio (linear regression: $y = 41.434+0.572x$, $R^2 = 0.313$; d.f. = 1, $F = 4.548$, $P = 0.059$, one-way ANOVA) (Figures 7.7B and 7.7C, respectively).

7.4 DISCUSSION

7.4.1 Contaminating Effect of Crystalline Epicuticular Waxes on Insect Attachment Devices

Similar to other studies on insect attachment to pruinose substrates [8,11–13,16,18,19, 22,23,25–27], our previous experiments with *C. fastuosa* showed that these insects were unable to attach properly to all tested plant surfaces having wax bloom [28]. Most pruinose surfaces did not influence the attachment ability of beetles in further experiments on the smooth glass plate used as a reference substrate. It was observed that after having contact with pruinose plant surfaces, insects usually cleaned their attachment organs from probable wax contamination and, within a short time, recovered their attachment ability. Only stem of *A. negundo* had a strong disabling effect on insect attachment, but recovery of the attachment ability was relatively fast. It was assumed that wax crystals of this plant can probably temporarily impair the function of the tenent setae in *C. fastuosa*.

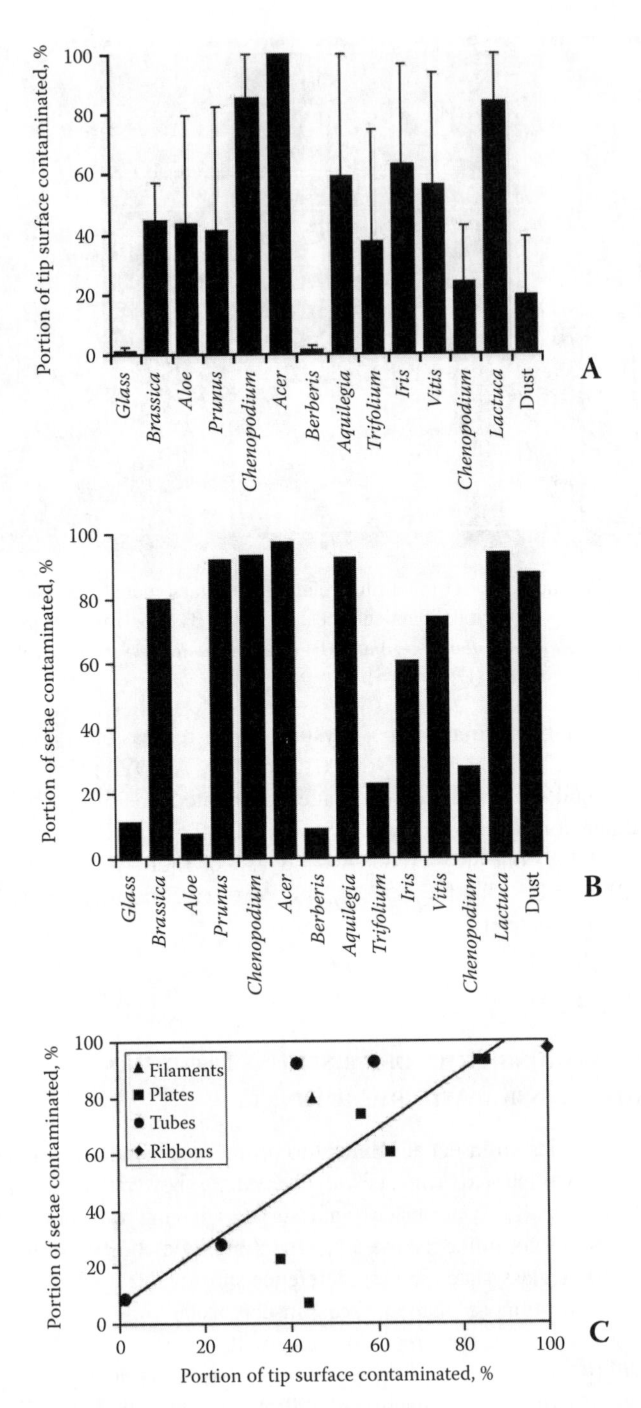

FIGURE 7.6 (A) The portion of the area of the setal tips covered with contaminating particles, (B) the portion of setae contaminated with wax crystals and dust particles, and (C) the relationship between these two parameters.

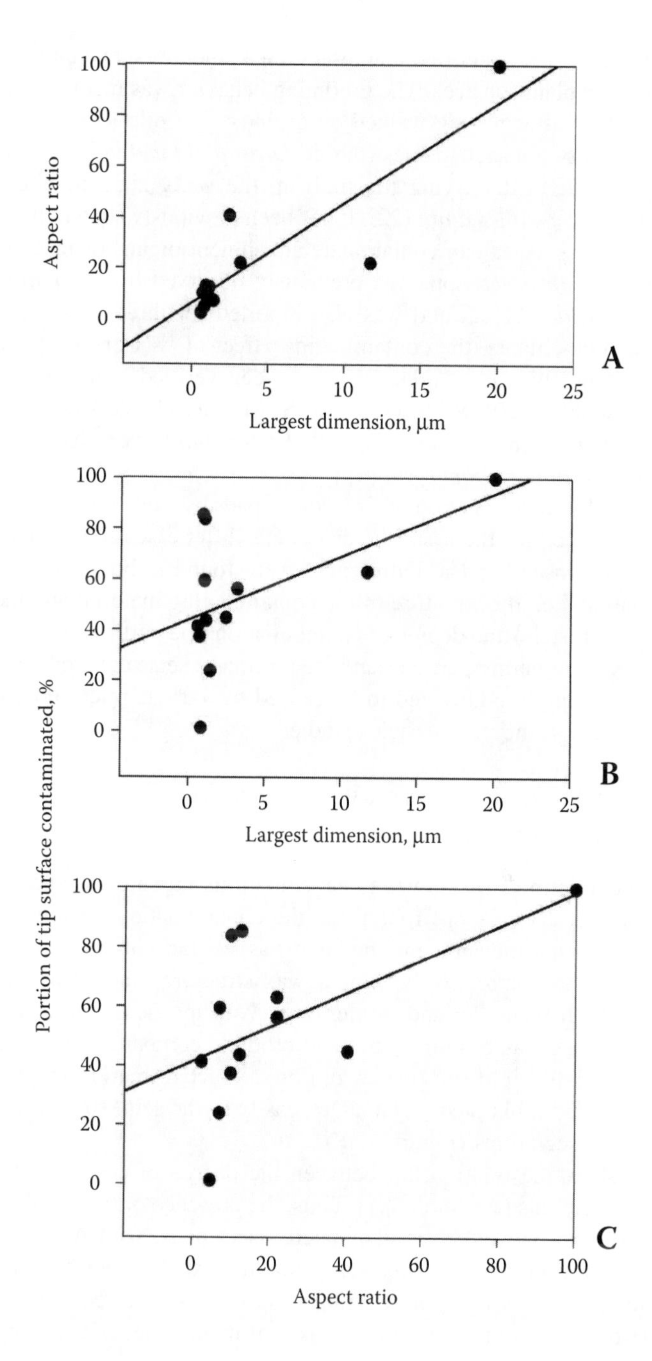

FIGURE 7.7 (A) The relationships between the largest aspect ratio and the largest dimension of wax crystals, (B) between the portion of setal tip surface contaminated and the largest dimension, and (C) between the portion of setal tip surface contaminated and the aspect ratio.

It has been previously reported that insects clean their tarsi and pads after having contact with waxy plant surfaces. The grooming behavior was recorded in the beetle *Paropsis charybdis* after having walked on pruinose juvenile leaves of *Eucalyptus nitens* [13] and in two insect species (the fly *Drosophila melanogaster* and the ant *Iridomyrmex humilis*) after trying to climb up the waxy zone of pitchers in the carnivorous plant *Nepenthes alata* [22]. It has been previously shown that crystalline wax of some plant species can contaminate attachment organs of insects. Contamination by wax of *Brassica* spp. was previously observed by SEM in the beetles *Phaedon cochleariae* [11,12] and was also reported for three generalist predators [35]. In carnivorous plants, the contaminating effect of wax crystals in *Nepenthes* sp. on pads of the fly *Musca domestica* has been verified with the transmission electron microscopy [5,20]. With the use of SEM, wax platelets of carnivorous *N. ventrata*, *Brocchinia reducta*, and *Catopsis berteroniana* were found attached to pads of the fly *Calliphora vomitoria* [23].

Our data on the contamination of *C. fastuosa* pads by wax crystals of 12 pruinose plant surfaces showed for the first time that plants differ essentially in their contaminating effects on insect's pads. Differences were found in both quality (nature of the contamination, i.e., the structure of the contaminating material and the presence of recognizable crystals) and degree of contamination (the portion of setal tip surface covered with contaminating particles and the portion of setae covered with the wax). These differences are hypothesized to be caused by various micromorphologies of epicuticular waxes in the plant species studied.

7.4.2 Dependence of Pad Contamination on Wax Micromorphology

Analyzing the relationship between the contaminating ability and geometrical parameters of wax crystals, we found that the contamination positively correlated with both the largest dimension and the largest aspect ratio of crystals. Thus, larger, relatively longer, and/or relatively thinner wax structures should adhere to insect pads more easily than smaller and shorter ones. We suppose that large crystals with high aspect ratio are much more brittle and erodible compared with compact ones and can break readily into tiny pieces during contact formation (Figure 7.8B vs. 7.8E). These hydrophobic pieces can attach easily to the setal surface covered with hydrophobic pad secretions (Figure 7.8F).

Our data show the relationship between the degree of contamination and the type of wax projections (see Table 7.1). Thus, the presence of the tubulelike crystals led to almost clean or weakly contaminated insect pads, with the only exception being the very short tubules in *P. domestica* that nevertheless cause relatively strong contamination. After having contact with the substrates, covered with wax rodlets, pads were strongly contaminated. In the cases of membraneous and irregular platelets, strong or very strong contamination was recorded.

The crystals or at least their contours were clearly observed on the contaminated insect surfaces only in the case of platelets (except *A. vera*); they were not recognizable for other wax crystals such as tubules and rodlets. The homogeneous structure of the wax contamination and the absence of the crystals in the contaminating

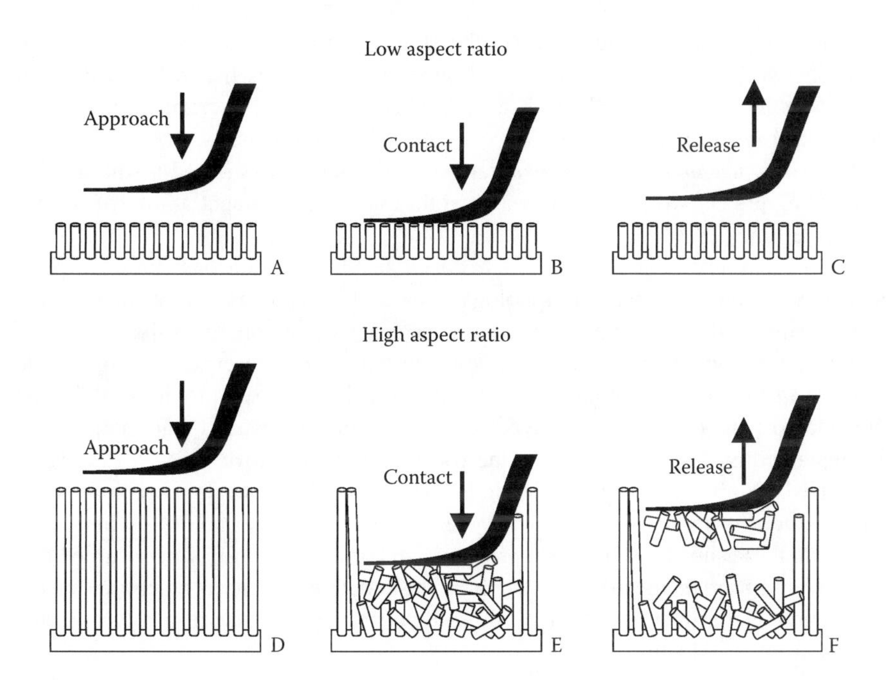

FIGURE 7.8 Diagram explaining the effect of the aspect ratio of wax crystals on the degree of pad contamination. (A) to (C) Low aspect ratio. (D) to (F) High aspect ratio.

material may be explained by the wax dissolving in the secretory fluid produced by tenent setae (wax-dissolving hypothesis). It has been previously found that insect pads contain oily substances [36–39]. Since only very nonpolar solvents, such as benzene, hexane, and chloroform, have been previously reported to be able to dissolve epicuticular waxes of most plants [40], experiments focusing on the dissolution of various plant waxes in the pad secretion or its oily components should be performed to prove the wax-dissolving hypothesis.

It can be assumed that not only the type of the wax projection and the geometry of crystals, but also other characteristics of the wax coverage, may influence the pad contamination. When the plant surface is not uniformly covered with the crystalline wax and there are gaps free of crystals, weak contamination of insect adhesive pads is expected. However, this effect was observed only in *B. vulgaris*, whereas in most other cases, the contamination appeared to be strong (*T. montanum* and *A. vulgaris*) or even very strong (*V. vinifera* and *L. serriola*). On the contrary, when the wax bloom is very dense and consists of several layers of crystals, relatively strong contamination can be expected. Since crystals from an upper layer of the composite wax system presumably detach more easily than those in direct contact with the subjacent surface, composite wax systems should lead to stronger contamination. Such an effect has been previously considered for the upper layer of wax platelets in pitchers of carnivorous plants from the genus *Nepenthes* to explain the antiadhesive properties of the waxy zone [5,20]. The compact network of crystals, especially when they have a high aspect ratio, should also cause a strong contaminating effect.

The delicate mechanical removal (by brushing) of threadlike crystals composing dense networks, and strong contamination of adhesive pads in the fly *Calliphora vomitoria* by these waxes have been previously described in carnivorous plants *Brocchinia reducta* and *Catopsis berteroniana* [23]. In our study, ribbonlike terete rodlets of *A. negundo* forming relatively compact networks showed a similar contaminating effect. Moreover, this wax can disable the attachment ability of beetles for some time [28].

Thus, our experiments with *C. fastuosa* beetles and various pruinose plant substrates differing in wax morphology showed that wax crystals in many plant species are easily detachable structures that can readily adhere to the surface of attachment organs in insects and contaminate them. The presence of recognizable crystals in the contaminating material found for some waxes provides additional evidence for the contamination hypothesis. The homogeneous contamination may be explained by wax dissolution in the pad secretion. To prove the wax-dissolving hypothesis, further experiments should be carried out. In a few plant species, waxes showed no contaminating effect; however, a decrease in the ability of insects to attach on these substrates has been previously recorded [28]. It may be assumed that in these cases other mechanisms, such as the reduction of the contact area or the absorption of the pad secretion by crystalline wax coverage, are involved in the adhesion force reduction.

ACKNOWLEDGMENTS

Constant support by Jürgen Berger (Electron Microscopy Unit at the MPI of Developmental Biology, Tübingen, Germany) is greatly acknowledged. Florian Haas (University of Heidelberg, Germany) helped with data quantification. This project was supported by the Federal Ministry of Education, Science and Technology, Germany, to SNG (Project BioFuture 0311851).

REFERENCES

1. Barthlott, W. and Ehler, N., Raster-Elektronenmikroskopie der Epidermisoberflächen von Spermatophyten, *Trop. Subtrop. Pflanzenwelt*, 19, 367, 1977.
2. Barthlott, W. and Wollenweber, E., Zur Feinstruktur, Chemie und taxonomischen Signifikanz epikutikularer Wachse und ähnliche Sekrete, *Trop. Subtrop. Pflanzenwelt*, 32, 7, 1981.
3. Barthlott, W., Scanning electron microscopy of the epidermal surface in plants, in *Scanning Electron Microscopy in Taxonomy and Functional Morphology*, Claugher, D., Ed., Clarendon, Oxford, 1998, p. 69.
4. Jeffree, C.E., The cuticle, epicuticular waxes and trichomes of plants, with references to their structure, functions and evolution, in *Insects and the Plant Surface*, Juniper, B.E. and Southwood, T.R.E., Eds., Edward Arnold, London, 1986, p. 23.
5. Juniper, B.E., Robins, R.J., and Joel, D.M., *The Carnivorous Plants,* Academic Press, London, 1989.
6. Harley, R., The greasy pole syndrome, in *Ant-Plant Interactions*, Huxley, C.R. and Cutler, D.E., Eds., Oxford University Press, Oxford, 1991, p. 430.

7. Juniper, B.E., Waxes on plant surfaces and their interactions with insects, in *Waxes: Chemistry, Molecular Biology and Functions*, Hamilton, R.J., Ed., Oily, West Ferry, Dundee, 1995, p. 157.
8. Eigenbrode, S.D., Plant surface waxes and insect behaviour, in *Plant Cuticles – An Integral Functional Approach*, Kerstiens, G., Ed., BIOS, Oxford, 1996, p. 201.
9. Anstey, T.H. and Moore, J.F., Inheritance of glossy foliage and cream petals in green sprouting broccoli, *J. Heredity*, 45, 39, 1954.
10. Way, M.J. and Murdie, G., An example of varietal resistance of Brussel sprouts, *Ann. Appl. Biol.*, 56, 326, 1965.
11. Stork, N.E., Role of wax blooms in preventing attachment to brassicas by the mustard beetle, *Phaedon cochleariae*, *Entomol. Exp. Applicata*, 28, 100, 1980.
12. Stork, N.E., The form of plant waxes: a means preventing insect attachment? in *Insects and the Plant Surface*, Juniper, B.E. and Southwood, T.R.E., Eds., Edward Arnold, London, 1986, p. 346.
13. Edwards, P.B., Do waxes of juvenile *Eucalyptus* leaves provide protection from grazing insects? *Austral. J. Ecol.,* 7, 347, 1982.
14. Edwards, P.B. and Wanjura, W.J., Physical attributes of eucalypt leaves and the host range of chrysomelid beetles, *Symp. Biol. Hung.,* 39, 227, 1990.
15. Stoner, K.A., Glossy leaf wax and host-plant resistance to insects in *Brassica oleracea* L. under natural infestation, *Environ. Entomol.*, 19, 730, 1990.
16. Bodnaryk, R.P., Leaf epicuticular wax, an antixenotic factor in Brassicaceae that affects the rate and pattern of feeding in flea beetles, *Phyllotreta cruciferae* (Goeze), *Can. J. Plant Sci.*, 72, 1295, 1992.
17. Eigenbrode, S.D. and Espelie, K.E., Effects of plant epicuticular lipids on insect herbivores, *Annu. Rev. Entomol.*, 40, 171, 1995.
18. Federle, W. et al., Slippery ant-plants and skillful climbers: selection and protection of specific ant partners by epicuticular wax blooms in *Macaranga* (Euphorbiaceae), *Oecologia*, 112, 217, 1997.
19. Eigenbrode, S.D. and Kabalo, N.N., Effects of *Brassica oleracea* waxblooms on predation and attachment by *Hippodomia convergen, Entomol. Exp. Applicata*, 91, 125, 1999.
20. Juniper, B.E. and Burras, J.K., How pitcher plants trap insects, *New Scientist*, 269, 75, 1962.
21. Martin, J.T. and Juniper, B.E., *The Cuticle of Plants*, Edward Arnold, London, 1970.
22. Gaume, L., Gorb, S., and Rowe, N., Function of epidermal surfaces in the trapping efficiency of *Nepenthes alata* pitchers, *New Phytol.*, 156, 476, 2002.
23. Gaume, L. et al., How do plant waxes cause flies to slide? Experimental tests of wax-based trapping mechanisms in the three pitfall carnivorous plants, *Artropod Str. Develop.*, 33, 103, 2004.
24. Riedel, M., Eichner, A., and Jetter, R., Slippery surfaces of carnivorous plants: composition of epicuticular wax crystals in *Nepenthes alata* Blanco pitchers, *Planta*, 218, 87, 2003.
25. Eigenbrode, S.D., Kabalo, N.N., and Stoner, K.A., Predation, behavior, and attachment by *Chrysoperla plorabunda* larvae on *Brassica oleracea* with different surface waxblooms, *Entomol. Exp. Applicata*, 90, 225, 1999.
26. Eigenbrode, S.D. et al., Effects of wax bloom variation in *Brassica oleracea* on foraging by a vespid wasp, *Entomol. Exp. Applicata*, 97, 161, 2000.
27. Federle, W., Rohrseitz, K., and Hölldobler, B., Attachment forces of ants measured with a centrifuge: better "wax-runners" have a poorer attachment to a smooth surface, *J. Exp. Biol.*, 203, 505, 2000.

28. Gorb, E.V. and Gorb, S.N., Attachment ability of the beetle *Chrysolina fastuosa* on various plant surfaces, *Entomol. Exp. Applicata*, 105, 13, 2002.
29. Gorb, S.N., *Attachment Devices of Insect Cuticle*, Kluwer Academic Publishers, Dordrecht, 2001.
30. Ramires, V.B., Can dusts be used against *Varroa jacobsoni*? *Apiacta*, 24, 2, 1989.
31. Dixon, A.F.G., Croghan, P.C., and Gowing, R.P., The mechanism by which aphids adhere to smooth surfaces, *J. Exp. Biol.*, 152, 243, 1990.
32. Barthlott, W. et al., Classification and terminology of plant epicuticular waxes, *Bot. J. Linn. Soc.*, 126, 237, 1998.
33. Beutel, R. and Gorb, S.N., Ultrastructure of attachment specializations of hexapods (Arthropoda): evolutionary patterns inferred from a revised ordinal phylogeny, *J. Zool. Syst. Evol. Res.*, 39, 177, 2001.
34. Stork, N.E., A scanning electron microscope study of tarsal adhesive setae in the Coleoptera, *Zool. J. Linn. Soc.*, 68, 173, 1980.
35. Eigenbrode, S.D. et al., Mobility of three generalist predators is greater on cabbage with glossy leaf wax than on cabbage with a wax bloom, *Entomol. Exp. Applicata*, 81, 335, 1999.
36. Ishii, S., Adhesion of a leaf feeding ladybird *Epilachna vigintioctomaculta* (Coleoptera: Coccinellidae) on a vertically smooth surface, *Appl. Entomol. Zool.*, 22, 222, 1987.
37. Kosaki, A. and Yamaoka, R., Chemical composition of footprints and cuticula lipids of three species of lady beetles, *Jap. J. Appl. Entomol. Zool.*, 40, 47, 1996.
38. Eisner, T. and Aneshansley, D.J., Defence by foot adhesion in a beetle (*Hemisphaerota cyanea*), *Proc. Natl. Acad. USA*, 97, 6568, 2000.
39. Vötsch, W. et al., Chemical composition of the attachment pad secretion of the locust *Locusta migratoria, Insect Biochem. Molec. Biol.*, 32, 1605, 2002.
40. Hallam, N.D., Fine structure of the leaf surface and the origin of leaf waxes, in *The Plant Cuticle*, Cutler, D.E., Alvin, K.L., and Price, C.E., Eds., Academic Press, London, 1982, p. 197.

8 Ecology and Biomechanics of Slippery Wax Barriers and Wax Running in *Macaranga*–Ant Mutualisms

Walter Federle and Tanja Bruening

CONTENTS

8.1 INTRODUCTION

Most terrestrial ecosystems have been shaped by interactions between plants and insects. Insects are involved in a variety of interactions with plants, not only as pollinators and herbivores, but also as predators of herbivores, seed dispersers, and as prey of insectivorous plants. The mechanisms giving rise to and maintaining these interactions are often chemical in nature. Plant secondary compounds may have direct effects on herbivores (e.g., feeding inhibitors or toxic compounds) or act as signals (semiochemicals) perceived by insects. The chemical ecology of insect–plant interactions has been studied extensively over the past decades (see, e.g., [1–3]). However, insect–plant interactions are not only determined by chemical factors, but also by physical factors. For example, mechanical interactions are well-known from specialized flower morphologies tailored for specific pollinators [4,5], and mechanical factors play a significant role in the context of herbivory (see, e.g., [6–8]). Despite their ecological importance, surprisingly little work has focused on mechanical interactions between insects and plants (but see Chapters 6, 7, and 9 of this book).

A mechanical factor that represents a fundamental prerequisite for the existence of insect–plant interactions is surface attachment. To hold onto and maneuver on smooth plant cuticles, insects have evolved specialized adhesive pads on their legs [9]. To keep away insects with adhesive organs, many plant species form surfaces that inhibit insect attachment or hamper insect locomotion. Some insects in turn have developed adaptations to circumvent these barriers (see, e.g., [10–14]). The evolution of mechanical barriers and of mechanisms to circumvent them is analogous to the evolution of chemical plant defenses and insect mechanisms to detoxify or sequester them. Such "coevolutionary" processes (in the sense of "escape-and-radiate" or reciprocal adaptation) have probably played a central role in the evolution of plants and insects [15,16].

One of the clearest examples of the importance of biomechanical factors for the ecology and evolution of insect–plant interactions is represented by the protective "wax barriers" in the ant-plant genus *Macaranga* [17]. The aim of the present chapter is to give a survey of the ecological and evolutionary implications of wax barriers and "wax running" capacity in *Macaranga*–ant mutualisms. We combine this review with new findings on the biomechanics of wax running in *Crematogaster* (*Decacrema*) ants.

8.2 ECOLOGY AND EVOLUTION OF WAX BARRIERS IN THE ANT-PLANT GENUS *MACARANGA*

8.2.1 Protection of Specific Ant Partners against Generalist Ants

Macaranga (Euphorbiaceae) is a large paleotropical pioneer tree genus consisting of approximately 280 species [18]. The majority of species are myrmecophilic, i.e., they offer nectar and food bodies on the plant surface to attract generalist ants, which provide protection against herbivory. Only 29 *Macaranga* species in southeast Asia are myrmecophytic, i.e., obligately associated with specialized plant-ants. These ants (belonging to the genera *Crematogaster* and *Camponotus*) exclusively nest in the hollow (or hollowed-out) stems of their host trees.

In 14 *Macaranga* ant-plant species, the stems are densely covered by blooms of epicuticular wax crystals. These waxy surfaces are very slippery for most generalist insects. A test of the climbing capacity of 17 generalist ant species on waxy *Macaranga* stems showed that ants often fell down from the plant or moved forward very slowly. This barrier effect of the wax crystal surfaces is only based on reduced physical adhesion; no evidence of chemical repellence was found [17]. In striking contrast to generalist ants, the specialized ant partners of waxy *Macaranga* hosts had no difficulty running on the slippery stems. Because of their exceptional wax running capacity, these ants escape predation and competition by generalist ants. In fact, *Crematogaster* (*Decacrema*) wax runners were clearly inferior to most other ants in direct encounters, so that their survival depended on the presence of vertical waxy stem sections preventing generalist ants from invading their nests [17]. The barrier effect of waxy stems is enhanced by the whorled growth form and the extreme epidermis longevity of many *Macaranga* ant-plants, which increase the length of slippery, waxy stem sections that have to be traversed to reach the upper parts of a tree [17].

8.2.2 Effect of Wax Barriers on Host Specificity

Macaranga wax barriers not only act as an exclusion filter favoring specialists over generalists, but also as an ecological isolation mechanism between different ant partners of *Macaranga*. Most of the obligate plant-ants associated with *Macaranga* are members of the *Crematogaster* subgenus *Decacrema*. The *Crematogaster* (*Decacrema*) complex currently represents a group of nine similar morphospecies (msp.), many of which can only be identified by morphometry of queen head dimensions [19]. A formal description of the recognized morphospecies colonizing *Macaranga* and a taxonomic revision of the *Decacrema* subgenus are still to come.

Crematogaster (*Decacrema*) ants colonize both waxy and nonwaxy *Macaranga* host plants. Running tests in several *Crematogaster* (*Decacrema*) morphospecies showed that ants associated with nonwaxy *Macaranga* hosts had great difficulty climbing on slippery waxy stems [17,20], suggesting that only certain *Crematogaster* (*Decacrema*) morphospecies are able to colonize waxy *Macaranga* hosts. The host plant spectrum of *Crematogaster* (*Decacrema*) morphospecies surveyed by

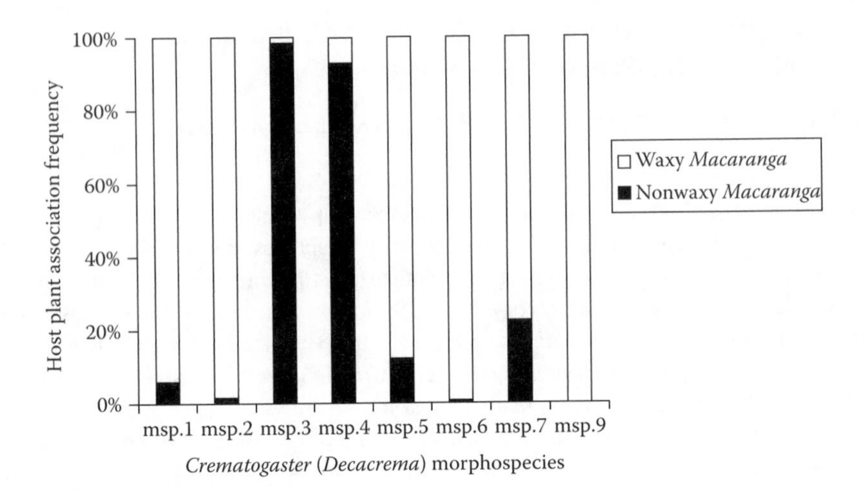

FIGURE 8.1 Association frequency of eight *Crematogaster* (*Decacrema*) morphospecies with waxy vs. nonwaxy *Macaranga* host plants. The data is from Table 2A in Fiala et al., *Biol. J. Linn. Soc.*, 66, 305, 1999; of the plants identified as *M. indistincta* by Fiala et al., 28 were *M. glandibracteolata* trees (B. Fiala, personal communication) and are included among the waxy host plants here. *Crematogaster* msp.8 is not shown here because it is not a member of the *Crematogaster*-subgenus *Decacrema*.

Fiala et al. [19] clearly confirms that *Macaranga* wax barriers act as an ecological isolation mechanism between different ant associates of *Macaranga*. Of the eight *Crematogaster* (*Decacrema*) morphospecies recognized by Fiala et al. [19], six morphospecies are "wax runners" and two are "non–wax runners" (Figure 8.1). A recently discovered new *Crematogaster* (*Decacrema*) morphospecies is also associated with nonwaxy hosts (H. Feldhaar, unpublished results). This specificity pattern becomes even more clear-cut when only adult colonies (and no founding queens) are considered ([19]; B. Fiala, personal communication).

While it is evident that *Crematogaster* (*Decacrema*) non–wax runners can hardly be successful as colonizers of waxy *Macaranga* hosts, it is less obvious why wax runners almost never colonize nonwaxy trees. Even though *Crematogaster* (*Decacrema*) wax runners adhere less well to smooth substrates than non–wax runners (suggesting the existence of a trade-off between the ants' ability to walk on waxy vs. smooth stems [20]), their attachment forces to smooth surfaces are not smaller than those of other arboreal ants and do not seem to represent any limitation. The absence of wax runners on nonwaxy *Macaranga* hosts is rather based on interspecific competition by non–wax runners (see Section 8.2.4).

The only apparent exception from the pattern of specificity conferred by wax barriers (Figure 8.1) is the association between the waxy *M. glandibracteolata* and two morphospecies of non–wax runners in Sabah (Borneo). The proximate factors explaining the existence of this association are still unclear. Even though the wax crystal surfaces of *M. glandibracteolata* appear to be less slippery than those of other waxy *Macaranga* species, the resident ant colonies show less vitality and growth rate than the same morphospecies living sympatrically on nonwaxy *Macaranga* hosts (H. Feldhaar, unpublished results).

Specialization of *Crematogaster* (*Decacrema*) clades (identified from mitochondrial cytochrome oxidase subunit I [COI] gene sequences) to host plants with vs. without wax barriers has also been confirmed by Quek et al. [21], but their study suggested that several clades occurred both on waxy and nonwaxy hosts. However, due to interspecific hybridization and mtDNA gene flow, mtDNA patterns of variation can be misleading about species relationships in closely related species groups (see, e.g., [22, 23]). In fact, recent work by H. Feldhaar has revealed evidence for COI pseudogenes and mtDNA introgression in the *Crematogaster* (*Decacrema*) complex ([24], H. Feldhaar, unpublished results), which may explain the conflicting clade boundaries derived from mtDNA gene sequence data and morphology [25].

8.2.3 Evolution of *Macaranga* Wax Barriers

Wax crystal surfaces occur frequently among myrmecophytic but only rarely among myrmecophilic species of the genus *Macaranga* [17]. In the few waxy, nonmyrmecophytic *Macaranga* species, the waxy stem surface is interspersed with hairs, which facilitate access for generalist ants and reduce the effectiveness of the wax barriers. Even though myrmecophytism and the presence of waxy stems in *Macaranga* exhibit phylogenetic autocorrelation [26,27], the more frequent occurrence of wax barriers among ant-plants is not solely a consequence of phylogenetic history. Inspection of the distribution of gains and losses of slippery wax barriers (i.e., waxy stems without hairs) across a phylogenetic tree of *Macaranga* confirmed that they are functionally linked with myrmecophytism [27], consistent with our earlier conclusions from a phylogenetically uninformed analysis [17].

Although many *Macaranga* species benefit from the presence of generalist ant visitors, there are several reasons why it is beneficial for *Macaranga* ant-plants to promote and protect only one or a few specialized plant-ant species:

1. Obligate ant associations can provide advantages over myrmecophilic relationships. *Macaranga* trees are protected against overgrowth by the pruning behavior of *Crematogaster* (*Decacrema*) ants, which is absent in generalist ants [28,29]. Moreover, specific ant partners permanently nesting on a tree can defend herbivores more effectively than generalist visitors [30].
2. Once an obligate ant associate is present, myrmecophytes can become more dependent on the ant partners' presence because of an evolutionary loss of plant defensive compounds ([31]; for *Macaranga*, see [32,33]). In fact, *Macaranga* ant-plants are unable to survive in the absence of their specific ant partners [30]. For this reason, invasions of generalist ants, which compete for host plant food resources or behave aggressively against the ant partners, can be costly for the host plant.

Selection may have forced *Macaranga* ant-plants to evolve traits that minimize predation and competition for their specific ant partners. There are two alternative ways of how this can be achieved. When plant-ants primarily feed on host plant food resources — as is the case in *Crematogaster* (*Decacrema*) ants — plants can increase the ants' defensive capability by supplying them with more food, which results in larger colonies with more competitive workers. Alternatively, ant-plants can evolve exclusion filters that restrict access to the mutualists. Both "strategies" are found in *Macaranga*–ant associations (see Section 8.2.4.).

8.2.4 Adaptive Syndromes of Ant Associations in Waxy and Nonwaxy *Macaranga* Ant-Plants

The presence or absence of protective wax barriers has important ecological implications for *Macaranga*–ant associations. Despite its effectiveness, this exclusion filter is only present in 14 out of 29 *Macaranga* ant-plant species. This raises the question of why and how non–wax runners are able to compete and survive despite the absence of the protective wax barriers. Our work has revealed that these associations compensate for the absence of wax barriers by a variety of other mechanisms, which involve both ant and host plant traits.

8.2.4.1 Ant Traits

8.2.4.1.1 Pruning Behavior

Crematogaster (*Decacrema*) non–wax runners prune neighboring vegetation more intensely than their congeners inhabiting waxy *Macaranga* hosts [29]. Stronger pruning not only more effectively prevents host plants from being overgrown by climbers, but also provides a better protection against invasions by foreign ants [29,34].

8.2.4.1.2 Aggressiveness

The greater intensity of pruning behavior might reflect a generally higher level of aggressiveness by *Crematogaster* (*Decacrema*) non–wax runners [29]. Moreover, ants from the nonwaxy *M. trachyphylla* were reported to be more aggressive than ants from the waxy *M. beccariana* [35].

8.2.4.1.3 Colony Density

Data from young *Macaranga* trees (1.80 to 3.00 m tall) indicated that colonies of non–wax runners have a higher number of workers per twig length [29], which probably result in a greater defensive capability. However, a different pattern was found in mature *Macaranga* trees [36], suggesting that other factors (e.g., food supply) contribute to interspecific variation of colony density.

8.2.4.2 Host Plant Traits

8.2.4.2.1 Hidden Food Presentation

In many *Macaranga* host plants devoid of wax crystals, food bodies are presented in a secluded space under the recurved, succulent stipules, whereas waxy hosts offer them openly accessible on the plant surface. Similarly, extrafloral nectaries (EFN) are largely absent in nonwaxy *Macaranga* hosts but not in many waxy members of the genus [27]. The correlation of open food presentation and wax barriers is supported both by traditional and phylogenetically based comparative methods (independent contrasts [37]). Thus, the hidden food presentation in nonwaxy *Macaranga* host plants probably represents an adaptation to protect the specific ant partners against competition [27].

8.2.4.2.2 Prostomata

The prostoma, a preformed thin-walled zone of the stem where ants preferably chew their entrance holes, represents a further host plant trait that correlates with the presence and absence of wax barriers. In most *Macaranga* myrmecophytes devoid of wax crystals, the hollow internodes possess a longitudinal prostoma above the leaf insertion [38]. Many waxy host plants, however, lack prostomata and their domatium walls are uniformly thin on all sides of the internode. Absence of prostomata and presence of wax barriers were found to be significantly correlated when *Macaranga* ant-plant species were treated as independent samples [38]. When the distribution of prostomata is mapped on a phylogenetic tree of *Macaranga*, however, it can be seen that most prostoma-containing species in the genus section *Pachystemon* may have originated from a single ancestor. Even though prostomata are therefore unlikely to have evolved as an adaptation to the absence of wax barriers, they could function as filters promoting only the specialized *Crematogaster* (*Decacrema*) ants capable of finding the prostoma. *Crematogaster* (*Decacrema*) workers recognize the position of the prostoma from inside the domatia by using incident daylight as an orientation cue, a capacity that may be absent in many generalist stem-nesting ants that chew entrance holes into hollow twigs [38].

Exclusion filters such as wax barriers, hidden food presentation, or prostomata have also been reported from other myrmecophytic systems. For example, the round or flattened prostomata in the ant-plant genus *Leonardoxa* facilitate hole boring for specific ants [39]; long and dense trichomes in a variety of ant-plant genera inhibit the movements of larger ants [40]. The selective effect of all these plant traits is based on more or less specialized behaviors or morphological properties of the ant partners.

8.2.4.3 Evolution of Adaptive Syndromes

The ant and plant characteristics listed in the previous section correlate with the ants' ability or inability to run on waxy stems and the presence or absence of wax crystals. The combinations of characters may thus be considered adaptive syndromes. Both ant and host plant traits may maximize the ecological benefits and minimize the ecological costs of each stem wax type. For the host plants, the evolutionary outcome may be governed by trade-offs between costs of producing morphological exclusion filters and costs of sustaining highly competitive ant colonies. Similar trade-offs may be associated with the ants' capacity of wax running.

Reconstructions of ancestral states in *Macaranga* ant-plants are ambiguous with regard to the question of whether species devoid of wax crystals have evolved from waxy ancestors [21,27]. However, it is likely that wax barriers have been lost once at the base of a large clade in the genus section *Pachystemon*. Thus, *Macaranga* host plants have possibly undergone a change of adaptive syndrome, followed by radiation.

Even though the existing phylogenetic information on the *Macaranga*-associated *Crematogaster* (*Decacrema*) species group is still uncertain [21,25], it suggests that wax running capacity represents the ancestral condition in this clade and that non–wax runners have originated through several independent losses [25]. The evolutionary origin of wax running capacity may thus have occurred before the radiation of this plant-ant clade.

To summarize, the slippery wax barriers and the ants' capacity to run on them represent a mechanical factor that plays a central role in the biology of *Macaranga*–ant associations and has multiple implications on traits of both host plants and ant partners. An essential prerequisite for clarifying the mechanisms of speciation and the extent of coevolution vs. coadaptation (i.e., the match of characters with independent origins) in this ant-plant system is a knowledge of the mechanical basis of wax barriers and of wax running behavior. A biomechanical analysis of wax running is necessary to identify traits and adaptations involved in the capacity to colonize waxy hosts. By mapping these traits on the phylogeny of *Crematogaster* (*Decacrema*) ants, it will be possible to study the evolution of this biomechanical capacity and to compare it with the evolutionary history of host plant adapations.

8.3 BIOMECHANICS OF WAX RUNNING IN *CREMATOGASTER* (*DECACREMA*) ANTS

8.3.1 Tarsal Attachment Devices in Ants

The pretarsus of arthropods contains two sickle-shaped claws that are used to cling to rough-textured surfaces ([41,42] (Figure 8.2A). Attachment to smooth plant substrates is accomplished by adhesive organs, which are structurally diverse in different insect orders. In ants and bees (Hymenoptera), the adhesive organ (arolium) is an unfoldable pad located between the claws. The adhesive surface of the arolium is relatively smooth and is typically wetted by a fluid secretion, as in many other insects [43,44].

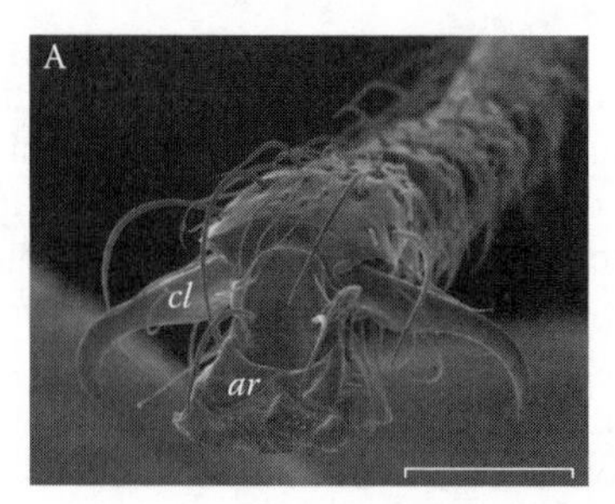

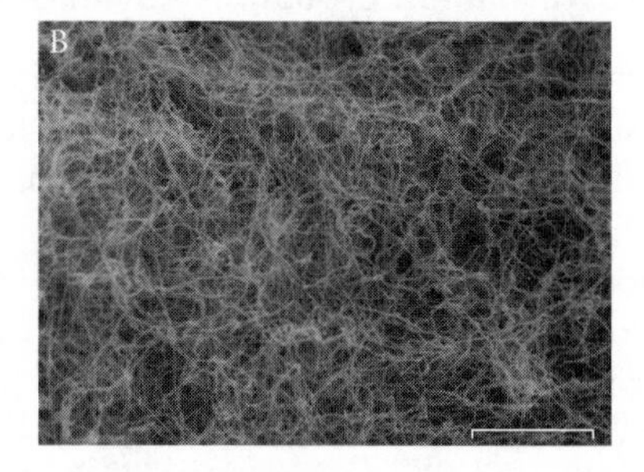

FIGURE 8.2 (A) Pretarsus of *Crematogaster* (*Decacrema*) msp.2 with claws (*cl*) and arolium (*ar*). (B) Wax crystal surface on *M. hypoleuca* stem. Scale bars: 50μm (A) and 5μm (B).

8.3.2 Mechanisms of Slipperiness

Epicuticular wax crystals are typically formed by pure lipid compounds when they are present as the predominant component in the wax mixture. In *Macaranga*, the threadlike crystals (Figure 8.2B) are composed of triterpenoids [45]. Plant surfaces covered with wax crystals are mostly slippery for insects (see, e.g., [46–52]). Several hypothetical mechanisms have been proposed to explain this phenomenon.

1. Wax crystals form substrates with a microscopic surface roughness. Surface roughness may effectively prevent attachment of insect tarsi if (a) the roughness height and width are too large for surface minima to be filled completely with adhesive secretion, (b) the roughness width is too small to allow the soft adhesive pad cuticle to deform and make sufficient contact between asperities (which leads to a strongly reduced area of real contact), and (c) the surface texture is too fine to allow the claws to interlock with surface protrusions [42]. These criteria are met for intermediate surface roughnesses in the micron and submicron range, which correspond to the order of magnitude found in natural waxy plant surfaces. Evidence in favor of microscopic roughness reducing insect attachment forces is given by Gorb [41].
2. Wax crystals can be easily detached so that insect tarsi become contaminated when walking on a waxy plant surface [47,48,50,53–55]. Wax crystal particles not only contaminate insect attachment structures, but they can form an amorphous substance on the surface of adhesive pads when partly dissolved by insect adhesive secretions [49,50]. Both forms of contamination impede attachment.
3. The hydrophobic microrough wax crystal substrates may absorb and deplete insect adhesive secretions necessary for capillarity-based adhesion [50].

In the *Crematogaster–Macaranga* system, both (1) surface roughness and (2) wax crystal detachment and pad contamination contribute to the slipperiness of waxy *Macaranga* stem surfaces. We measured in-plane detachment forces of *Crematogaster* (*Decacrema*) msp.2 wax runners on nonexfoliating aluminium oxide substrates of different roughness using the centrifuge method [56] (Figure 8.3). Forces were large on smooth and on coarse rough surfaces, but were minimal on the intermediate microrough substrate (0.05-μm particle size, Figure 8.3B). Second, tarsi of ants freshly collected after locomotion on waxy *Macaranga* stems were clearly contaminated with wax crystals (Figure 8.4). Similar to the interaction between fly tarsi and wax crystal surfaces of the carnivorous plants *Nepenthes ventrata* (Nepenthaceae) and *Brocchinia reducta* (Bromeliaceae) [49], the crystal material appeared to be partly dissolved and amorphous on the surface of the arolia but less so on the claws and other parts of the tarsus (Figure 8.4).

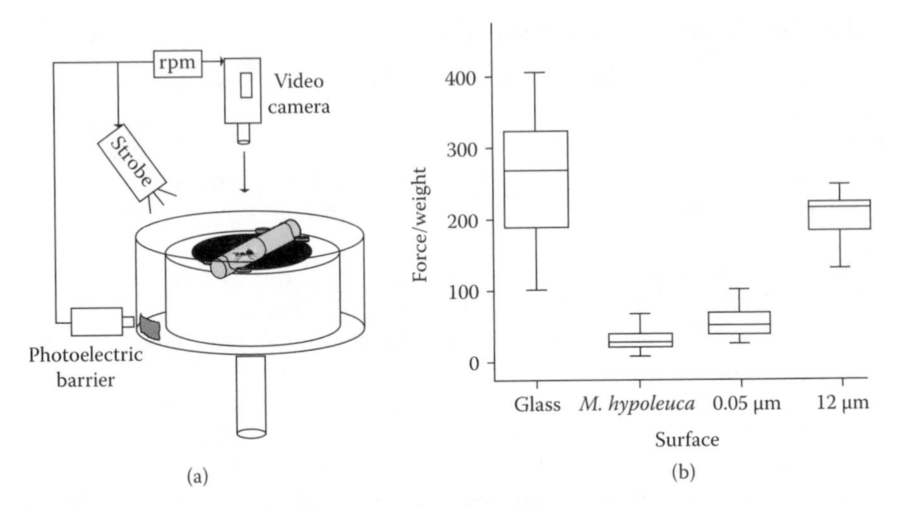

FIGURE 8.3 (A) Centrifuge method (Federle et al., *J. Exp. Biol.*, 203, 505, 2000) used to measure in-plane attachment forces of ants on artificial substrates and waxy *Macaranga* stems. (B) Attachment forces of *Crematogaster* (*Decacrema*) mspp.2 and 4 ants on substrates of different surface roughness. For each ant, the mean of three consecutive force measurements was used (n = 16 to 41 ants per species and substrate). The labels "0.05 μm" and "12 μm" on the x axis denote rough aluminium oxide substrates with the given particle sizes.

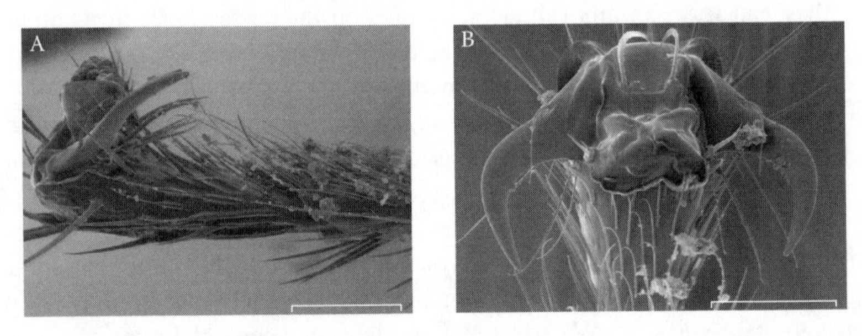

FIGURE 8.4 Tarsus of *Crematogaster* (*Decacrema*) msp.4 after climbing on a vertical waxy *M. hypoleuca* stem for 5–8 min. The ants were anaesthetized by cooling at 4°C and prepared for scanning electron microscopy (SEM) by freezing and drying overnight in a lyophilizer. (A) Lateral view of the tarsus. (B) Frontal view of the pretarsus. Note the threadlike wax crystal material typical of *Macaranga* (Figure 8.2B) on the claws and on the hairs at the ventral side of the tarsus. In contrast, the crystal particles are apparently dissolved by adhesive secretion on the arolium surface. Scale bars: 50 μm.

8.3.3 Mechanisms of Wax Running

The complex of nine closely related *Crematogaster* (*Decacrema*) morphospecies represents an excellent model system to study the function of wax barriers and the biomechanics of wax running. Even though *Crematogaster* (*Decacrema*) ants are very similar in morphology [19,25], wax runners and non–wax runners differ strikingly in their running performance. *Crematogaster* (*Decacrema*) non–wax runners placed on vertical waxy *Macaranga* stems are at best capable of walking very slowly and often fall down [17]. Our initial study revealed a weak tendency for smaller ant species to be better wax runners, which may be explained by scaling laws (greater surface area to weight ratio in smaller ants). However, this effect was not significant, and many ant species even smaller than the specialized ant partners had problems climbing the waxy stems [17]. In a subsequent study [20], we disproved the hypothesis that wax running is simply based on larger adhesive pads relative to ant body size. Comparison of adhesive forces on a smooth Plexiglass substrate between wax runners and non–wax runners indicated even an inverse effect with wax runners clinging *less* well to the smooth substrate [20]. Wax running, however, is not necessarily a consequence of special adaptations of the tarsal adhesive system. The capacity to climb slippery stems could also involve behavioral and locomotory adaptations and other morphological traits. In the following section, we present a preliminary analysis of wax running in *Crematogaster* (*Decacrema*) ants by addressing three questions: (1) Is wax running capacity based on greater attachment or superior locomotion? (2) Are there any relevant morphological differences between *Crematogaster* (*Decacrema*) wax runners and non–wax runners? (3) How do the kinematics of climbing differ between wax runners and non–wax runners?

8.3.3.1 Attachment Force vs. Climbing Performance: Is Wax Running Capacity Based on Greater Attachment or Superior Locomotion?

To investigate whether the wax running capacity of *Crematogaster* (*Decacrema*) ants is based on special adaptations of the attachment system to waxy substrates, we compared the ants' climbing performance with actual attachment forces generated on waxy *Macaranga* stems. Climbing performance was quantified by placing workers of *Crematogaster* (*Decacrema*) msp.2 (wax runners) and *Crematogaster* (*Decacrema*) msp.4 (non–wax runners) on vertical waxy *M. hypoleuca* stems (n = 34 workers from 2 colonies of each morphospecies; 2 stems of 11 mm diameter) and determining the proportion of ants capable of walking to one of two finishing lines 5 cm above and below the release point within 10 min [17]. Attachment was measured in the same ants as the centrifugal shear force required to detach them from waxy *M. hypoleuca* stems (13 stems with diameters between 8 and 13 mm). Surprisingly, we found that attachment forces per body weight were not significantly different between species, with a trend toward greater forces in the *non–wax runners* (two-tailed U-test, $U = 483.5$, $P = 0.092$) (Figure 8.5B). This is in striking contrast to the clearly superior climbing performance of wax runners (Fisher's exact test, $P < 0.001$) (Figure 8.5A). To test whether this unexpected result was caused by artifacts due to ill-defined surfaces of *Macaranga* trees kept in the greenhouse, we repeated the same set of experiments using a standardized, artificial substrate and found exactly the same effect (unpublished results). Our finding thus indicates that the difference in wax running capacity between *Crematogaster* (*Decacrema*) morphospecies is not the result of superior adhesion but rather due to behavioral and/or locomotory adaptations. These locomotory adaptations could be based both on

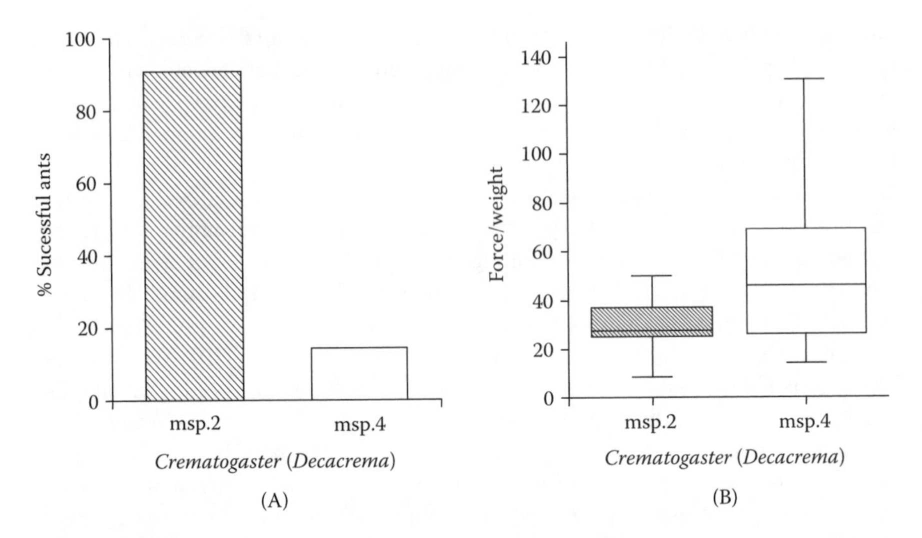

FIGURE 8.5 Climbing performance vs. attachment force of *Crematogaster* (*Decacrema*) wax runners (msp.2) and non–wax runners (msp.4) on waxy stems of *M. hypoleuca*. (A) Climbing capacity; ants that walked a stretch of 5 cm upward or downward along the stem were classified as "successful." (B) Maximum in-plane attachment force measured on the same substrate, using the centrifuge method (Figure 8.3A).

morphological characters and on kinematic variables, which will be considered in the following sections. It should be noted that our result does not indicate that adhesion is irrelevant for wax running. The much poorer wax running performance of many generalist ants [17] may also be caused by weaker attachment forces to the wax crystal substrates.

8.3.3.2 Comparative Morphometry of Wax Runners and Non–Wax Runners

Workers of the different *Crematogaster* (*Decacrema*) morphospecies associated with *Macaranga* are closely related and morphologically similar. We did not find any qualitative difference in leg and tarsus structure between wax running and non–wax running morphospecies. Quantitative morphometry, however, revealed strong interspecific variation of the total leg length, with wax runners (*Crematogaster* [*Decacrema*] mspp. 2 and 6) having longer legs than the non–wax runners (*Crematogaster* [*Decacrema*] mspp. 3 and 4) (Figure 8.6). To test this effect for significance, we performed analyses of covariance (ANCOVA) on log-transformed data with body mass as the covariate. Because the ANCOVA assumption of parallel regression lines was violated for the middle and hind legs ($P < 0.01$), we performed analogous analyses with log (pronotum width) instead of log (body mass) as the covariate (regression slopes not significantly different from each other for the front and middle legs, $F_{3,181} < 2.6$, $P > 0.05$, but for the hind legs: $P = 0.025$). All the three leg pairs yielded significant effects ($F_{3,184} \geq 115.1$; $P < 0.001$). Interspecific *post hoc*

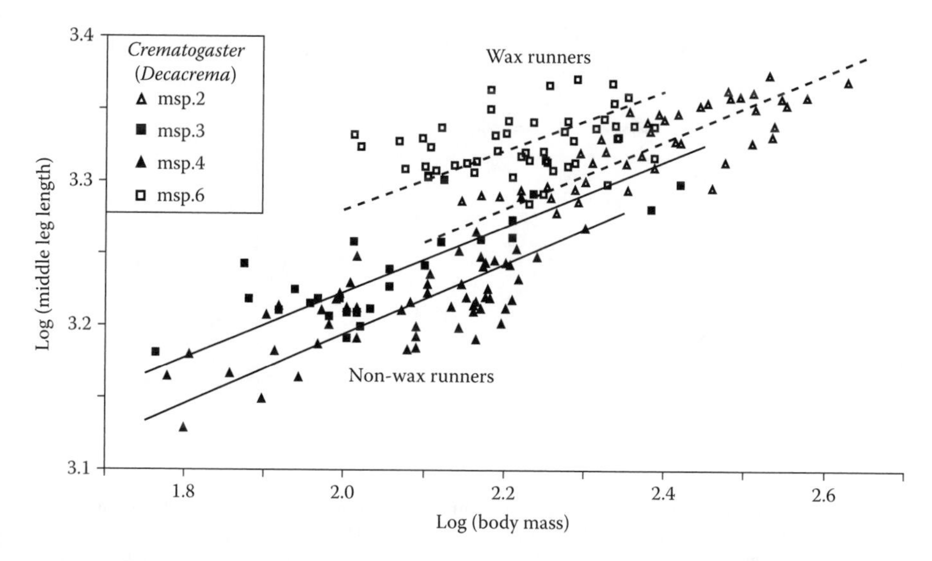

FIGURE 8.6 Double logarithmic plot of middle leg length vs. dry body mass in two wax running species (*Crematogaster* [*Decacrema*] msp.2 and 6) and two species of non–wax runners (*Crematogaster* [*Decacrema*] msp.3 and 4). Data represent a total of 205 ant workers from 29 colonies ($n > 32$ ants and $n > 4$ colonies in each morphospecies); lines show model II (reduced major axis) regressions.

TABLE 8.1
Results of ANCOVA Comparing Leg Length between Four *Crematogaster* (*Decacrema*) morphospecies with Pronotum Width as the Covariate

Crematogaster (*Decacrema*)	Front Legs [µm]		Middle Legs [µm]		Hind Legs [µm]	
msp.2 (wax runner)	1700	a	1997	a	2351	a
msp.3 (non–wax runner)	1465	b	1690	b	1975	b
msp.4 (non–wax runner)	1445	b	1628	c (b)	1915	c (b)
msp.6 (wax runner)	1700	c	2181	a	2599	d (a)

Note: Numbers indicate the back-transformed adjusted means of leg length, and the letters indicate the Scheffé grouping. Different letters denote significantly different leg lengths at the 5% level, and letters in parentheses denote significantly different leg lengths at the 0.1% level.

comparisons revealed highly significant differences in leg length between *Crematogaster* (*Decacrema*) mspp.2, 6 and mspp.3, 4 but weaker or no effects within each group (Table 8.1). Thus, the investigated ant morphospecies can be grouped into wax runners and non–wax runners based solely on their leg lengths.

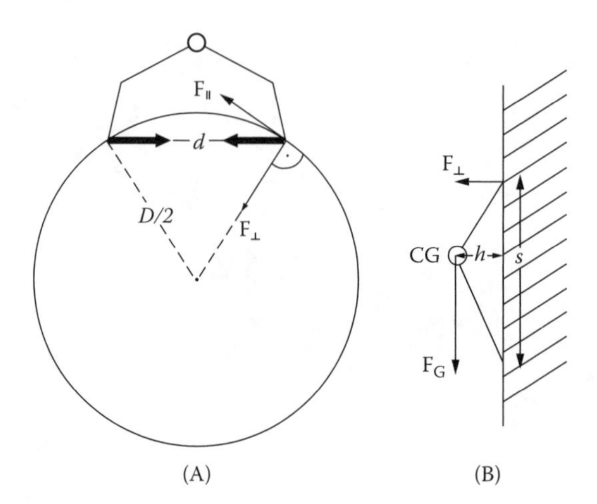

FIGURE 8.7 Two-leg model of ant climbing up a vertical stem. (A) Stem cross section (i.e., view from above). (B) Side view. D: stem diameter; d: lateral distance between tarsi of a leg pair; $F_\perp$: force perpendicular to stem surface; $F_\parallel$: force parallel to stem surface; F_G: gravity force; s: (vertical) distance between front and hind legs; CG: ant's center of gravity; h: height of center of gravity above stem.

8.3.3.3 Mechanical Benefit of Long Legs for Climbing Ants

Longer legs may convey two mechanical benefits to *Crematogaster* (*Decacrema*) ants climbing on waxy *Macaranga* stems, which are illustrated by a simple two-leg model in Figure 8.7:

1. Longer-legged ants could achieve a greater normal force $F_\perp$ by grasping around the stem (Figure 8.7A).
2. The perpendicular detachment force acting on the front legs is smaller in longer-legged ants because of the longer lever arm (Figure 8.7B).

Insects climbing up a cylindrical stem can hold on more firmly by drawing each foot toward the other, which increases the tarsal ground reaction force $F_\perp$ (Figure 8.7A). However, leg contraction is limited by the concomitantly growing shear forces $F_\parallel$, which will cause the legs to slide on larger diameter stems. The proportion of tangential to normal force gained by leg adduction depends on the distance d between the tarsi and the stem diameter D:

$$\frac{F_\parallel}{F_\perp} = \sqrt{\left(\frac{D}{d}\right)^2 - 1} \tag{8.1}$$

To prevent slipping, this proportion should not exceed the static friction coefficient μ (i.e., the proportion of friction force to normal load) of the tarsus on the waxy stem surface:

$$\sqrt{\left(\frac{D}{d}\right)^2 - 1} < \mu \text{ and thus, } d > \frac{D}{\sqrt{\mu^2 + 1}} \tag{8.2}$$

Assuming that ants rely wholly on friction in clinging to a stem with typical friction coefficient, $\mu < 1$, Equation 8.2 suggests that in the absence of adhesive forces, climbing ants can only enhance attachment by grasping around the stem if $d > 0.71 \times D$, i.e., if the legs subtend at least one quarter of the stem circumference. This situation, however, would be different if the ants were able to dig the tips of their claws into the stem surface. When claws interlock with the substrate, they generate a new contact surface that is more nearly perpendicular to the adduction force. This would have the effect that the ants can cling to the stems even if $d < 0.71 \times D$ [57].

The second possible advantage of longer legs is related to the increased distance between front and hind leg footfall positions. The perpendicular force acting on the front legs tending to topple the ant from the stem during upward climbing depends on the height h of the center of gravity above the stem surface and the distance s between front and hind footfall positions (Figure 8.7B):

$$F_{\perp} = -\frac{h \cdot F_G}{s} \tag{8.3}$$

Assuming a constant height of the center of gravity above the stem, longer legs would reduce the detachment force acting on the upper legs of climbing ants. Unlike the increase of normal force achieved by grasping around the stem, this effect would not only be beneficial on a cylindrical twig but also on a plane vertical substrate.

8.3.3.4 Kinematics of Climbing in *Crematogaster* (*Decacrema*) Wax Runners and Non–Wax Runners

To search for possible locomotory adaptations involved in wax running and to evaluate the predictions of the above climbing model (Figure 8.7), we recorded the ants' climbing behavior on waxy *M. hypoleuca* ($n = 6$ stems; mean diameter of 9.2 mm) and on 10-mm diameter glass rods by comparing *Crematogaster* (*Decacrema*) msp.2 wax runners ($n = 21$ and 24 on glass and wax, respectively) and *Crematogaster* (*Decacrema*) msp.4 non–wax runners ($n = 19$ and 24 on glass and wax, respectively). Climbing ants were recorded using two synchronized digital high-speed video cameras (Redlake PCI 1000 B/W). The cameras were oriented at 90° to each other and perpendicular to the stem surface so that dorsal and lateral views could be recorded. We only analyzed continuous runs in which the ants climbed upward in the middle line of the stem in dorsal view (cutoff for lateral excursions: 15% of the stem diameter). To evaluate gait parameters and body posture, body coordinates and footfall positions were digitized and further analyzed in MATLAB (The MathWorks, Inc.).

Both for wax runners and non–wax runners, the gait strongly differed between runs on glass rods and on waxy *Macaranga* stems. Ants climbed faster and assumed a typical tripod gait when climbing on glass (*Crematogaster* (*Decacrema*) msp.2: duty factor = 0.74, mean velocity = 11.5 mm/sec), whereas slower and "safer" gaits with more legs on the ground at any one time were used on the waxy stems (*Crematogaster* (*Decacrema*) msp.2: duty factor = 0.88, mean velocity = 2.2 mm/sec). Wax running typically involved a much greater number of steps in the front legs (*Crematogaster* (*Decacrema*) msp.2: stepping frequency = 6.6 sec^{-1}) than in the middle and hind legs (stepping frequency of middle legs = 1.7 sec^{-1}, of hind legs = 2.1 sec^{-1}). The ants repeatedly put down their front legs and pulled them toward the body, apparently to achieve sufficient traction on the waxy substrate.

Crematogaster (*Decacrema*) msp.2 wax runners not only climbed faster and almost never fell off the *M. hypoleuca* stems, but they also assumed a different body posture than *Crematogaster* (*Decacrema*) msp.4 non–wax runners (Figure 8.8). We measured relative leg distances between footfall positions by dividing the absolute footfall distances by the total length of the corresponding legs. Relative lateral distances (tangential to the stem) between front, middle, and hind feet were significantly greater in the *Crematogaster* (*Decacrema*) msp.2 wax runners (Figure 8.8; two-tailed U-tests; front legs, U = 145.0, P = 0.003; middle legs, U = 116.0, $P < 0.001$; hind legs, U = 104.0, $P < 0.001$). Thus, *Crematogaster* (*Decacrema*) msp.2 wax runners not only have longer legs but also spread them out more than the non–wax runners. This finding corroborates the functional significance of long legs for wax running. Despite the conspicuous sprawled posture of the middle and hind legs during wax running (Figure 8.8), the legs subtended in all cases considerably less than 0.71 times the stem diameter (median ratio of front, middle, and hind leg distance to stem diameter = 0.18, 0.42, and 0.42, respectively, with maxima = 0.24, 0.53, and 0.53, respectively). According to the prediction derived from Equation 8.2, the wider spreading of the legs in wax runners should thus not convey any mechanical advantage if leg contacts were entirely based on friction. It would also be conceivable that the wax runners' wider stance improves attachment by lowering the height h of the ant's center of gravity according to Equation 8.3. Because of the extension of leg joints, however, d can vary independently of h. Preliminary data have not shown any evidence that climbing *Crematogaster* (*Decacrema*) wax runners hold their center of gravity closer to the stem than non–wax runners (unpublished results). However, leg spreading would have a positive mechanical effect if some interlocking of the claw tips with the substrate is involved (see Section 8.3.3.3).

In contrast to the lateral footfall distances, the relative distance between front and hind tarsi (measured in the vertical direction) did not differ significantly between wax runners and non–wax runners (Figure 8.8). As a better estimate of the effective lever arm, we used the distance between front and hind tarsi divided by pronotum width as an arbitrarily chosen body length variable. Here, the distance between front and hind tarsi in the wax runner *Crematogaster* (*Decacrema*) msp.2 significantly exceeded that of *Crematogaster* (*Decacrema*) msp.4 (two-tailed U-test, U = 158.0, P = 0.012). Thus, even though the ants do not appear to extend their legs much further in the fore–aft direction, the greater leg length of the wax runners by itself

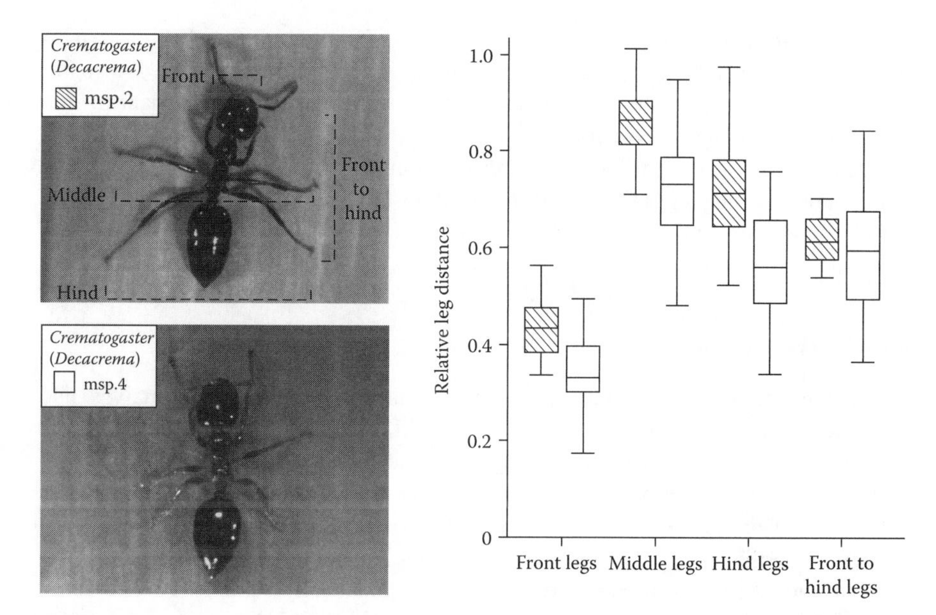

FIGURE 8.8 Climbing kinematics of *Crematogaster* (*Decacrema*) wax runners (msp.2) and non–wax runners (msp.4) on waxy *M. hypoleuca* stems. The relative leg distance corresponds to the distance between footfall positions divided by the actual length of the (two) legs. Lateral distances between front, middle, and hind legs were measured perpendicular to the stem, with the distance between front and hind legs parallel to the stem axis. Each raw data point represents the mean distance of one running sequence.

leads to a longer lever arm that may be mechanically advantageous as shown in Figure 8.7B.

Longer legs appear to be important not only for *Macaranga*-inhabiting ants, but also for other insects having to cope with waxy plant surfaces. The aphid *Brevicoryne brassicae* was found to walk more effectively on waxy *Brassica* leaves than *Lipaphis erysimi* [58]; this difference was explained by the greater leg length (relative to body length) of *B. brassicae* [59].

8.4 CONCLUSIONS

The mutualism between *Macaranga* trees and their specialized *Crematogaster* (*Decacrema*) ant partners illustrates how strongly insect–plant interactions can be determined and shaped by biomechanical factors. The slippery wax barriers on the stems of *Macaranga* myrmecophytes not only keep away generalist ants and thus protect the resident wax running ant partners against predation and competition, but they also act as an ecological isolation mechanism between different *Crematogaster* (*Decacrema*) ant associates. Numerous differences between associations of *Crematogaster* (*Decacrema*) wax runners and non–wax runners demonstrate that this mechanical factor has important ecological implications for *Macaranga*–ant mutualisms.

Our preliminary study into the biomechanics of wax running behavior surprisingly suggests that the difference in wax running capacity between *Crematogaster* (*Decacrema*) species is not caused by superior adhesion but by mechanical and locomotory adaptations. We found that *Crematogaster* (*Decacrema*) wax runners not only have longer legs but also spread them out more during climbing. The combination of long legs and a more sprawled posture is mechanically advantageous for climbing ants and may partly explain the exceptional wax running capacity of *Macaranga* ant partners.

ACKNOWLEDGMENTS

We wish to thank Patrick Drechsler for his assistance in the kinematic analysis and the development of video analysis routines in MATLAB. Sebastian Busch helped in the ant morphometry and motion analysis. Heike Feldhaar and Brigitte Fiala provided unpublished information on the phylogeny and distribution of *Crematogaster* (*Decacrema*) morphospecies. We are grateful to Werner Baumgartner, Bert Hölldobler, Reinhard Jetter, Ulrich Maschwitz, Frank Rheindt, and Markus Riederer for support and fruitful discussions. Our study was financially supported by research grants of the Deutsche Forschungsgemeinschaft (SFB 567/A4 (C6) and Emmy-Noether fellowship FE 547/1-3 to WF).

REFERENCES

1. Rosenthal, G.A. and Berenbaum, M.R., *Herbivores: Their Interactions with Secondary Plant Metabolites, Vol. I: The Chemical Participants,* Academic Press, San Diego, 1991.
2. Bernays, E.A. and Chapman, R.F., *Host-Plant Selection by Phytophagous Insects*, Chapman & Hall, New York, 1994.
3. Schoonhoven, L.M., Jermy, T., and Van Loon, J.J.A., *Insect-Plant Biology: From Physiology to Evolution*, Academic Press, London, 1998.
4. Kerner von Marilaun, A., *Flowers and Their Unbidden Guests*, Kegan Paul Co., London, 1878.
5. Proctor, M., Yeo, P., and Lack, A., *The Natural History of Pollination*, Harper Collins, London, 1996.
6. Sanson, G., et al., Measurement of leaf biomechanical properties in studies of herbivory: Opportunities, problems and procedures, *Austral. Ecology*, 26, 535, 2001.
7. Nichols-Orians, C.M. and Schultz, J.C., Interactions among leaf toughness, chemistry, and harvesting by attine ants, *Ecol. Entomol.*, 15, 311, 1990.
8. Wright, W. and Vincent, J.F.V., Herbivory and the mechanics of fracture in plants, *Biol. Rev.*, 71, 401, 1996.
9. Gorb, S.N. and Beutel, R.G., Evolution of locomotory attachment pads of hexapods, *Naturwiss.*, 88, 530, 2001.
10. Eisner, T. and Shepherd, J., Caterpillar feeding on a sundew plant, *Science,* 150, 1608, 1965.
11. Kennedy, C.E., Attachment may be a basis for specialization in oak aphids, *Ecol. Entomol.*, 11, 291, 1986.

12. Ellis, A.G. and Midgley, J.J., A new plant-animal mutualism involving a plant with sticky leaves and a resident hemipteran insect, *Oecologia*, 106, 478, 1996.
13. Southwood, T.R.E., Plant surfaces and insects — An overview, in *Insects and the Plant Surface*, Juniper, B.E. and Southwood, T.R.E., Eds., Edward Arnold, London, 1986, p. 1.
14. Clarke, C.M. and Kitching, R.L., Swimming ants and pitcher plants: A unique ant-plant interaction from Borneo, *J. Trop. Ecol.*, 11, 589, 1995.
15. Futuyma, D.J., Some current approaches to the evolution of plant–herbivore interactions, *Plant Species Biol.*, 15, 1, 2000.
16. Thompson, J.N., *The Coevolutionary Process*, Chicago University Press, Chicago, 1994.
17. Federle, W., et al., Slippery ant-plants and skillful climbers: Selection and protection of specific ant partners by epicuticular wax blooms in *Macaranga* (Euphorbiaceae), *Oecologia*, 112, 217, 1997.
18. Whitmore, T.C., *Macaranga* Thou., in *The Euphorbiaceae of Borneo*, Airy-Shaw, H.K., Ed., HMSO (Her Majesty's Stationery Office), London, 1975, p. 140.
19. Fiala, B., et al., Diversity, evolutionary specialisation and geographic distribution of a mutualistic ant-plant complex: *Macaranga* and *Crematogaster* in South East Asia, *Biol. J. Linn. Soc.*, 66, 305, 1999.
20. Federle, W., Rohrseitz, K., and Hölldobler, B., Attachment forces of ants measured with a centrifuge: better "wax-runners" have a poorer attachment to a smooth surface, *J. Exp. Biol.*, 203, 505, 2000.
21. Quek, S.-P., et al., Codiversification in an ant-plant mutualism: stem texture and the evolution of host use in *Crematogaster* (Formicidae: Myrmicinae) inhabitants of *Macaranga* (Euphorbiaceae), *Evolution*, 58, 554, 2004.
22. Shaw, K.L., Conflict between nuclear and mitochondrial DNA phylogenies of a recent species radiation: What mtDNA reveals and conceals about modes of speciation in Hawaiian crickets, *Proc. Natl. Acad. Sci.*, 99, 16122, 2002.
23. Sota, T. and Vogler, A.P., Incongruence of mitochondrial and nuclear gene trees in the carabid beetles *Ohomopterus*, *Syst. Biol.*, 50, 39, 2001.
24. Feldhaar, H., Fiala, B., and Gadau, J., Is ecological speciation the driving force in the radiation of the *Macaranga*-associated *Crematogaster* (*Decacrema*) plant-ants? Abstract 16.7, *IX Congress of the European Society for Evolutionary Biology (ESEB)*, Leeds, UK, 2003.
25. Feldhaar, H., et al., Molecular phylogeny of *Crematogaster* subgenus *Decacrema* ants (Hymenoptera: Formicidae) and the colonization of *Macaranga* (Euphorbiaceae) trees, *Mol. Phyl. Evol.*, 27, 441, 2003.
26. Davies, S.J., Systematics of *Macaranga* Sects. *Pachystemon* and *Pruinosae* (Euphorbiaceae), *Harvard Pap. Bot.*, 6, 371, 2001.
27. Federle, W. and Rheindt, F., *Macaranga* ant-plants hide food from intruders: correlation of food presentation and presence of wax barriers analysed using phylogenetically independent contrasts, *Biol. J. Linn. Soc.*, 84, 177, 2005.
28. Fiala, B. et al., Studies of a South East Asian ant-plant association: protection of *Macaranga* trees by *Crematogaster borneensis*, *Oecologia*, 79, 463, 1989.
29. Federle, W., Maschwitz, U., and Hölldobler, B., Pruning of host plant neighbors as defence against enemy ant invasions: *Crematogaster* ant partners of *Macaranga* protected by "wax barriers" prune less than their congeners, *Oecologia*, 132, 264, 2002.
30. Heil, M. et al., On benefits of indirect defence: Short- and long-term studies of antiherbivore protection via mutualistic ants, *Oecologia*, 126, 395, 2001.

31. Janzen, D.H., Coevolution of mutualism between ants and acacias in Central America, *Evolution*, 20, 249, 1966.
32. Heil, M., et al., Reduced chemical defence in ant-plants? A critical re-evaluation of a widely accepted hypothesis, *Oikos*, 99, 457, 2002.
33. Eck, G. et al., Trade-off between chemical and biotic antiherbivore defense in the South East Asian plant genus *Macaranga*, *J. Chem. Ecol.*, 27, 1979, 2001.
34. Davidson, D.W., Longino, J.T., and Snelling, R.R., Pruning of host plant neighbors by ants: An experimental approach, *Ecology*, 69, 801, 1988.
35. Itioka, T. et al., Difference in intensity of ant defense among three species of *Macaranga* myrmecophytes in a Southeast Asian dipterocarp forest, *Biotrop.*, 32, 318, 2000.
36. Feldhaar, H. et al., Patterns of the *Crematogaster–Macaranga* association: The ant partner makes the difference, *Insectes Soc.*, 50, 9, 2003.
37. Felsenstein, J., Phylogenies and the comparative method, *Am. Nat.*, 125, 1, 1985.
38. Federle, W. et al., Incident daylight as orientation cue for hole-boring ants: Prostomata in *Macaranga* ant-plants, *Insectes Soc.*, 48, 165, 2001.
39. Brouat, C. et al., Plant lock and ant key: Pairwise coevolution of an exclusion filter in an ant-plant mutualism, *Proc. R. Soc. Lond. B*, 268, 2131, 2001.
40. Davidson, D.W., Snelling, R.R., and Longino, J.T., Competition among ants for myrmecophytes and the significance of plant trichomes, *Biotropica*, 21, 64, 1989.
41. Gorb, S., *Attachment Devices of Insect Cuticle*, Kluwer Academic Publishers, Dordrecht, Boston, 2001.
42. Dai, Z., Gorb, S.N., and Schwarz, U., Roughness-dependent friction force of the tarsal claw system in the beetle *Pachnoda marginata* (Coleoptera, Scarabaeidae), *J. Exp. Biol.*, 205, 2479, 2002.
43. Federle, W. et al., An integrative study of insect adhesion: Mechanics and wet adhesion of pretarsal pads in ants, *Integ. Comp. Biol.*, 42, 1100, 2002.
44. Vötsch, W. et al., Chemical composition of the attachment pad secretion of the locust *Locusta migratoria*, *Insect Biochem. Mol. Biol.*, 32, 1605, 2002.
45. Markstädter, C. et al., Chemical composition of the slippery epicuticular wax blooms on *Macaranga* (Euphorbiaceae) ant-plants, *Chemoecology*, 10, 33, 2000.
46. Kerner von Marilaun, A., *Pflanzenleben, Band 2*, Bibliographisches Institut, Leipzig, 1891.
47. Knoll, F., Über die Ursache des Ausgleitens der Insektenbeine an wachsbedeckten Pflanzenteilen, *Jahrb. wiss. Bot.*, 54, 448, 1914.
48. Stork, N.E., Role of wax blooms in preventing attachment to brassicas by the mustard beetle, *Phaedon cochleariae*, *Entom. Exp. Appl.*, 28, 100, 1980.
49. Gaume, L. et al., How do plant waxes cause flies to slide? Experimental tests of wax-based trapping mechanisms in three pitfall carnivorous plants, *Arthropod Struct. Dev.*, 33, 103, 2004.
50. Gorb, E.V. and Gorb, S.N., Attachment ability of the beetle *Chrysolina fastuosa* on various plant surfaces, *Entom. Exp. Appl.*, 105, 13, 2002.
51. Eigenbrode, S.D. and Jetter, R., Attachment to plant surface waxes by an insect predator, *Integ. Comp. Biol.*, 42, 1091, 2002.
52. Eigenbrode, S.D. and Kabalo, N.N., Effects of *Brassica oleracea* waxblooms on predation and attachment by *Hippodamia convergens*, *Entomol Exp. Applic.*, 91 (April), 125, 1999.
53. Juniper, B.E. and Burras, J.K., How pitcher plants trap insects, *New Scientist*, 13, 75, 1962.

54. Eigenbrode, S.D. et al., Mobility of three generalist predators is greater on cabbage with glossy leaf wax than on cabbage with a wax bloom, *Entom. Exp. Appl.*, 81, 335, 1996.
55. Stork, N.E., The form of plant waxes: A means of preventing insect attachment? in *Insects and the Plant Surface*, Juniper, B.E. and Southwood, T.R.E., Eds., Edward Arnold, London, 1986, p. 346.
56. Federle, W., Baumgartner, W., and Hölldobler, B., Biomechanics of ant adhesive pads: Frictional forces are rate- and temperature-dependent, *J. Exp. Biol.*, 207, 67, 2004.
57. Cartmill, M., Climbing, in: *Functional Vertebrate Morphology*, Hildebrand, M. et al., Eds., The Belknap Press, Cambridge, 1985, p. 73.
58. Åhman, I., Plant-surface characteristics and movements of two Brassica-feeding aphids, *Lipaphis erysimi* and *Brevicoryne brassicae*, *Symp. Biol. Hung.*, 39, 119, 1990.
59. Eigenbrode, S.D., Plant surface waxes and insect behaviour, in *Plant Cuticles — An Integrated Functional Approach*, Kerstiens, G., Ed., Bios Scientific Publishers, Oxford, 1996, p. 201.

9 Nectar Feeding in Long-Proboscid Insects

Brendan J. Borrell and Harald W. Krenn

CONTENTS

9.1 INTRODUCTION

That [bees] and other insects, while pursuing their food in the flowers, at the same time fertilize them without intending and knowing it and thereby lay the foundation for their own and their offspring's future preservation, appears to me to be one of the most admirable arrangements of nature.

Sprengel [1]

Although Sprengel, writing in 1793, may not have recognized the evolutionary implications of his life's work on plant–pollinator interactions, he was among the first to relate the morphological features of flowering plants to those of nectar-feeding animals. Indeed, the early evolution and diversification of angiosperms have frequently been attributed to an "arrangement" between plants and their pollinators, but how "admirable" such relationships often are remains questionable [2]. Darwin postulated that extended corollas of certain flowers represent the outcome of an evolutionary arms race between plants and their pollinators [3], with plants evolving to match, in depth, mouthpart lengths of pollinating taxa [4–7]. Consequently, the rise of flowering plants in the late Cretaceous also corresponded with a period of rapid diversification in insect feeding strategies, including the evolution of the famously elongate mouthparts associated with nectar feeding in certain Lepidoptera, Diptera, and Hymenoptera [8,9].

Although many nectar-feeding insects consume floral nectars with short mouthparts, the benefits nectar feeders derive from their long proboscides are clear: exclusive access to deep flowers, providing copious amounts of nectar [10–13]. In fact, long-proboscid insects are able to capitalize on a wider diversity of resources than their short-proboscid counterparts as they frequent any flowers from which they can physically extract nectar whether deep or shallow [11,14–16]. Such advantages lead to the fundamental questions: Do insect nectarivores incur a cost to having such long mouthparts? If so, how can we measure these costs? What are the functional requirements of elongate mouthparts and how might they influence pollinator behavior? Clearly, a long proboscis can be unwieldy [17,18]; the control, extension, and retraction of the proboscis requires specialized machinery [19–23], and imbibement of a viscous fluid through such a slender duct entails a whole other set of biomechanical problems [24–26]. The goal of the present chapter is to examine the functional morphology and biomechanics of nectar feeding with elongate mouthparts and to explore how physical constraints may have shaped feeding ecology and plant–pollinator relationships over evolutionary time.

9.2 FUNCTIONAL DIVERSITY OF LONG MOUTHPARTS

9.2.1 Evolution of Suction Feeding

The first fluid-feeding insects employed a lapping or sponging mechanism to imbibe their liquid meals. This modality, which uses capillary forces for fluid uptake, is widespread among insects, including those that specifically visit plants to consume floral nectars [27]. The elongation of mouthparts is derived and enables insects to develop a pressure gradient along the food canal, allowing them to consume nectar from the concealed nectaries found in long, tubular corollas (Figure 9.1). This type of proboscis, termed a "concealed nectar extraction apparatus" by Jervis [28], often matches or exceeds the body length in holometabolous insects (Endopterygota) and other nectar feeders (Table 9.1 and Figure 9.1). At 280 mm, a tropical sphingid holds the record for mouthpart length in absolute terms [29]. Relative to body length, however, record holders are South African nemestrinid flies (Figure 9.1C) whose proboscides may be over four times the length of their bodies [15]. A number of disparate evolutionary pathways have preceded the development of these long, suctorial mouthparts in various taxa (Table 9.2).

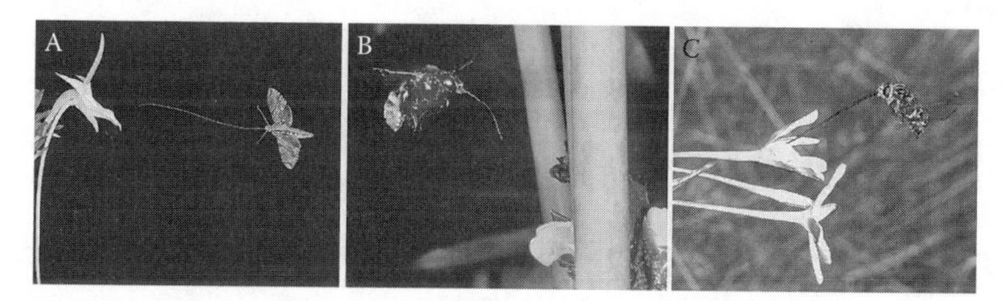

FIGURE 9.1 (A) Hawkmoth *Xanthopan* (Sphingidae) approaching the long-spurred blossom of an *Angraecum* orchid; proboscis length approximately 220 mm (photo with permission of L.T. Wasserthal). (B) Orchid bee, *Eulaema meriana*, departing from a *Calathea* inflorescence (photo with permission of G. Dimijian). (C) Long-proboscid fly *Moegistorhynchus longirostris* (Nemestrinidae) at a flower of *Ixia* (photo with permission of S. Johnson).

Many taxa within Hymenoptera have evolved elongate mouthparts in the context of nectar feeding [28,30]. Many of these feed on nectar using a lapping and sucking mode, but the Euglossini (orchid bees) and long-tongued Masarinae (pollen wasps) have shifted to pure suction feeding [31,32]. In other cases, a suctorial mode of feeding is suggested from the length and general composition of the mouthparts (e.g., some species of Tenthredinidae, Eumenidae, and Sphecidae [27,28,30]).

Suctorial nectar feeding via an elongate proboscis has arisen multiple times in Diptera [33]. Suction feeding in hoverflies (Syrphidae) [34] and beeflies (Bombyliidae) [19,35] likely evolved from unspecialized flower-visiting ancestors employing a sponging feeding mode on floral and extrafloral nectar and pollen. Specialized nectar feeding in the Culicidae and Tabanidae evolved from hematophagous ancestors [36]. While both sexes of the tropical culicid genus *Toxorhychites* shifted entirely to floral nectar, female horseflies in the genus *Corizoneura* are equipped with both a short proboscis (10 mm) for piercing and sucking blood, and a long proboscis (50 mm) for nectar feeding [37]. In addition, nectar-feeding flies belonging to the Empitidae (dance flies) are derived from predatory insect feeders [36].

Even though generalized feeding on petals, nectar, and pollen is frequent among adult beetles, only two taxa of blister beetles (Meloidae) have independently shifted to specialized nectar feeding via an elongate proboscis [36,38].

Ancestors of butterflies and moths fed on nonfloral plant fluids with a simply formed, coilable proboscis. The proboscides of all nectar-feeding Lepidoptera exhibit the same set of derived features, suggesting that nectar feeding evolved only once in a taxon of glossatan Lepidoptera known as the Eulepidoptera [39,40].

9.2.2 Anatomical Considerations

Mouthpart elements that make up the proboscis vary considerably among insect taxa. In Hymenoptera, where nectar feeding has evolved independently multiple times, proboscis morphology is similarly diverse. Most frequently, the hymenopteran proboscis is formed by basally linked maxillary and/or labial components, known as the labiomaxillary complex. In the "long-tongued" bees (Apidae + Megachilidae), the proboscis is composed of the elongated galeae and labial palps that together form the food canal surrounding the long and hairy glossa (Figure 9.2) [41]. In some

TABLE 9.1
Principal Composition and Maximal Reported Proboscis Length of the Proboscides of Selected Nectar Feeders

Taxon	Proboscis Components	Length (mm)	Ref.
Coleoptera			
Meloidae (blister beetles) Nemognathinae[a]	Galeae or maxillary palps	10	132
Hymenoptera			
Apidae Bombini (bumblebees) *Bombus hortorum*	Galeae, glossa, labial palps	19	2
Euglossini (orchid bees) *Eufriesea ornata*	Galeae, glossa, labial palps	41	133
Colletidae ("short-tongued" bees) *Niltonia virgili*	Labial palps	9	43
Vespidae Masarinae (pollen wasps) *Ceramius metanotalis*	Glossa	6.2	134
Lepidoptera			
Sphingidae (hawkmoths) *Amphimoea walkeri*[b]	Galeae	280	29
Riodinidae (metalmark butterflies) *Eurybia lycisca*	Galeae	45	H.W. Krenn, unpublished
Diptera			
Tabanidae (horseflies) *Corizoneura longirostris*	Labrum/epipharynx, hypopharynx, mandible stylets, lacinia, labium; distally labium alone[c]	50	37
Nemestrinidae (tangle-veined flies) *Moegistorynchus longirostris*	Labrum/epipharynx, hypopharynx, lacinia, labium; distally labium alone	90	15
Bombyliidae (beeflies) *Bombylius major*	Labrum/epipharynx, hypopharynx, maxillary structures, labium	12.5	19
Syrphidae (hoverflies) *Rhingia campestris*	Labrum/epipharynx, hypopharynx, maxillary structures, labium	10.5	135
Chiroptera			
Phyllostomidae (leaf-nosed bats) *Choeronycteris mexicana*	Tongue	77	94

(continued)

TABLE 9.1 (CONTINUED)
Principal Composition and Maximal Reported Proboscis Length of the Proboscides of Selected Nectar Feeders

Taxon	Proboscis Components	Length (mm)	Ref.
Aves			
Trochilidae (hummingbirds)			
Ensifera ensifera	Mandibles and tongue	91[d]	136

[a] No detailed studies are available.
[b] World record holder in proboscis length.
[c] Piercing blood feeding and nectar feeding in females.
[d] Functional proboscis length may exceed reported bill length.

TABLE 9.2
Evolutionary Transitions to Specialize Suction Feeding in Some Nectar-Feeding Insect

Taxon	Ancestral Feeding Mode	Derived Taxon	Ref.
Coleoptera			
Meloidae	Biting/chewing on various floral food sources	*Nemognatha, Leptopalpus*	36
Hymenoptera			
Apidae	Lapping nectar feeding	Euglossini	31
Vespidae	Lapping nectar feeding	Masarinae	32
Lepidoptera			
Glossata	Suction feeding of nonfloral plant fluid	Eulepidoptera[a]	39, 137
Diptera			
Culicidae	Piercing blood feeding females	*Toxorhynchites*	36
Nemestrinidae	Unknown	Nemestrinidae[b]	36
Tabanidae	Piercing blood feeding females	*Corizoneura*[c]	37
Bombyliidae	Mopping up fluid feeding	*Bombylius*	19, 35
Empididae	Predatory insect feeding	*Empis*	2, 36
Syrphidae	Nectar and pollen feeding	*Rhingia*	34

[a] Secondarily nonfeeding in several taxa.
[b] Unknown whether all are suction-feeding flower visitors.
[c] Proboscis of females specialized to both nectar and blood feeding.

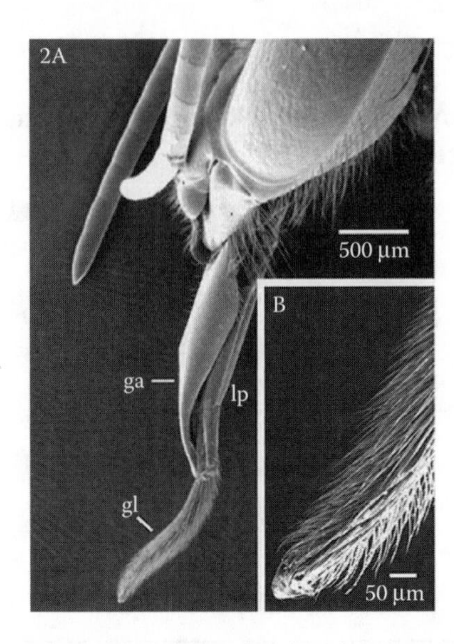

FIGURE 9.2 (A) Head and extended proboscis of *Melipona* sp. (Hymenoptera: Apidae); proboscis consists of galeae (ga), labial palps (lp), and glossa (gl). (B) Close up of the glossal tip.

"long-tongued" bees, even basal elements of the mouthparts have a significant influence on a bee's functional tongue length [42]. Remarkably, one group of "short-tongued" bees (Colletidae, *Niltonia*), which feeds on deep *Jacaranda* flowers in the New World tropics, has a proboscis that approaches its body length but is composed of the labial palps alone [43]. Another group of colletid bees has a proboscis formed mostly from the concave maxillary palps [27,44]. In long-tongued pollen wasps (Vespidae: Masarinae), the proboscis and food canal are formed from the glossa alone [36]. There are many other compositions found in various groups of Hymenoptera, including Braconidae, Sphecidae, and even in Tenthredinoidea. Overviews on the occurrence and principal compositions are given in Jervis [28], Jervis and Vilhelmsen [30], and Krenn, Plant, and Szucsich [27].

In contrast to mouthpart diversity exhibited by Hymenoptera, the proboscides of all "higher" Lepidoptera consist only of the two maxillary galeae enclosing the food canal (Figure 9.3) [20,39,40].

Most Diptera have sponging and sucking mouthparts that are similar in composition but with highly variable lengths. Their proboscis is complex, consisting of an elongated labrum–epipharynx unit and a hypopharynx, which, sometimes together with rodlike maxillary structures, form the food canal and are enclosed by the gutter-shaped labium. The paired labellae (a homologue to the labial palps of other insects) at the apical end protrude from the proboscis (Figure 9.4) [41]. Adaptations to nectar feeding include elongation of the whole functional unit, a simplified composition of the food canal formation, and a slender labellae [27,34].

The long suctorial proboscis of the typical nectar-feeding insect is characterized by a tightly sealed food canal (Figures 9.5A, 9.5B, and 9.5C), a specialized tip region

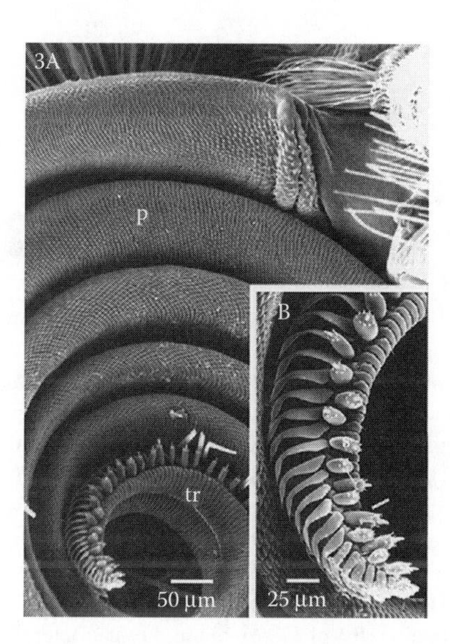

FIGURE 9.3 (A) Spirally coiled proboscis (p) of *Vanessa cardui* (Lepidoptera: Nymphalidae) in lateral view; tip region (tr). (B) Proboscis tip slits into food canal formed by extended galeal-linking structures; sensilla styloconica (s) are characteristic sensory organs of the lepidopteran proboscis.

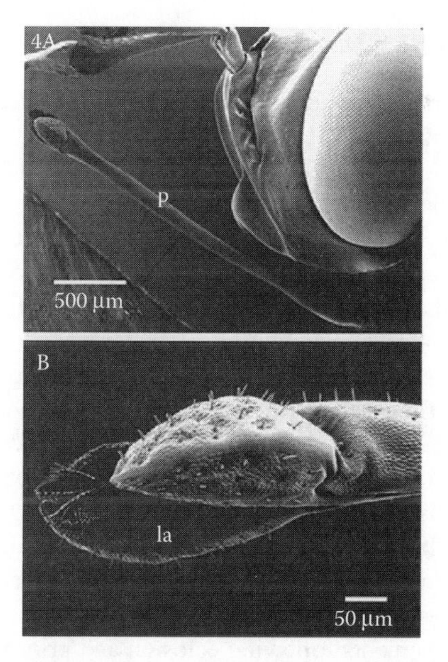

FIGURE 9.4 (A) Head of *Physocephala rufipes* (Diptera: Conopidae) with proboscis (p) tip projecting forward in resting position. (B) Labella (la) of proboscis tip.

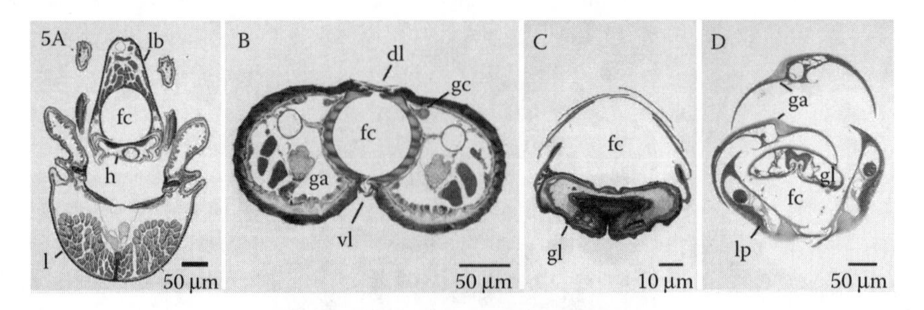

FIGURE 9.5 Cross-sections of the feeding canals (fc) of some nectar feeding insects. (A) In *Volucella bombylans* (Diptera: Syrphidae), food canal is formed by groove and tongue junction of labrum–epipharynx unit (lb) and the hypopharynx (h); labium (l) surrounds the other proboscis components. (B) In *Pieris brassicae* (Lepidoptera: Pieridae) the galeae (ga) interlock on the dorsal and ventral margins to enclose the central food canal. Dorsal linkage (dl) consists of overlapping platelets sealed by gland cell (gc) substances; ventral linkage (vl) is formed by cuticular hooks. (C) Overlapping cuticular structures of the glossa (gl) form the food canal in *Ceramius hispanicus* (Hymenoptera: Vespidae: Masarinae). (D) Food canal is formed from the galeae (ga) and labial palps (lp) in *Euglossa* sp. (Hymenoptera: Apidae: Euglossini), and is disengaged in the resting position.

(Figures 9.2B, 9.3B, and 9.4B), and a powerful suction pump (Figure 9.6 and Figure 9.7). These features are integral to the functioning of the proboscis and must be considered in detail before biomechanical generalizations can be developed.

9.2.2.1 Proboscis-Sealing Mechanisms

One to five individual parts interlock to form a fluid-tight suction tube (Figure 9.5). Various modes of interlocking exist: Individual components can be interlocked by tongue and groove junctions, e.g., bees and flies (Figure 9.5A), or by a series of overlapping cuticle plates and hook-shaped structures, e.g., Lepidoptera (Figure 9.5B) [23,39,45]. When a single component forms the food canal (e.g., long-tongued pollen wasps), overlapping cuticle plates shape the food tube (Figure 9.5C) [32]. In long-proboscid flies, the distal region of the food tube is formed by the strongly arched labium, the margins of which interlock to form the tube (Figure 9.5D) [36]. In butterflies, epidermal gland cells in the galeal lumen may produce substances that help seal the linkage of the galeae (Figure 9.5B) [20].

In long-tongued bees, the food canal is assembled anew each time the proboscis is extended for feeding (Figure 9.5D). During folding and extension, the components of the dipteran proboscis remain interlocked, but tongue and groove junctions permit sliding movements of the components against each other [35]. The butterfly proboscis is assembled once during pupal emergence and remains permanently interlocked. In pupae, the two galeae develop separately and can only interlock by a distinct sequence of galeae movements following eclosion and prior to cuticular sclerotization. For nymphalid butterflies, interlocking of the galeae is an irreversible and indispensable process that occurs only once during a short time interval following eclosion [46].

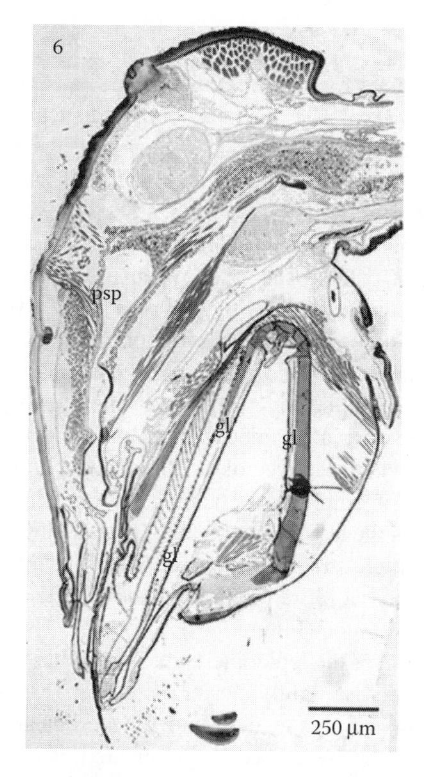

FIGURE 9.6 Sagittal section of the head of *Ceramius hispanicus* (Hymenoptera: Vespidae: Masarinae); pharyngeal suction pump (psp) enlargeable and contractable by pumping musculature; and glossa (gl) in retracted position inside the labium.

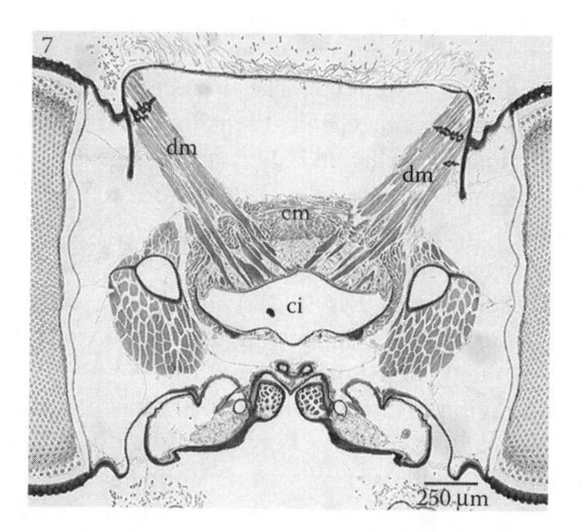

FIGURE 9.7 Cross section of the head of *Heliconius melpomene* (Lepidoptera: Nymphalidae); large dilator muscles (dm) can expand the cibarial suction pump; and circular musculature (cm) can compress the cibarium (ci) for swallowing (images with permission of S. Eberhard).

9.2.2.2 Tip Region

The presence of a fluid-tight food tube requires a specially adapted tip, which must interact with the fluid surface. The tips of lapping and sucking mouthparts of many Hymenoptera are characterized by their hairy glossae (Figure 9.2A). In some long-tongued bees, the glossa is extended just beyond the food canal, and nectar is loaded between extendible hairs by capillary forces (see Section 9.3.2). The lapping movement of the glossa is mediated by muscles that originate on the basal sclerites of the labium and insert at the glossal base. When these muscles relax, the glossa extends because of the elasticity of the glossal rod [42,47,48]. Contraction of these muscles draws the proximal end of the glossal rod into an S-shaped position. As a result, the glossa retracts between the galeae and the labial palps [42]. It is unknown whether nectar is unloaded either by "squeezing" the glossa [49,50] or via suction pressure generated in the cibarial chamber [25]. For suction-feeding euglossine bees, the glossa no longer plays an active role in fluid transport [31]. In short-tongued pollen wasps, the glossa is employed in lapping, whereas in long-tongued taxa, the modified glossa serves as the actual suction tube (Figure 9.5C) [32]. In long-tongued pollen wasps, arched cuticle structures form an incomplete food canal in the bifurcated tip region of the glossa. More proximally, these flattened structures overlap to form a tightly closed food tube (Figure 9.5B) [32].

The flexible tip region of the lepidopteran proboscis has been modified to permit fluid uptake into the otherwise tightly closed food tube. Terminal ends of the galeae are characterized by rows of slits leading into the food canal (Figure 9.3B). There, the galeal-linking structures are arched and elongated, not tightly sealing the food canal; instead, they interlock only at their tips with those of the opposite galea. Because of their curved and extended shape, a slit is formed between consecutive structures. These slits are found on the dorsal side of the proboscis tip in a region that makes up 5 to 20% of the total proboscis length [39,51–53]. Because there is no apical opening into the food canal, the intake slits of the tip region must be immersed into the fluid prior to sucking. The tip region is further characterized by rows of combined contact chemomechanical sensilla [54–56]. Each of these sensilla consists of a variably shaped stylus and short apical sensory cone (Figure 9.3B). Their shape and arrangement are correlated to some extent with butterfly feeding ecology [51,53,57]. When the butterfly feeds from a surface, the fluid adheres to these structures, forming a droplet that is then ingested [58]. In Lepidoptera with particularly long proboscides (e.g., *Papilio* and *Sphinx*), these sensillae are short and barely extend over the surface [51], suggesting that they are adapted to work within the narrow confines of the tubular flowers these insects visit.

The proboscis tip region of brachyceran Diptera has paired movable and variously shaped labellae [34,59] that contact nectar on their inner surface; that surface is equipped with an elaborate system of tiny cuticular channels known as the pseudotracheae (Figure 9.4B). Pseudotracheae distribute saliva over the labellae [60], helping to dissolve nutrients and dilute dried up nectar (see Section 9.3.3). In unspecialized flies, labellae tend to be broad and cushionlike, equipped with a comblike arrangement of pseudotracheae [34,59]. In nectar-feeding hoverflies and beeflies, the labellae are slender and elongate, and the number of pseudotracheal channels is reduced [19,34]. In other nectar-feeding flies (e.g., Conopidae), they are also short and slender, not exceeding the diameter of the labium (Figure 9.4) [59]. In all, the pseudotracheal system forms an extension of the food canal, and pure suction feeding is likely in all those that feed from tubular flowers where spreading of the labellae is impaired.

9.2.2.3 Fluid Pumps

Fluid pumps (Figures 9.6 and 9.7) create the pressure gradient required for imbibing nectar through the slender proboscis. In series with the food canal, these pumps are located in the head and are formed mainly by the cibarium. In Diptera, however, fluid feeding involves an interplay of successive suction pumps that enlarge subsequent sections of the food pathway through the mouthparts and the foregut [19,60–62]. Fluid pumps are not restricted to obligatory nectar-feeding insects because all fluid-feeding insects possess similar pump organs to consume liquid nutrients.

The functional anatomy of suction pumps has been studied in detail in butterflies (Figure 9.7) [20,63]. Contractions of dilator muscles enlarge the cibarium, and at the same time, a ring of muscles in the foregut closes the connection into the pharynx. When the pump lumen is enlarged, nectar is drawn in from the food tube. Subsequently, the entrance of the pump is sealed by a flaplike valve structure, and circularly arranged muscles, which form the wall of the cibarial pump, contract, thus forcing fluid into the opened pharynx. Based on video analysis of air bubbles in the food canal, the dilation–contraction cycle in a pierid butterfly occurs approximately once per second [64]. In addition, electrophysiological measurements have shown that contraction frequencies range from 4 Hz in the nectar-feeding ant, *Camponotus mus* [65], to 6 Hz in a hematophagous bug, *Rhodnius prolixus* [66].

9.3 FEEDING MECHANICS AND FORAGING ECOLOGY

One general conclusion of optimal foraging studies has been that animals seek to maximize their rate of energy intake [67]. Indeed, floral features that influence the rate of energy intake of pollinators have been shown to affect patterns of flower visitation and specificity of pollinators [17,68–72]. Although the utility of energy intake rate has been called into question by some authors [73–76], apparent violations of this rule may result from a misunderstanding of an animal's "temporal scale of optimization" [77]. For a nectarivorous animal, the rate of energy intake can be measured over the timescale of feeding, over a single flower visit, or over an entire foraging bout. In the following sections, we partition functional aspects of nectar feeding into several phases of a flower visit: proboscis extension, floral probing, fluid feeding, and proboscis retraction.

9.3.1 Proboscis Mobility and Floral Handling

The insect proboscis is a deployable structure. During nectar feeding, the position of the proboscis ranges from being directed anteriorly or held perpendicular to the main body axis (Figure 9.1). When not in use, the proboscis is stowed, probably to reduce body drag during flight and possible force asymmetries generated during flight maneuvers (Table 9.3; Figures 9.3A, 9.4A, and 9.6). In many Diptera and Hymenoptera, the proboscis is flexed under the head and body where the tip projects anteriorly or posteriorly. In most taxa, this flexion is accompanied by partial or complete retraction of the proboscis into the labium or head capsule. A number of unique resting positions correspond with these myriad proboscis morphologies. Long-tongued pollen wasps have evolved a rather unique and extreme solution to the problem of proboscis storage. In contrast to short-tongued pollen wasps where the glossa is flexed outside and in front of the head, long-tongued pollen wasps possess a modified basal glossa joint, which allows a double 90° flexion, effectively retracting the glossa in a backward loop under the basal labium sclerite (Figure 9.6). This strongly arched mouthpart sclerite forms a pouchlike formation wherein the folded glossal rod fits and structures forming the food canal are retracted. In extremely long-tongued pollen wasps, the labium actually forms a saclike protrusion posterior to the head wherein the retracted glossa lies [36]. The spirally coiled resting position of the lepidopteran proboscis (Figure 9.3A) is unique among nectar-feeding insects. This space-saving posture may be one reason why the longest proboscides evolved in this group. Recoiled primarily by intrinsic galeal musculature [21, 22], the proboscis fits under the head and between the labial palps, where it locks itself

TABLE 9.3
Resting Positions in Selected Nectar-Feeding Insects with Long Proboscides

Resting Position of Proboscis	Representative Taxa	Ref.
Flexed under body, tip pointing backward	*Nemognatha, Leptopalpus* (Coleoptera: Meloidae)	36, 38
Flexed under body and partly retracted, tip pointing backward	Long-tongued Apoidea (Hymenoptera)	42, 47
	Prosoeca (Diptera: Nemestrinidae)	N.U. Szucsich, personal communication
Flexed under head, tip pointing forward	*Corizoneura* (Diptera: Tabanidae)	37
Folded under the head and partly retracted, tip pointing forward	*Rhingia* (Diptera: Syrphidae)	34, 60
	Bombyliidae (Diptera)	35
	Conopidae (Diptera)	59
	Tachinidae (Diptera)	59
Fully retracted loop in labium, tip pointing forward	Masarina (Hymenoptera: Vespidae)	32
Spiral of three to seven coils under head	Glossata (Lepidoptera)	45

using the elasticity of the spirally coiled galeae without the need of further muscle action [45].

The time an insect spends deploying the proboscis and handling floral structures decreases foraging profitability, and a number of adaptations allow nectar feeders to minimize floral-handling time. Hummingbirds, nectar-feeding bats, and certain insects frequently hover when probing flowers, probably reducing floral access times [78] while simultaneously reducing possible predation risks [79]. Many long-proboscid insects partially extend their proboscis before landing, but others extend it after landing, thus making proboscis extension a rather cumbersome process. In bees, cranial muscles of the labiomaxillary complex unfold the proboscis by moving basal components anteriorly [80], a design that requires a substantial amount of space. In bumblebees, long proboscides may be a hindrance owing to the need to rear the head backward prior to proboscis insertion into the corollae [81]. In long-tongued euglossine bees, this process reaches comical proportions as they fumble to extend their ungainly tongues while barely hanging onto the petals of a *Costus* flower. By contrast, long-tongued pollen wasps are able to immediately extend their proboscis into narrow corolla tubes after landing since the glossa is propelled forward from its internally looped resting position [32].

Proboscis movements are well-studied in butterflies. After uncoiling the proboscis with a hydraulic mechanism [45,82,83], the proboscis assumes a flexed position during feeding that permits easy adjustment to various corolla lengths. Probing movements are controlled by this hydraulic mechanism in addition to high cuticular flexibility, proboscis musculature, and accompanying sensory equipment [45,55]. Elevation of the entire proboscis, combined with extension and flexion of the distal parts, leads to rapid and precise probing movements without whole body movements. These probing movements are likely to be advantageous in handling inflorescences [45,64].

The comparison of bombyliid flies with short and long proboscides indicates that the same principal mechanisms govern their proboscis movements. One remarkable innovation in long-proboscid bombyliid species is their ability to take up nectar from laterally open flowers with the proboscis directed anteriorly but without fully extending it or spreading the labellae [19,35].

Nectar-feeding insects are typically generalist pollinators, and there is little evidence to support the partitioning of floral resources on the basis of proboscis length alone [11,14–16]. Not surprisingly, animals with longer mouthparts are able to access deeper flowers, but the specificity of these relationships often depends on other aspects of plant and pollinator morphology [84–86]. In hummingbirds, foraging efficiency is influenced by the match between corolla and bill morphologies [70–72], and in bumblebees, there is some evidence to suggest that efficiency is maximized when foragers visit flowers matching their tongue length [14,17,18]. Unfortunately, because of a lack of comparative foraging studies, there are few data to address the relationship between handling time, feeding modality, and proboscis length in other insects. However, because insects with long proboscides tend to follow foraging traplines on a few nectar-rich resources [87], fluid-handling times may be more significant than probing times.

9.3.2 Factors Influencing Fluid Handling

The rate and efficiency with which an insect can transport nectar from the floral nectar reservoir and through its proboscis depends on the physical properties of the nectar solution, the modality of fluid feeding, the geometry of the feeding apparatus, and the dynamics of muscle contraction [25]. Betts [88] was the first to recognize the importance of viscosity in limiting nectar ingestion rates in honeybees, and Baker [89] hypothesized that similar biophysical constraints may have influenced the evolution of the dilute nectars found in hummingbird flowers. Early biomechanical analyses [26,90] employed the Hagen–Poiseiulle relation to describe how the rate of nectar intake, Q, varies with viscosity, μ, proboscis length, L, food canal radius, R, and the driving pressure gradient, P:

$$Q = R^4P/(8\mu \mathrm{L}) \tag{9.1}$$

One prediction derived from Equation 9.1 is that the nectar intake rate declines linearly as proboscis length increases. Thus, based on this simple analysis, an obvious disadvantage to a long proboscis may be a slower nectar intake rate. Alternatively, long-proboscid insects may compensate for this handicap by developing proportionally larger pump muscles and/or increasing the radius of their food canal. Presently, no published studies have addressed these possibilities, but preliminary data from 33 species of euglossine bees suggest that nectar intake rates decline with tongue length after the confounding effects of body size have been removed [91].

In seeking to maximize their rate of energy intake, insect nectarivores must select from a variety of floral resources. One constraint faced by these foragers is that nectar viscosity increases exponentially with sucrose concentration, and Equation 9.1 tells us that nectar intake rate declines with viscosity. Thus, the rate of energy intake will be maximized at some intermediate concentration (Figure 9.8). Because the pressure drop P varies with fluid properties [92], the position of this optimal nectar concentration will depend on the precise mechanism of force production.

Researchers have identified two primary mechanisms of fluid transport during nectar loading: capillary-based lapping and suction feeding (see Section 9.2). Lapping insects such as ants (on extrafloral nectars [48,93]), bees [42,48–50,93], hummingbirds (Trochilidae), and nectar-feeding bats (Phyllostomidae: Glossophaginae) dip their hairy tongues (or glossae in insects) into the nectar solution whereupon liquid is drawn up via capillary forces and subsequently unloaded internally via "squeezing" or suction from the cibarial pump [25,26,49,94]. Suction feeding, which depends solely on a pressure gradient generated by fluid pumps in the head and along the intestinal tract, occurs primarily in the Lepidoptera, Diptera, and some Hymenoptera (Table 9.2). Many flies use a primitive sponging mode of nectar feeding where nectar is first taken up by the spread labella and later sucked into the food canal. The loading phase of sponging likely depends on both capillary forces and suction pressure generated by the spreading labella.

These two mechanisms of feeding lead to different predictions regarding the value of the optimal nectar sugar concentration [25]. Daniel et al. [24] used A.V.

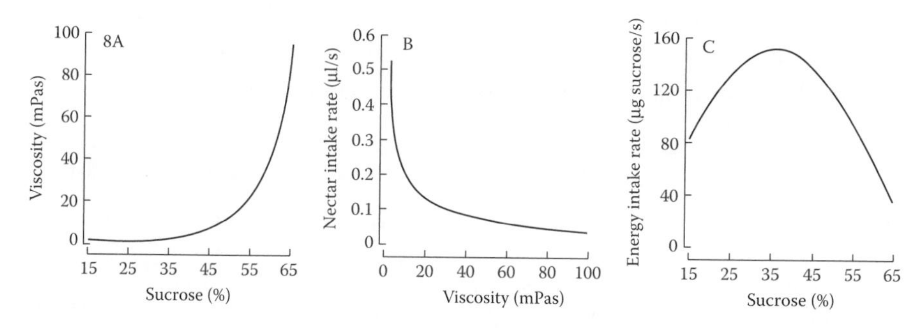

FIGURE 9.8 Relationships between energy intake rate, nectar intake rate, viscosity, and sucrose concentration. Because viscosity increases exponentially with sucrose concentration (A) and volumetric nectar intake rate declines with viscosity (B), energy intake rates will be maximized at intermediate sugar concentrations (C). Graphs are calculated for a 150-mg insect using the suction feeding model of Daniel et al., *Oecologia*, 79, 66, 1989.

Hill's classic model of muscle contraction dynamics to describe the behavior of the cibarial pump musculature in butterflies. This model predicted an optimal range of sucrose concentrations between 31 and 39% (sugar weight to total weight) depending on parameter estimates. Empirical studies with eight lepidopteran species and numerous other insects have largely confirmed these predictions (Table 9.4). Remarkably, although proboscis length influences the absolute rate of energy intake for suction feeders (see above), the sugar concentration that maximizes energy flux is predicted to be independent of proboscis length [24].

Using a capillary pressure term to examine the mechanics of lapping by bees, Kingsolver and Daniel [25] predicted that optimal nectar sugar concentrations for lappers should be greater than those for suction feeders. Indeed, maximal energy intake rates for lapping bees and ants are at sugar concentrations nearly 15% (w/w) higher than those for suction-feeding insects (Table 9.4). Because the frequency and amplitude of glossal extension in hymenopterans relies on passive mechanical properties [48], Borrell [31] suggested that as tongue length increases, lapping ceases to be an effective mechanism of fluid transport. One consequence of the evolution of greatly elongated proboscides in the Diptera and Hymenoptera may have been a downward shift in the sugar concentration that maximizes the rate of energy intake.

9.3.3 Environmental Influences on Floral Nectar Constituents

Although laboratory feeding experiments have been largely confined to nectar intake rates on pure sucrose solutions (but see [95–97]), floral nectars in nature are often composed of a suite of sugars in various proportions along with small concentrations of amino acids and other compounds [98]. These chemical constituents influence both the physical properties of nectar [26] and its energetic value to a given pollinator [96,98]. Fructose and glucose, for instance, which are found in moderate concentrations in insect flowers, are both less viscous than sucrose at the same concentration [26]. However, in choice tests, pure sucrose is preferred over either of these sugars

TABLE 9.4
Optimal Nectar Sugar Concentrations (% w/w) Reported for Some Nectar Feeders

Common Name	Genus	Feeding Mode	Optimal %	Ref.
Ponerine ant	*Pachycondyla*	Lapping	50	93
Ponerine ant	*Rhytidoponera*	Lapping	50	93
Bumblebee	*Bombus*	Lapping	55	49
Honeybee	*Apis*	Lapping	55	101
Stingless bee	*Melipona*	Lapping	60	101
Leaf-nosed bat	*Glossophaga*	Lapping	60	118
Rufous hummingbird	*Selasphorus*	Lapping	50	123, 138
Honeyeater bird	*Various*[a]	Lapping	40	115
Leafcutter ant	*Atta*	Suction	30	93
Carpenter ant	*Camponotus*	Suction	40	93
Orchid bee	*Euglossa*	Suction	35	31
Fritillary butterfly	*Agraulis*	Suction	40	92
Sulphur butterfly	*Phoebis*	Suction	35	92
Fritillary butterfly	*Speyeria*	Suction	35	139
Skipper butterfly	*Thymelicus*	Suction	40	126
Painted lady butterfly	*Vanessa*	Suction	40	140
Armyworm noctuid moth	*Pseudaletia*	Suction	40	126
Hummingbird hawkmoth	*Macroglossum*	Suction	35	97
Tobacco hawkmoth	*Manduca*	Suction	30	141
Human	*Homo*	Suction	40	126
Blowfly	*Phormia*	Sponging	35	95
Mean(±95% C.I.) for lappers			**50.5 ± 5.1**	
Mean(±95% C.I.) for suction feeders[b]			**36.2 ± 2.7**	

Note: In general, animals were timed while feeding from large volumes of aqueous sucrose solution and the volume or mass change of the solution was recorded upon completion of the feeding bout.

[a] *Anthochaera* (45%), *Phylidonyris* (45%), and *Acanthorhynchus* (35%)

[b] Not including humans.

[96,99,100], perhaps because of its ease of assimilation. Nectar viscosity increases with the addition of amino acids [26], and although they are ubiquitous at low concentrations in floral nectars [98], their significance to nectar-feeding insects has yet to be convincingly demonstrated [99,101,102].

In general, insect-pollinated flowers tend to be sucrose dominant [98], and nectar intake rates observed in the laboratory provide a window to understanding the mechanics of nectar ingestion at real flowers. Sucrose concentrations of nectars in insect-pollinated flowers vary widely, ranging from just a few percent to a high of 88% in the crystallized nectar of one Mediterranean shrub [98]. High concentrations are typically diluted with saliva prior to ingestion, and under these conditions salivation rate may even be a limiting factor in foraging efficiency. Diurnally pollinated flowers normally exhibit a single peak in nectar production in the midmorning, whereas flowers that are pollinated at night exhibit this peak shortly after dusk [11,103,104]. Although available nectar volumes change over the course of the day, nectar sugar concentrations are relatively stable in most flower species [11,12, 105–107]. In fact, flowers with long corollas and concealed nectaries are less affected by evaporation or dilution by rain than flowers with open nectaries [103].

Studies isolating the effects of environmental factors such as temperature, humidity, water stress, and atmospheric carbon dioxide on nectar sugar concentrations have produced mixed results [108–111]. Similarly, the heritability of nectar sugar concentration appears to vary by species and environment, making generalizations difficult at the present time [108,112]. It is important to note, however, that whereas interindividual variation in nectar volume can be quite large, variation in nectar sugar concentration tends to be rather low [113]. Patterns of high interspecific and low intraspecific variation in sugar concentration are at least suggestive of strong stabilizing selection.

9.3.4 Have Nectar Sugar Concentrations Evolved to Match Pollinator Preferences?

The cost of producing nectar can be substantial, and at least in environments of low water stress, this cost can be directly related to sugar content [114]. Thus, if total sugar mass is held constant, it costs a plant the same amount to offer a pollinator 35% sugar as it does 65% sugar [115]. If nectar-feeding insects seek to maximize their rate of energy intake during feeding, then they should prefer to visit plants that provide nectars matching their optimal sugar concentration. Consequently, flowers specializing on a particular pollinator may be expected to evolve sugar concentrations that match pollinator preferences.

Euglossine bees are derived suction feeders with an optimal nectar sugar concentration that falls between 30 and 40% sucrose [31]. We compiled data on the nectar sugar concentrations recorded from flowers in 28 species in 9 families that euglossine bees are known to visit and categorized these flowers as euglossine specialists or generalists (Table 9.5). Overall, we found a close match between optimal nectar sugar concentrations and the concentrations found in specialist flowers. More significantly, however, we observed lower variance in sugar concentrations in specialist as compared to generalist flowers, but we caution that verifying this trend requires additional data and phylogenetic controls. In comparison with sympatric bees that lap nectars, euglossine bees also tend to forage from flowers with more dilute rewards (Table 9.6). Other analyses of floral nectars have supported partitioning of pollinator guilds on the basis of sugar concentration [89,98,116], but as is evident from Table 9.6, feeding biomechanics is clearly only one factor influencing these trends. Opposing physiological pressures to minimize water loads in flight [117] or obtain dietary water [89,118,119] may also influence choice behavior by nectarivores and the evolution of nectar sugar concentrations in flowers. Additionally, floral generalization, recent pollinator shifts, and phylogenetic inertia may contribute to the mismatch between sugar concentration and feeding mechanics in some taxa.

One method for assessing how the biomechanics of nectar ingestion has influenced nectar constituents of flowers is to evaluate choice behavior of nectarivorous

TABLE 9.5
Mean Nectar Sugar Concentrations (% w/w) for Some Flowers Visited by Euglossine Bees

Family	Genus (Species *N*)	Specialist	Sucrose (%)	Ref.
Apocynaceae	*Stemmadenia* (1)	N	34	142
Apocynaceae	*Thevetia* (1)	N	32	142
Bignoniaceae	*Jacaranda* (1)	N	15	B.J. Borrell, unpublished
Bignoniaceae	*Tabebuia* (3)	N	39	142
Convolvulaceae	*Ipomoea* (1)	N	31	143
Gesneriaceae	*Drymonia* (2)	N	34	12, 144
Gesneriaceae	*Sinningea* (2)	N	26	145
Mimosaceae	*Inga* (2)	N	26	105, 146
Passifloriaceae	*Passiflora* (2)	N	40	107
Costaceae	*Costus* (4)	Y	36	12, 106
Costaceae	*Dimerocostus* (1)	Y	35	B.J. Borrell, unpublished
Gesneriaceae	*Sinningea* (2)	Y	34	145
Lecythidaceae	*Coratari* (1)	Y	39	147
Lecythidaceae	*Eschweilera* (2)	Y	36	147
Marantaceae	*Calathea* (3)	Y	38	148–150
Mean (± 95% C.I.) for generalist flowers			**31 ± 5.7**	
Mean (± 95% C.I.) for euglossine specialists			**36 ± 1.7**	

Note: In general, nectars were extracted from new flowers during times of pollinator visitation, and the equivalent sucrose concentration was measured using a handheld refractometer. Designation of flowers as euglossine specialists was based on visitation frequency data reported by the authors, not taking into account pollinator efficiency.

TABLE 9.6
Mean Nectar Sugar Concentrations (% w/w) of Flowers Visited by Different Animal Taxa in a Variety of Habitats

Common Name	Feeding Mode	Sucrose (%)	Habitat Type	References
Bumblebee	Lapping	44	Temperate meadow	151
Centridine bee	Lapping	48	Tropical forest	102
Stingless bee	Lapping	44	Tropical forest	102
Hummingbird	Lapping	22	Tropical wet forest	89
Leaf-nosed bat	Lapping	14	Tropical wet forest	152
Orchid bee	Suction	36	Tropical wet forest	See Table 9.5
Long-proboscid fly	Suction	26	Mediterranean shrub	15
Hawkmoth	Suction	22	Tropical dry forest	11
Butterfly	Suction	25	Temperate	26

insects in laboratory studies. Numerous investigations have measured visitation rates of nectarivores to effectively infinite volume sucrose solutions and concluded that these animals prefer the most concentrated solutions offered them [96,120–122]. One problem with this approach is that it confounds nectar sugar concentration with total meal energy [75,123]. The more relevant question is how much water should a plant add to a fixed quantity of sugar in order to maximize attractiveness to pollinators [115]. Furthermore, behavioral studies should use realistic nectar volumes and monitor transport costs to and from nectar sources so that the data may be analyzed for a variety of timescales [26,77]. Roberts' exemplary study of hummingbird foraging [123] analyzed concentration preferences at different timescales but employed an equal volume rather than an equal sugar design. Hainsworth and Hamill [75] conducted the only published sugar choice experiment we know of by offering the butterfly *Vanessa cardui* a choice between feeding from a 70% solution for 30 sec or a 35% solution for 20 sec. In spite of the decline in energy intake rate, these authors found that butterflies still preferred the more concentrated solution. One caveat with interpreting these results is that butterflies were not freely foraging but were captured and hand fed upon landing at color-coded feeding sites. The euglossine bee *Euglossa imperialis* does not discriminate between 35% (feeding time [FT] = 9 sec) and 55% (FT = 15 sec) solutions offered in an equal sugar design; *Euglossa imperialis* does, however, show a slight but significant preference for 35% (FT = 9 sec) sucrose over 60% (FT = 30 sec; B.J. Borrell, unpublished). Neither B.J. Borrell (unpublished) nor Hainsworth and Hamill [75] monitored transport costs, which when taken into account, predict preferences for more concentrated nectars than consideration of feeding costs alone [26].

An alternative route of investigation has been to augment the viscosity of pure sucrose solutions using small quantities of polymers such as tylose or methyl cellulose [97,124,125]. Hummingbirds do not distinguish between 20% sucrose solution and a 20% sucrose solution with the viscosity increased to that of a 40% solution [125]. However, the bee *Euglossa imperialis* shows a strong preference for low viscosity nectars in choice experiments (B.J. Borrell, unpublished).

9.3.5 Temperature and Optimal Nectar Foraging

Environmental temperature and nectar sugar concentration interact to influence both the energetic costs or foraging and the rate of energy intake during feeding. Nectar viscosity increases at colder temperatures, and the dependence of viscosity on temperature increases with increasing sugar concentration [26]. Consequently, one general prediction is that nectar intake rates should decline at cooler temperatures, a prediction that has been confirmed in experiments with both butterflies [126] and euglossine bees (B.J. Borrell, unpublished). Thus, foraging insects would do well to forage in sunny patches [26,111,127] or at inflorescences with endogenous heat sources [128]. The relevant behavioral experiment would involve independently controlling nectar temperature and air temperature to partition thermoregulatory costs from feeding costs.

Some researchers have argued that nectars are less concentrated at high elevations owing to temperature effects on viscosity [89,103]. However, Heyneman [26] showed that optimal concentrations should shift no more than 1 to 2% for a 10°C decrease in air temperature. Indeed, for the butterfly *Thymelicus lineola*, optimal nectar sugar concentrations lie at approximately 40% sucrose at both 25 and 35°C. At the cooler temperature, however, energy intake rate exhibited a less well-defined peak, remaining equally rewarding between 25 and 45% sucrose [126].

For endotherms such as hummingbirds, hawkmoths, or large bees, temperature can also have a direct influence on the energetic cost of foraging. The energetic cost of preflight warm-up and shivering during flower visits is substantially higher at colder temperatures [69]. As noted above, endothermic flowers have the potential to offset these costs by providing pollinators a heat reward [128]. Neotropical euglossine bees are known to regulate heat production during flight: A 10°C decline in air temperature results in a 30% increase in metabolic power requirements [129]. Consequently, an increase in transport costs at lower temperatures may have a greater effect on optimal nectar sugar concentrations than changes in nectar physical properties [26]. Contrary to this hypothesis, Borrell [91] found that euglossine bees harvest nectars of the same concentration in both dry and wet forests in both the lowlands and highlands of Costa Rica. For hummingbirds, Tamm [130] demonstrated a preference for more concentrated nectars as transport costs increased, and it would be interesting to see if the same relation holds true for temperature-mediated changes in flight costs. One final note is that the metabolic cost of warming nectar on a cold day cannot be ignored in examining thermal effects on foraging choice [131].

9.4 CONCLUDING REMARKS

In this review, we have endeavored to synthesize functional morphology, biomechanics, and behavioral ecology to develop an integrative view of the interactions between flowering plants and nectar-feeding animals. Proboscides exceeding body length have arisen multiple times among nectar-feeding taxa, and although the morphological composition of these proboscides vary widely, all of these insects share several key attributes, including the possession of a fluid-tight food canal, a specialized tip region, and one or more fluid pumps. These insects have overcome functional problems of proboscis control, storage, and extension to maximize profitability of nectar-foraging activities. The rate of fluid flow in an insect's proboscis depends on the modality of fluid feeding, the morphology of the feeding apparatus, and the chemistry of floral nectars. Optimal nectar-foraging strategies may also be influenced by environmental temperatures and the distribution of nectar resources. Future studies should aim to test proposed links between morphology and ecology to further our understanding of the evolution of long proboscides.

ACKNOWLEDGMENTS

We thank G. Byrnes, C. Clark, R. Dudley, R. Hill, S. Horisawa, and two anonymous reviewers for comments and discussions which greatly improved this manuscript. The SEM micrographs were prepared with the help of the electron microscopy lab in the Institute of Zoology at the University of Vienna. B.J.B. was supported by a graduate research fellowship from the U.S. National Science Foundation.

REFERENCES

1. Sprengel, C.K., Discovery of the secret nature in the structure and fertilization of flowers, in *Floral Biology: Studies on Floral Evolution in Animal-Pollinated Plants*, Lloyd, D.G. and Barrett, S.C. H., Eds., Chapman and Hall, New York, 1996, p. 65.
2. Proctor, M., Yeo, P., and Lack, A., *The Natural History of Pollination*, Timber Press, Portland, Oregon, 1996, p. 479.
3. Darwin, C., *On the Various Contrivances by which British and Foreign Orchids Are Fertilized by Insects*, Murray, London, 1862.
4. Nilsson, L.A., The evolution of flowers with deep corolla tubes, *Nature*, 334, 147, 1988.
5. Alexandersson, R. and Johnson, S.D., Pollinator-mediated selection on flower-tube length in a hawkmoth pollinated *Gladiolus* (Iridaceae), *Proc. R. Soc. Lond. B*, 269, 631, 2002.
6. Johnson, S.D. and Steiner, K.E., Long-tongued fly pollination and evolution of floral spur length in the *Disa draconis* complex (Orchidaceae), *Evolution*, 51, 45, 1997.
7. Schemske, D.W. and Horvitz, C.C., Temporal variation in selection on a floral character, *Evolution*, 43, 461, 1989.
8. Labandeira, C.C., Insect mouthparts: Ascertaining the paleobiology of insect feeding strategies, *Annu. Rev. Ecol. Syst.*, 28, 153, 1997.
9. Ren, D., Flower-associated Brachycera flies as fossil evidence for Jurassic angiosperm origins, *Science*, 280, 85, 1998.
10. Ackerman, J.D., Euglossine bees and their nectar hosts, in *The Botany and Natural History of Panama*, D'Arcy, W.G. and Correa, M.D., Eds., Missouri Botanical Garden, St. Louis, MO, 1985, p. 225.
11. Haber, W.A. and Frankie, G.W., A tropical hawkmoth community: Costa Rican dry forest Sphingidae, *Biotropica*, 21, 155, 1989.
12. Grove, K.F., *Reproductive Biology of Neotropical Wet Forest Understory Plants*, University of Iowa, Iowa City, 1985, p. 187.
13. Heinrich, B., Resource partitioning among some eusocial insects: Bumblebees, *Ecology*, 57, 874, 1976.
14. Ranta, E. and Lundberg, H., Resource partitioning in bumblebees: The significance of differences in proboscis length, *Oikos*, 35, 298, 1980.
15. Manning, J.C. and Goldblatt, P., The *Moegistorhynchus longirostris* (Diptera: Nemestrinidae) pollination guild: long-tubed flowers and a specialized long-proboscid fly pollination system in southern Africa, *Plant Syst. Evol.*, 206, 51, 1997.
16. Borrell, B.J., Long tongues and loose niches: Evolution of euglossine bees and their nectar flowers, *Biotropica*, 37, 664, 2005.
17. Inouye, D.W., The effect of proboscis and corolla tube lengths on patterns and rates of flower visitation by bumblebees, *Oecologia*, 45, 197, 1980.
18. Harder, L.D., Flower handling efficiency of bumble bees: Morphological aspects of probing time, *Oecologia*, 57, 274, 1983.
19. Szucsich, N.U. and Krenn, H.W., Flies and concealed nectar sources: Morphological innovations in the proboscis of Bombyliidae (Diptera), *Acta Zool.*, 83, 183, 2002.
20. Eastham, L.E.S. and Eassa, Y.E.E., The feeding mechanism of the butterfly *Pieris brassicae* L., *Phil. Trans. R. Soc. Lond. B*, 239, 1, 1955.
21. Wannenmacher, G. and Wasserthal, L.T., Contribution of the maxillary muscles to proboscis movement in hawkmoths (Lepidoptera: Sphingidae) — An electrophysiological study, *J. Insect Physiol.*, 49, 765, 2003.

22. Krenn, H.W., Proboscis musculature in the butterfly *Vanessa cardui* (Nymphalidae, Lepidoptera): Settling the proboscis recoiling controversy, *Acta Zool.*, 81, 259, 2000.
23. Hepburn, H.R., Proboscis extension and recoil in Lepidoptera, *J. Insect Physiol.*, 17, 637, 1971.
24. Daniel, T.L., Kingsolver, J.G., and Meyhofer, E., Mechanical determinants of nectar-feeding energetics in butterflies: Muscle mechanics, feeding geometry, and functional equivalence, *Oecologia*, 79, 66, 1989.
25. Kingsolver, J.G. and Daniel, T.L., Mechanics of food handling by fluid-feeding insects, in *Regulatory Mechanisms in Insect Feeding*, Chapman, R.F. and, de Boer, G. Eds., Chapman & Hall, New York, 1995, p. 32.
26. Heyneman, A.J., Optimal sugar concentrations of floral nectars: Dependence on sugar intake efficiency and foraging costs, *Oecologia*, 60, 198, 1983.
27. Krenn, H.W., Plant, J.D. and, Szucsich, N.U., Mouthparts of flower-visiting insects, *Arthropod Struct. Dev.*, 34, 1, 2005.
28. Jervis, M., Functional and evolutionary aspects of mouthpart structure in parasitoid wasps, *Biol. J. Linn. Soc.*, 63, 461, 1998.
29. Amsel, H.G., *Amphimoea walkeri* Bsd., der Schwärmer mit dem längsten Rüssel! *Entomol Rundsch*, 55, 165, 1938.
30. Jervis, M. and, Vilhelmsen, L., Mouthpart evolution in adults of the basal, "symphytan," hymenopteran lineages, *Biol. J. Linn. Soc.*, 70, 121, 2000.
31. Borrell, B.J., Suction feeding in orchid bees (Apidae: Euglossini), *Proc. R. Soc. Lond. B*, 271, S164, 2004.
32. Krenn, H.W., Mauss, V., and, Plant, J., Evolution of the suctorial proboscis in pollen wasps (Masarinae, Vespidae), *Arthropod Struct. Dev.*, 31, 103, 2002.
33. Gilbert, F. and Jervis, M., Functional, evolutionary and ecological aspects of feeding-related mouthpart specializations in parasitoid flies, *Biol. J. Linnean Soc.*, 63, 495, 1998.
34. Gilbert, F.S., Foraging ecology of hoverflies: Morphology of the mouthparts in relation to feeding on nectar and pollen, *Ecol. Entomol.*, 1981, 245, 1981.
35. Szucsich, N.U. and Krenn, H.W., Morphology and function of the proboscis in Bombyliidae (Diptera, Brachycera) and implications for proboscis evolution in Brachycera, *Zoomorphology*, 120, 79, 2000.
36. Schremmer, F., Morphologische Anpassungen von Tieren — insbesondere Insekten — an die Gewinnung von Blumennahrung, *Verh. Deutschen Zoologischen Ges Saarbrücken*, 1961, 375, 1961.
37. Dierl, W., Zur Nahrungsaufnahme von *Corizoneura longirostris* (Hardwicke) (Diptera: Tabanidae), *Khumbu Himal*, 3, 76, 1968.
38. Handschin, E., Ein neuer Rüsseltyp bei einem Käfer. Biologische und morphologische Beobachtungen an *Leptopalpus rostratus* F., *Z. Morphol. Ökologie Tiere*, 14, 513, 1928.
39. Krenn, H.W. and Kristensen, N.P., Early evolution of the proboscis of Lepidoptera (Insecta): External morphology of the galea in basal glossatan moths lineages, with remarks on the origin of pilifers, *Zool. Anz.*, 239, 179, 2000.
40. Kristensen, N.P., *Lepidoptera: Moths and Butterflies 2. Handbook of Zoology IV/36*, Walter De Gruyter, New York, 2003.
41. Smith, J.J. B., Feeding mechanisms, in *Comprehensive Insect Physiology, Biochemistry and Pharmacology*, Kerkut, G.A. and Gilbert, L.I., Eds., Pergamon Press, New York, 1985, p. 64.
42. Harder, L.D., Functional differences of the proboscides of short- and long-tongued bees (Hymenoptera, Apoidea), *Can. J. Zool.*, 61, 1580, 1982.

43. Laroca, S., Michener, C.D., and Hofmeister, R.M., Long mouthparts among short-tongued bees and the fine structure of the labium in *Niltonia* (Hymenoptera, Colletidae), *J. Kans. Entomol. Soc.*, 62, 400, 1989.
44. Houston, T.F., An extraordinary new bee and adaptation of palpi for nectar-feeding in some Australian Colletidae and Pergidae (Hymenoptera), *J. Aust. Entomol. Soc.*, 22, 263, 1983.
45. Krenn, H.W., Functional morphology and movements of the proboscis of Lepidoptera (Insecta), *Zoomorphology*, 110, 105, 1990.
46. Krenn, H.W., Proboscis assembly in butterflies (Lepidoptera): A once in a lifetime sequence of events, *Eur. J. Entomol.*, 94, 495, 1997.
47. Snodgrass, R.E., *Anatomy of the Honey Bee*, Comstock, Ithaca, 1956.
48. Paul, J., Roces, F., and Hölldobler, B., How do ants stick out their tongues? *J. Morphol.*, 254, 39, 2002.
49. Harder, L.D., Effects of nectar concentration and flower depth on flower handling efficiency of bumble bees, *Oecologia*, 69, 309, 1986.
50. Harder, L.D., Measurement and estimation of functional proboscis length in bumblebees (Hymenoptera: Apidae), *Can. J. Zool.*, 60, 1073, 1982.
51. Paulus, H.F. and Krenn, H.W., Morphologie des Schmetterlingsrüssels und seiner Sensillen — Ein Beitrag zur phylogenetischen Systematik der Papilionoidea (Insecta, Lepidoptera), *J. Zool. Syst. Evol. Res.*, 34, 203, 1996.
52. Krenn, H.W. and Penz, C.M., Mouthparts of *Heliconius* butterflies (Lepidoptera: Nymphalidae): A search for anatomical adaptations to pollen-feeding behavior, *Int. J. Insect Morphol. Embryol.*, 27, 301, 1998.
53. Krenn, H.W., Zulka, K.P., and Gatschnegg, T., Proboscis morphology and food preferences in nymphalid butterflies (Lepidoptera: Nymphalidae), *J. Zool.*, 254, 17, 2001.
54. Altner, H. and Altner, I., Sensilla with both terminal pore and wall pores on the proboscis of the moth, *Rhodogastria Bubo* Walker (Lepidoptera, Arctiidae), *Zool. Anz.*, 216, 129, 1986.
55. Krenn, H.W., Proboscis sensilla in *Vanessa cardui* (Nymphalidae, Lepidoptera): Functional morphology and significance in flower-probing, *Zoomorphology*, 118, 23, 1998.
56. Walters, B.D., Albert, P.J., and Zacharuk, R.Y., Morphology and ultrastructure of sensilla on the proboscis of the adult spruce budworm, *Choristoneura fumiferana* (Clem.) (Lepidoptera: Tortricidae), *Can. J. Zool.*, 76, 466, 1998.
57. Büttiker, W., Krenn, H.W., and Putterill, J.F., The proboscis of eye-frequenting and piercing Lepidoptera (Insecta), *Zoomorphology*, 116, 77, 1996.
58. Knopp, M.C.N. and Krenn, H.W., Efficiency of fruit juice feeding in *Morpho peleides* (Nymphalidae, Lepidoptera), *J. Insect Behav.*, 16, 67, 2003.
59. Elzinga, R.J. and Broce, A.B., Labellar modifications of Muscomorpha flies (Diptera), *Ann. Entomol. Soc. Am.*, 79, 150, 1986.
60. Schuhmacher, H. and Hoffmann, H., Zur Funktion der Mundwerkzeuge von Schwebfliegen bei der Nahrungsaufnahme (Diptera: Syrphidae), *Entomol. Gen.*, 7, 1982.
61. Schiemenz, H., Vergleichende funktionell-anatomische Untersuchungen der Kopfmuskulatur von *Theobaldia* und *Eristalis* (Diptera, Culicidae und Syrphidae), *Dtsch. Entomol. Z. N.F.*, 4, 268, 1957.
62. Bonhag, P.F., The skeleto-muscular mechanism of the head and abdomen of the adult horsefly (Diptera: Tabanidae), *Trans. Am. Entomol. Soc.*, 77, 131, 1951.
63. Eberhard, S.H. and Krenn, H.W., Salivary glands and salivary pumps in adult Nymphalidae (Lepidoptera), *Zoomorphology*, 122, 161, 2003.

64. Krenn, H.W., *Artogeia (Pieris) rapae* L. (Pier.) — *Russelbewegungen und Nahrungsaufnahme. – Film C 1819 des ÖWF*, Österreichisches Bundesinstitut für den wissenschaftlichen Film, Wien, 1985.
65. Josens, R.B., Nectar feeding and body size in the ant *Camponotus mus*, *Insectes Soc.*, 49, 326, 2002.
66. Smith, J.J.B., Effect of diet viscosity on the operation of the pharyngeal pump in the blood-feeding bug *Rhodnius prolixus*, *J. Exp. Biol.*, 82, 93, 1979.
67. Stephens, D.W. and Krebs, J.R., *Foraging Theory*, Princeton University Press, Princeton, 1987.
68. Whitham, T.G., Coevolution of foraging in *Bombus* and nectar dispensing in *Chilopsis*: A last dreg theory, *Science*, 197, 593, 1977.
69. Heinrich, B., Energetics of pollination, *Annu. Rev. Ecol. Syst.*, 6, 139, 1975.
70. Wolf, L.L., Hainsworth, F.R., and Stiles, F.G., Energetics of foraging: rate and efficiency of nectar extraction by hummingbirds, *Science*, 176, 1351, 1972.
71. Temeles, E.J. and Roberts, M.D., Effect of sexual dimorphism in bill length on foraging behavior: An experimental analysis of hummingbirds, *Oecologia*, 94, 87, 1993.
72. Temeles, E.J. et al., Evidence for ecological causation of sexual dimorphism in a hummingbird, *Science*, 289, 441, 2000.
73. Montgomerie, R.D., Eadie, J.M., and Harder, L.D., What do foraging hummingbirds maximize? *Oecologia*, 63, 357, 1984.
74. Hainsworth, F.R. and Wolf, L.L., Nectar characteristics and food selection by hummingbirds, *Oecologia*, 25, 101, 1976.
75. Hainsworth, F.R. and Hamill, T., Foraging rules for nectar: food choices by painted ladies, *Am. Nat.*, 142, 857, 1993.
76. Hainsworth, F.R., "Fast food" vs "haute cuisine": painted ladies, *Vanessa cardui* (L.), select food to maximize net meal energy, *Funct. Ecol.*, 3, 701, 1989.
77. Gass, C.L. and Roberts, M.D., The problem of temporal scale in optimization: Three contrasting views of hummingbird visits to flowers, *Am. Nat.*, 140, 829, 1992.
78. Hainsworth, F.R., Discriminating between foraging rules and why hummingbirds hover, *Anim. Behav.*, 41, 902, 1991.
79. Wasserthal, L.T., Swing-hovering combined with long tongue in hawkmoths, an antipredator adaptation during flower visits, in *Animal-Plant Interactions in Tropical Environments*, Barthlott, W., Ed., Museum Koenig, Bonn, 1993, p. 77.
80. Plant, J.D. and Paulus, H.F., Comparative morphology of the postmentum of bees (Hymenoptera: Apoidea) with special remarks on the evolution of the lorum, *Z. Zool. Syst. Evolutionsforsch.*, 25, 81, 1987.
81. Plowright, C.M.S. and Plowright, R.C., The advantage of short tongues in bumblebees (*Bombus*): Analysis of species distributions according to flower corolla depth, and of working speeds on white clover, *Can. Entomol.*, 129, 51, 1997.
82. Schmitt, J.B., The feeding mechanism of adult Lepidoptera, *Smithson. Misc. Coll.*, 97, 1, 1938.
83. Bänziger, H., Extension and coiling of the lepidopterous proboscis: A new interpretation of the blood-pressure theory, *Mitt. Schweiz. Entomol. Gest.*, 43, 225, 1971.
84. Temeles, E.J. et al., The role of flower width in hummingbird bill length-flower length relationships, *Biotropica*, 34, 68, 2002.
85. Inoue, T. and Kato, M., Inter- and intraspecific morphological variation in bumblebee species, and competition in flower utilization, in *Effects of Resource Distribution on Animal-Plant Interactions*, Hunter, M.D., Ohgushi, T., and Price, P.W. Eds., Academic Press, San Diego, 1992, p. 393.

86. Harder, L.D., Morphology as a predictor of flower choice by bumble bees, *Ecology*, 66, 198, 1985.
87. Janzen, D.H., Euglossine bees as long-distance pollinators of tropical plants, *Science*, 171, 203, 1971.
88. Betts, A.D., Das Aufnahmevermögen der Bienen beim Zuckerwasserfüttern, *Arch. Bienenkunde*, 10, 301, 1929.
89. Baker, H.G., Sugar concentrations in nectars from hummingbird flowers, *Biotropica*, 7, 37, 1975.
90. Kingsolver, J.G. and Daniel, T.L., On the mechanics and energetics of nectar feeding in butterflies, *J. Theor. Biol.*, 76, 167, 1979.
91. Borrell, B.J., Optimality and allometry in nectar foraging of orchid bees (Apidae: Euglossini), *Integr. Comp. Biol.*, 43, 869, 2003.
92. May, P.G., Nectar uptake rates and optimal nectar concentrations of two butterfly species, *Oecologia*, 66, 381, 1985.
93. Paul, J. and Roces, F., Fluid intake rates in ants correlate with their feeding habits, *J. Insect Physiol.*, 49, 347, 2003.
94. Winter, Y. and von Helversen, O., Operational tongue length in phyllostomid nectar-feeding bats, *J. Mammal.*, 84, 886, 2003.
95. Dethier, V.G., Evans, D.R., and Rhoades, M.V., Some factors controlling the ingestion of carbohydrates by the blowfly, *Biol. Bull.*, 111, 204, 1956.
96. Dethier, V.G. and Rhoades, M.V., Sugar preference-aversion functions for the blowfly, *J. Exp. Biol.*, 126, 177, 1954.
97. Josens, R.B. and Farina, W.M., Nectar feeding by the hovering hawk moth *Macroglossum stellatarum*: Intake rate as a function of viscosity and concentration of sucrose solutions, *J. Comp. Physiol. A*, 187, 661, 2001.
98. Baker, H.G. and Baker, I., A brief historical review of the chemistry of floral nectar, in *The Biology of Nectaries*, Bentley, B. and Elias, T., Eds., Columbia University Press, New York, 1983, p. 126.
99. Erhardt, A., Preferences and nonpreferences for nectar constituents in *Ornithoptera priamus* Poseidon (Lepidoptera, Papilionidae), *Oecologia*, 90, 581, 1992.
100. Wykes, G.R., The preferences of honeybees for solutions of various sugars which occur in nectar, *J. Exp. Biol.*, 29, 511, 1952.
101. Roubik, D.W. and Buchmann, S.L., Nectar selection by *Melipona* and *Apis mellifera* (Hymenoptera: Apidae) and the ecology of nectar intake by bee colonies in a tropical forest, *Oecologia*, 61, 1, 1984.
102. Roubik, D.W. et al., On optimal nectar foraging by some tropical bees (Hymenoptera: Apidae), *Apidologie*, 26, 197, 1995.
103. Cruden, R.W., Hermann, S.M., and Peterson, S., Patterns of nectar production and plant-pollinator coevolution, in *The Biology of Nectaries*, Bentley, B. and Elias, T., Eds., Columbia University Press, New York, 1983, p. 80.
104. Frankie, G.W. and Haber, W.A., Why bees move among mass-flowering neotropical trees, in *Handbook of Experimental Pollination Biology*, Jones, C.E. and Little, R.J., Eds., Scientific and Academic Editions, New York, 1983, p. 360.
105. Koptur, S., Flowering phenology and floral biology of *Inga* (Fabaceae: Mimosoideae), *Syst. Bot.*, 8, 354, 1983.
106. Schemske, D.W., Floral convergence and pollinator sharing in two bee-pollinated tropical herbs, *Ecology*, 62, 946, 1981.
107. Varassin, I.G., Trigo, J.R., and Sazima, M., The role of nectar production, flower pigments and odour in the pollination of four species of *Passiflora* (Passifloriaceae) in south-eastern Brazil, *Bot. J. Linn. Soc.*, 136, 139, 2001.

108. Carroll, A.B., Pallardy, S.G., and Galen, C., Drought stress, plant water status, and floral trait expression in fireweed, *Epilobium angustifolium* (Onagraceae), *Am. J. Bot.*, 88, 438, 2001.
109. Lake, J.C. and Hughes, L., Nectar production and floral characteristics of *Tropaeolum majus* L. grown in ambient and elevated carbon dioxide, *Ann. Bot. (Lond.)*, 84, 535, 1999.
110. Jakobsen, H.B. and Kristjansson, K., Influence of temperature and floret age on nectar secretion in *Trifolium repens* L, *Ann. Bot. (Lond.)*, 74, 327, 1994.
111. Corbet, S.A. and Willmer, P.G., The nectar of *Justicia* and *Columnea*: composition and concentration in a humid tropical climate, *Oecologia*, 51, 412, 1981.
112. Mitchell, R.J., Heritability of nectar traits: Why do we know so little? *Ecology*, 85, 1527, 2004.
113. Real, L.A. and Rathcke, B.J., Patterns of individual variablity in floral resources, *Ecology*, 69, 728, 1988.
114. Southwick, E.E., Photosynthate allocation to floral nectar: A neglected energy investment, *Ecology*, 65, 1775, 1984.
115. Mitchell, R.J. and Paton, D.C., Effects of nectar volume and concentration on sugar intake rates of Australian honeyeaters (Meliphagidae), *Oecologia*, 83, 238, 1990.
116. Perret, M. et al., Nectar sugar composition in relation to pollination syndromes in Sinningieae (Gesneriaceae), *Ann. Bot. (Lond.)*, 87, 267, 2001.
117. Bertsch, A., Foraging in male bumblebees (*Bombus lucorum* L.): Maximizing energy or minimizing water load? *Oecologia*, 62, 325, 1984.
118. Roces, F., Winter, Y., and Helverson, O.v., Nectar concentration preferences and water balance in a flower visiting bat, *Glossophaga soricina antillarum*, in *Animal-plant Interactions in Tropical Environments*, Barthlott, W., Ed., Museum Koenig, Bonn, 1993, p. 159.
119. Southwick, E.E. and Pimentel, D., Energy efficiency of honey production by bees, *Bioscience*, 31, 730, 1981.
120. Josens, R.B. and Farina, W.M., Selective choice of sucrose solution concentration by the hovering hawk moth *Macroglossum stellatarum*, *J. Insect Behav.*, 10, 651, 1997.
121. Kato, M., Roubik, D.W., and Inoue, T., Foraging behavior and concentration preference of male euglossine bees (Apidae: Hymenoptera), *Tropics*, 1, 259, 1992.
122. Waddington, K.D. and Kirchner, W., Acoustical and behavioral correlates of profitability of food sources in honeybee round dances, *Ethology*, 92, 1, 1992.
123. Roberts, W.M., Hummingbirds' nectar concentration preferences at low volume: The importance of time scale, *Anim. Behav.*, 52, 361, 1996.
124. Tezze, A.A. and Farina, W.M., Trophallaxis in the honeybee, *Apis mellifera*: The interaction between viscosity and sucrose concentration of the transferred solution, *Anim. Behav.*, 57, 1319, 1999.
125. Stromberg, M.R. and Johnsen, P.B., Hummingbird sweetness preferences: Taste or viscosity? *Condor*, 92, 606, 1990.
126. Pivnick, K.A. and McNeil, J.N., Effects of nectar concentration on butterfly feeding: Measured feeding rates for *Thymelicus lineola* (Lepidoptera: Hesperiidae) and a general feeding model for adult Lepidoptera, *Oecologia*, 66, 226, 1985.
127. Kevan, P.G., Sun-tracking solar furnaces in High Arctic flowers: Significance for pollination and insects, *Science*, 189, 723, 1975.
128. Seymour, R.S., White, C.R., and Gibernan, M., Heat reward for insect pollinators, *Nature*, 426, 243, 2003.

129. Borrell, B.J. and Medeiros, M.J., Thermal stability and muscle efficiency in hovering orchid bees (Apidae: Euglossini), *J. Exp. Biol.*, 207, 2925, 2004.
130. Tamm, S., Importance of energy costs in central place foraging by hummingbirds, *Ecology*, 70, 195, 1989.
131. Lotz, C.N., del Rio, C.M., and Nicolson, S.W., Hummingbirds pay a high cost for a warm drink, *J. Comp. Physiol. B*, 173, 455, 2003.
132. Lovell, J.H., The origin of anthophily among the Coleoptera, *Psyche*, 22, 67, 1915.
133. Roubik, D.W. and Hanson, P.E., *Orchid Bees of Tropical America: Biology and Field Guide*, Editorial InBio, San José, Costa Rica, 2004, 370.
134. Gess, S.K., *The Pollen Wasps: Ecology and Natural History of the Masarinae*, Harvard University Press, Cambridge, MA, 1996, p. 340.
135. Gilbert, F.S. et al., Morphological approaches to community structure in hoverflies (Diptera, Syrphidae), *Proc. R. Soc. London Series B-Biol. Sci.*, 224, 115, 1985.
136. Lindberg, A.B. and Olesen, J.M., The fragility of extreme specialization: *Passiflora mixta* and its pollinating hummingbird *Ensifera ensifera*, *J. Tropical Ecol.*, 17, 323, 2001.
137. Kristensen, N.P., Studies on the morphology and systematics of primitive Lepidoptera (Insecta), *Steenstrupia*, 10, 141, 1984.
138. Tamm, S. and Gass, C.L., Energy intake rates and nectar concentration preferences by hummingbirds, *Oecologia*, 70, 20, 1986.
139. Boggs, C.L., Rates of nectar feeding in butterflies: Effects of sex, size, age and nectar concentration, *Funct. Ecol.*, 2, 289, 1988.
140. Hainsworth, F.R., Precup, E., and Hamill, T., Feeding, energy processing rates and egg-production in painted lady butterflies, *J. Exp. Biol.*, 156, 249, 1991.
141. Stevenson, R.D., Feeding rates of the tobacco hawkmoth *Manduca sexta* at artificial flowers, *Am. Zool.*, 31, 57A, 1992.
142. Frankie, G.W. et al., Characteristics and organization of the large bee pollination system in the Costa Rican dry forest, in *Handbook of Experimental Pollination Biology*, Jones, C.E. and Little, R.J., Eds., Scientific and Academic Editions, New York, 1983, p. 411.
143. Real, L.A., Nectar availability and bee-foraging on *Ipomoea* (Convolvulaceae), *Biotropica*, 13, 64, 1981.
144. Steiner, K.E., The role of nectar and oil in the pollination of *Drymonia serrulata* (Gesneriaceae) by *Epicharis* bees (Anthophoridae) in Panama, *Biotropica*, 17, 217, 1985.
145. SanMartin-Gajardo, I. and Sazima, M., Non-euglossine bees also function as pollinators of *Sinningia* species (Gesneriaceae) in southeastern Brazil, *Plant Biol.*, 6, 506, 2004.
146. Koptur, S., Floral and extrafloral nectars of Costa Rican *Inga* trees: A comparison of their constituents and composition, *Biotropica*, 26, 276, 1994.
147. Knudsen, J.T. and Mori, S.A., Floral scents and pollination in neotropical Lecythidaceae, *Biotropica*, 28, 42, 1996.
148. Kennedy, H., *Calathea insignis*, in *Costa Rican Natural History*, Janzen, D.H., Ed., University of Chicago Press, Chicago, 1983, 204.
149. Ackerman, J.D. et al., Food-foraging behavior of male Euglossini (Hymenoptera: Apidae): Vagabonds or trapliners? *Biotropica*, 14, 241, 1982.
150. Kress, W.J., Moran, N., and Weiss, E., Ecological pressures on pollinator selection in *Calathea lutea*, in *OTS Tropical Biology: An Ecological Approach*, 1978, 425.

151. Harder, L.D. and Cruzan, M.B., An evaluation of the physiological and evolutionary influences of inflorescence size and flower depth on nectar production, *Funct. Ecol.*, 4, 559, 1990.
152. Tschapka, M., Energy density patterns of nectar resources permit coexistence within a guild of neotropical flower-visiting bats, *J. Zool.*, 263, 7, 2004.

10 Biomechanics and Behavioral Mimicry in Insects

Yvonne Golding and Roland Ennos

CONTENTS

10.1 INTRODUCTION

10.1.1 BATESIAN MIMICRY

Henry Walter Bates [1,2] was the first person to articulate a theory of mimicry from his detailed observations of insects in the Brazilian rainforest. While watching a day-flying moth mimicking a wasp, he wrote "the imitation is intended to protect the otherwise defenceless insect by deceiving insectivorous animals, which persecute the moth, but avoid the wasp." Bates applied this idea to his studies of ithomiine butterflies that exhibit red, yellow, and black aposematic coloration and pierid butterflies (Dismorphiinae); pierids are normally white or yellow, but some species, although palatable, exhibit the same warning coloration as the heliconiids. This has become known as *Batesian mimicry* and is generally defined as *the resemblance of a palatable animal (a mimic) to a distasteful or otherwise protected animal (a model) so that a predator is deceived or confused and protection is gained by the mimic* [3]. A mimic may employ visual, auditory, olfactory, or behavioral cues to aid in the deception or confusion that relies on the predator having already sampled the model and learned from the experience. The mimicry is most effective when (1) the mimic is rarer than the model, thereby increasing the chance that the model will be sampled more often than the mimic and when (2) the mimicry is accurate. However, there is some evidence that even very common and poor Batesian mimics may gain some protection by their mimicry [4,5]. Batesian mimicry has been described in vertebrates [6], invertebrates [7], and plants [8], but the overwhelming majority are described in tropical insects.

10.1.2 Müllerian Mimicry

Müllerian mimicry [9] differs from Batesian mimicry because it involves several organisms that are all toxic, distasteful, or protected to some degree, and resemble one another so that a predator avoids all of them. Consequently the mimicry is most effective when the component species are numerous. They are usually related species belonging to a broad taxonomic group, e.g., heliconiid butterflies or social wasps, unlike many Batesian mimics whose models can belong to different taxa, e.g., dipterans or coleopterans mimicking hymenopterans, or in the case of butterflies, belonging to two distinct families. In determining which type of mimicry we are dealing with, it is essential to determine the palatability status of a potential mimic. This can be problematic as suggested by Brower [10] and can lead to incorrect assumptions. Ritland [11] challenged a classical example of Batesian mimicry in temperate zone butterflies; the Florida viceroy butterfly (*Limenitis archippus floridensis*) was frequently quoted as being a Batesian mimic of the Florida queen (*Danaeus gilippus berenice*). However, experiments showed that both species were unpalatable, suggesting they were Müllerian mimics.

Batesian and Müllerian mimicry are fundamentally different; in Batesian mimicry deception is involved and the mimic benefits potentially at the expense of the predator and the model, but in Müllerian mimicry, all three species benefit. Therefore, as Fisher [12] first argued, natural selection would be expected to favor quite different adaptive strategies. It has been argued by some [13,14] that Müllerian mimicry cannot be regarded as a true type of mimicry because there is no deception involved.

10.2 MORPHOLOGICAL MIMICRY

Many accounts of mimicry in insects have concentrated on the morphological similarities, particularly the evolution of warning colors by palatable mimetic organisms to resemble their unpalatable or protected models with aposematic coloration. There are many well-studied examples, particularly in butterflies, which display their warning coloration on their large, conspicuous wings. The well-documented geographical correlations in color pattern between model and mimic species described by Bates and well illustrated by Moulton [15] have since been explored from a genetic perspective [16–20], and the potential for birds to act as selective agents of prey coloration and pattern, as suggested by Carpenter [21], has been verified experimentally for captive birds [4,22] and wild birds [23]. Many moths that have black and yellow banding on their body appear to be Batesian mimics of wasps [24].

Similarly, some beetles display black and yellow banding on the elytra [3]. Morphological mimicry in insects occurs widely throughout the tropics, but there are good examples occurring in temperate areas as well. Dipterans, including some asilids, conopids, tachinids, bombyliids, and most notably syrphids (hoverflies), mimic solitary and social wasps, honeybees, and bumblebees. The morphological similarities, particularly in color and markings, have been well-documented [25–28], and the abundance, distribution, and phenology of model and mimic species have also been intensively studied [29–31].

10.3 BEHAVIORAL MIMICRY

It has been widely acknowledged that behavior plays a major role in mimicry. Members of mimicry complexes dominated by unpalatable neotropical butterflies have been found to roost at similar heights in the canopy to their comimics [32] and to utilize host plants at similar heights [33]. It is also generally accepted that prey that are unprofitable because they are poisonous or unpalatable often exhibit slow and predictable movement; there is no selection pressure on them to adopt rapid movement to escape predators. Rather, conversely, there is selection pressure on such prey to advertise their defenses [34]. Bates [1] was probably the first person to observe that unpalatable or noxious butterflies flew slowly and deliberately, so that their warning coloration was easily visible, whereas palatable ones flew faster and more erratically. Aposematic beetles also adopt slow, sluggish behavior whereas palatable ones run quickly to avoid predatory birds [35].

On the other hand, prey may be unprofitable because they are simply hard to catch. Humphries and Driver [36] suggested that certain erratic behaviors shown by some prey animals when attacked by a predator were not accidental but specifically evolved as antipredator devices, confusing or disorientating the predator and thus increasing the prey's reaction time. Such behaviors, which seem to have no obvious aerodynamic or physiological function, appear highly erratic and include zigzagging, looping, and spinning. Driver and Humphries suggested this occurs in a wide range of animals, calling it *protean behavior* [37]. Examples include noctuid and geometrid moths, which show a bewildering range of seemingly unorientated maneuvers when exposed to the ultrasonics of hunting bats, a behavior that confers a 40% selective advantage for the moths [38]. Driver and Humphries [37] suggested that the behavior is advertising that the prey is difficult to catch and therefore unprofitable. This seems to suggest that a predator might not bother to attack, or another explanation is that such behavior could result in confusion, delaying an attack by a predator and allowing the prey to escape. Certainly erratic behavior is commonly observed in many insects including moths, orthopterans, dipterans, hemipterans, and homopterans [37], and Marden and Chai [39] described uncharacteristic upward movements shown by butterflies escaping predation.

Animals may even evolve morphological signals to reinforce or replace their behavioral ones that indicate they are hard to catch, which would help dissuade predators from attacking them. Of course, possession by one species of these signals can then lead to the evolution of such signals in other species, to produce what Srygley [40] has termed *escape mimicry*. There do not appear to be any clear examples of escape mimicry involving two or more species that are all hard to catch, though several candidates among tropical butterflies have been put forward [40]. However, an example in which an easy-to-catch mimic resembles a hard-to-catch model was given by Hespenheide [41]. He described an unusual and novel case of mimicry in which a group of Central American beetles, mostly weevils from the subfamily Zygopinae, mimic agile flies, notably robust-bodied species, such as tachinids, muscids, and tabanids. The weevils share common color patterns with the flies, which are unlike those of other beetles, none of which is considered distasteful. The weevils and flies share a behavioral characteristic that puts them in close association spatially; most perch on the same relatively isolated and exposed tree boles at midelevation in the canopy. Hespenheide [41] estimated that flies accounted for between 65 to 70% of flying insects in the area, and yet work carried out on the diet of neotropical birds found that flies, particularly robust-bodied species, formed a very small proportion of their diet. Hespenheide hypothesized, therefore, that the mimicry was based not on distastefulness but on the speed and maneuverability of the flies, which advertise they are difficult to catch. Gibson [42,43], in a series of experiments on captive birds, showed that escape mimicry is potentially plausible since over a period of several days, two species of birds both learned to avoid models of evasive prey and were also confused by escape mimics. However, Brower [10] wrote that erratic flight as an aversion tactic employed by insects and their Batesian mimics is unlikely to result in long-term learning by a predator, and so he was skeptical that such escape mimicry could evolve.

10.3.1 Behavioral Mimicry in Insects

Rettenmeyer [7] predicted that behavioral mimicry would be especially important among mimics of Hymenoptera, notably wasps and ants, because the behavior of their models is so conspicuous. Many mimetic invertebrates use *behavioral cues* to enhance their mimicry, and this is particularly remarkable when mimics and models are not closely related and have quite different morphologies. Good examples, which are included here because they mimic insects although they are not themselves insects, are ant-mimicking spiders, notably salticids of the genus *Myrmarachne* (Salticidae). They bring their front legs forward and wave them about to mimic the long antennae of ants (Hymenoptera: Formicidae) and thus also give the impression of having just six legs [3,13,44]. However, when alarmed, the spiders run off on all eight legs, so they retain full function of their front legs. The evolution of precise antlike behaviors in myrmecomorphic species might be predicted given that behavior is often identified as the most conspicuous feature of ants [7]. Some hoverflies (Diptera: Syrphidae), which mostly possess quite short antennae, also mimic the long antennae of social wasps (Hymenoptera: Vespidae) by bringing their front legs forward [45] while others (*Eristalis tenax*) mimic honeybees (Hymenoptera: Apidae) by dangling their legs in flight above flowers as if they are transferring pollen into pollen baskets (personal observation). Another example of leg-dangling behavior is shown by a syntomid moth (*Macrocneme*), mimicking the habit of its fossorial wasp model [46]. Carpenter [35] cited examples of flies that mimic the antennal behavior shown by stinging hymenoptera; they do this by waving the anterior pair of legs.

He suggested that the vibrating of antennae is part of an advertisement of aposematism by stinging insects. Some hymenopterans have white markings on the antennae that further highlight this warning behavior, and this is mimicked by syntomid moths. Cott [46] also described many examples of mimics that, when captured, behave as if they are likely to sting by curving the abdomen; these include forest dragonflies (*Microstigma maculatum*); a moth belonging to the genus *Phaegoptera*; a staphylinid beetle (*Xanthopugus*); and a longicorn beetle (*Dirphya*). The last example was described in detail by Carpenter and Poulton [47]; they commented that it was so impressive, they were reticent to handle the beetle!

10.3.2 Mimicry in Terrestrial Locomotion

Locomotory behavior is particularly effective in fooling or confusing predators, and again there are many anecdotal descriptions of across-taxa similarities. For example, although they look quite dissimilar, the Brazilian long-horned grasshopper *Scaphura nigra* mimics the fossorial wasp *Pepsis sapphires*; they both have the habit of running short distances with expanded wings [46]. The wasp adopts this behavior when hunting, but it is uncharacteristic behavior for a grasshopper. The grasshopper also has antennae modified to make them look shorter and more like those of the wasp [35]. Hoverflies of the genus *Xylota*, which resemble wasp species of the families Ichneumonidae and Pompilidae, also show similar running behavior when they are foraging on leaves; they both move in fits and starts with frequent changes in direction.

Nymphs of the bug *Hyalymenus* (Hemiptera: Alydidae) enhance their mimicry of ants by constantly agitating their antennae and adopting zigzag locomotion [48]. First instar nymphs of the stick insect *Extatasoma tiaratum* (Phasmidae) also adopt uncharacteristic behavior, running around very rapidly and looking very much like ants (personal observation).The ant-mimicking behavior of salticid spiders mentioned above, which only use six legs when running about with ants, has been the subject of some study [13], but the kinematics of the leg movements has not been investigated; it would be interesting to see if the gait of the two organisms is similar. Both clubionid and salticid spiders adopt a zigzag running gait to supplement their antennal deception [49], and some myrmecomorphic jumping spiders show a reluctance to jump unless seriously threatened [50]. Others show more specific mimetic behavior: *Synemosyna* spp. tend to walk on the outer edge of leaves like its model species of the genus *Pseudomyrmex* [51].

Wickler [13] described an example of superb morphological mimicry between a grasshopper and two beetles that requires the grasshopper to occupy two different niches at different stages in its life. *Tricondyla*, a genus of tiger beetles of varying size and with a powerful bite, scurry about on the forest floor in Borneo. The grasshopper *Condylodera trichondyloides* occurs in the same locations and looks very much like *Tricondyla* even in its mode of running. It seems to have compromised its jumping ability by evolving shorter hind legs, though once again the kinematics of its locomotion and the gait it adopts have not been investigated. Beetles pupate and thus do not alter their size in adulthood unlike grasshoppers, which pass through a number of moults, increasing in size each time. Young *Condylodera* grasshoppers are smaller than their model tiger beetle and do not live on the forest floor. Instead they live in the canopy, occurring in tree flowers along with another beetle *Collyris sarawakensis* that they resemble in size and color. So for many Batesian mimics, it is important that they are in the right place at the right time.

Many lycid beetles (Coleoptera: Lycidae) are considered to be models for Batesian mimics (although see the comments by Brower [10], who challenged the evidence for unpalatability). These have aposematic coloration on the elytra, gregarious habits, and sluggish behavior [52]. Two European wasp-beetles, *Strangalia* spp. and *Clytus arietis* are black and yellow, suggesting that they mimic social wasps. *Strangalia* is similar in color to a wasp but moves in a characteristically slow beetlelike way whereas *Clytus* differs in not having such a close resemblance to a wasp and adopts uncharacteristic active, jerky movements that are thought to resemble hunting wasps, suggesting that it is a Batesian mimic [3]. It may be that the mimic that is less convincing in terms of appearance is enhancing its mimicry by adopting wasplike behavior whereas the more-convincing mimic does not need to because it is convincingly unprofitable. The idea that "poor" mimics may enhance their mimicry by adapting their behavior has also been suggested for some hoverflies that mimic honeybees [53,54].

10.3.3 Flight Mimicry

Insect flight has been widely studied: Dudley [55] reviewed the biomechanics, Taylor [56] examined the control of insect flight, and Land [57] reviewed the visual control. Flight behavior of insects is commonly cited in early studies as being mimetic though these references are often anecdotal. For example, Opler [58] carried out an extensive study of the neotropical neuropteran *Climaciella brunnea* in Costa Rica. After studying their palatability, distribution, and markings, he concluded that five morphs of this harmless species were Batesian mimics of different species of polistine wasps. This was based on the wasps' palatability, distributions, and markings. However, he also suggested that their body posture and flight characteristics, which presumably were similar to those of the hymenopterans, were also evidence of the mimicry. Opler did not elaborate on the flight behavior, and so this is clearly an interesting research opportunity.

There are many other examples of anecdotal references to mimetic flight behavior. Dressler [59] described a Müllerian mimicry complex in bees of the genus *Eulaema* and in passing mentions two asilid flies that "in flight at least, mimic *Eulaema* quite accurately," thus suggesting that they are Batesian mimics. The implication is that they are not particularly similar in their morphology.

Carpenter [35] observed longicorn beetles that mimic wasps of the family Braconidae, describing them as indistinguishable in flight. Other coleopterans show remarkable flight adaptations. The fore wings of beetles are hardened into protective cases, the elytra. In flight the elytra are normally held out to the sides with the beetle relying on rapid beating of the hind wings for propulsion. *Acmaeodera* species, however, fly with the elytra lying in place over the abdomen, so that the warning coloration is still visible, giving the impression of a hymenopteran in flight. To achieve this, the elytra have evolved a special modification ("emargination") that allows free motion of the hind wings at the base. This uncharacteristic beetle behavior is clearly a mimetic adaptation, and because captive birds readily eat the beetle, it is undoubtedly a Batesian mimic [60].

However, in only two groups of insects has flight mimicry been studied in any detail and with any attempt at quantification: (1) the Müllerian mimicry of tropical butterflies and their Batesian mimics, and (2) the Batesian mimicry by hoverflies of the family Syrphidae of various members of the hymenoptera.

10.3.3.1 The Mimetic Flight Behavior of Butterflies

Much of the modern work on flight mimicry in tropical butterflies has been carried out in Central America starting with Chai's [22] study of butterfly predation by the specialist feeder, the rufous-tailed jacamar (*Galbula ruficauda*). These birds feed exclusively on flying insects; in fact, they do not recognize prey that does not fly. Chai established that captive jacamars fed on a wide range of butterflies and that they could distinguish between palatable species (*Papilio*; *Morpho*; Charaxinae; Brassolinae; Satyrinae and most Nymphalinae) and the unpalatable species (*Battus* and *Parides* [Papilionidae]; *Diathreia* and *Callicore* [Nymphalinae]; Heliconiinae; Acraeinae; Ithomiinae; Danainae and some Pieridae), most of which were members of Müllerian mimicry groups. Most of these were sight rejected. The birds were so adept that they could even distinguish between the very similar color patterns of some Batesian mimics and their models, although some mimics such as *Papilio anchisiades* were never taken by jacamars in feeding trials. Chai suggested that the birds made these assessments based both on the warning coloration on the wings and on the flight behavior of the butterflies. He stated that many unpalatable butterflies flew with "slow and fluttering wingbeats" and in a "regular path" that would enable them to display their warning colors. The flight of the Batesian mimics was described as being similar to that of their models, whereas other palatable butterflies flew faster and more erratically, making them harder to catch. Importantly, Chai established that jacamars could memorize the palatability of a large variety of butterflies, suggesting that insectivorous birds, such as jacamars, were likely to play a major role in the evolution of neotropical butterfly mimicry. Interestingly, Kassarov [61] suggested that the aerial hawking birds recognize butterflies by their flight pattern rather than by details of aposematic or mimetic coloration. Evidence has been presented that insectivorous birds can perceive motion two to four times faster than humans [62]; they also have superior color vision and are able to detect UV markings that humans cannot see [63].

The qualitative flight observations were later backed up by more quantitative studies of the flight behavior, body temperature, and body morphology of the butterflies [64–66]. Unpalatable Müllerian species did indeed fly more slowly and more regularly than palatable species when filmed flying in an insectary [64]. They were also able to fly at lower ambient temperatures and had lower thoracic temperatures when caught. Srygley and Chai [65] suggested that these differences could also be related to the contrasting body morphology of the two groups. The palatable, fast-flying butterflies had relatively wider thoraxes that could house the more massive flight muscles they would need for fast speed flight and rapid acceleration. To achieve this, however, they would have to have higher thoracic temperatures and so would be restricted to flying in warmer ambient temperatures. In contrast the unpalatable butterflies, with their slower, more economical flight, could fly at lower temperatures and divert more of their resources into a larger abdomen. This idea is incompatible, however, with a later idea of Srygley [67] about noncheatable signals. Here he suggested that flight mimicry might actually impose an aerodynamic cost, so that Müllerian and Batesian mimics would need to develop *more* power to fly than palatable butterflies. To test this idea, Srygley analyzed films of the flight of two species of palatable butterfly, four species of Müllerian mimics, and two Batesian mimics, and calculated the power required using the quasisteady analysis method of Ellington [68]. The Batesian mimics did have higher weight-specific power, but the power requirements of the other two groups showed extensive overlap. Theoretically, it seems unlikely in any case that an unpalatable butterfly would choose to fly in an uneconomical way. The matter is complicated because butterflies, like other insects, make extensive use of nonsteady aerodynamics [69], which will affect their power requirements. Clearly more research using additional species and examining the actual oxygen uptake of the insects rather than modeling the power is needed to settle this matter.

The morphological differences between palatable and unpalatable butterflies and their consequences were later examined in more detail [66,70]. These studies showed that in unpalatable Müllerian mimics, the wing center of mass was further from the body and the center of mass of the body was further behind the base of the wings than in palatable species. Both of these would make the insect less maneuverable, but give it smoother flight because of the increased moment of inertia [68], whereas the reverse was true for the palatable species. Srygley [10,70] therefore suggested that similarities in morphology in these insects lead almost automatically to "locomotor" mimicry in which "adaptive convergence of physiological and morphological features result in similar flight biomechanics and behaviour." In the Batesian mimic *Consul fabius*, although the center of mass of the wing was far from the body, as in the unpalatable Müllerian mimics, the center of mass of the body was near the wing base, as in other palatable species. Srygley [70] suggested that this intermediate morphology would enable this butterfly to fly rapidly and unpredictably if disturbed, like other palatable species, so it could escape if its mimicry proved unsuccessful.

In this context, it is interesting that the majority of Batesian mimetic butterflies are female [1,71], which typically have larger abdomens for the development of eggs and, as a consequence, relatively smaller thoraxes than males [64,72]. This would make them less maneuverable and hence more vulnerable to predation [55]. Since females are also longer-lived, there would thus be strong selection pressure for them to evolve the warning coloration of unpalatable butterflies to gain greater protection from birds. Females would also be more easily able to adopt the slow, regular flight of these species. Confirming this supposition, Ohsaki [73] found that in general female butterflies were attacked more frequently than males but that Batesian mimetic females (protected by their mimicry), males (protected by fast erratic flight if palatable), and the unpalatable models were attacked less than nonmimetic females. Ohsaki suggested that when the predation by avian predators is female biased, female-limited mimicry will be favored even if the costs of mimicry are the same for both sexes. The case is different in Müllerian mimicry in which unpalatable species resemble each other; all individuals of both sexes will usually become mimetic.

The locomotor mimicry that Srygley described has been most clearly demonstrated in studies of the flight kinematics of four mimetic butterflies of the genus *Heliconius* [74–76]. These make up two Müllerian mimicry pairs in which each species is more closely related to a member of the other mimicry pair than to the insect it mimics. Using multivariate statistics, Srygley [76] was able to separate the effects of evolutionary convergence from those of phylogeny. It was found that the mimics were more similar to each other than to the other closely related species in their wing-beat frequency, the degree of asymmetry in wing motion, and in their transport costs. It has been suggested that this is the first clear example of a mimetic behavioral signal for a flying insect [74–76]. However, since the morphological mimetic signal is displayed by the organs of locomotion, the wings, it might in any case be expected that flight mimicry would occur if the wings showed convergence in form; this would greatly constrain the kinematics and aerodynamics. It should also be emphasized that the research on flight mimicry in butterflies has been based on the analysis of a very few flights made often by only a single member of particular species. The films, moreover, are of captive individuals flying in artificial conditions. More film of free flight in the field might help show other more subtle aspects of flight mimicry and flight behavior.

10.3.3.2 The Mimetic Flight Behavior of Hoverflies

Most examples of mimetic insects occur in the tropics, but temperate Europe and United States. are home to many species of hoverflies (Diptera: Syrphidae) that are thought to be Batesian mimics of wasps and bees [28,77]. Some are black and yellow or red, resembling social and solitary wasps; others are large and hairy, resembling bumblebees (some are polymorphic, mimicking different species); while others, notably droneflies of the genus *Eristalis*, resemble honeybees. Some hoverfly mimics appear to closely resemble their models in morphology, while others are only superficially similar. Dipterans and hymenopterans, although both flying insects, have quite different ecologies; hymenopterans are often social insects that forage on flowers for nectar and pollen for the colony, or in the case of solitary species, for provisioning their nest. Hoverflies are always solitary animals that do not exhibit parental care. However, both spend much of their time foraging on flowers, during which time they are particularly obvious and vulnerable to predation by birds [78]. In a study of foraging behavior, Golding and Edmunds [53] found that droneflies often spent a similar amount of time as their honeybee models, both feeding on individual flowers and flying between them, when foraging on the same patch. Because they are seeking different rewards from the flowers and in different quantities, the most likely explanation is that this is a case of behavioral mimicry; the hoverflies, which are unprotected insects, are adapting their behavior to appear more like their model hymenopterans.

There have also been many suggestions that hoverflies show flight mimicry of hymenopterans. Different species have been referred to anecdotally as having beelike flight [13], bumblebeelike flight [79], and lazy wasplike flight [80]. One of us (Golding) has also observed the hoverfly *Xanthogramma pedissequum*, a black and yellow wasp mimic, adopting an uncharacteristic flight behavior; it flew about 30 cm above low-growing vegetation in a zigzag fashion very similar to the behavior of a hunting wasp (personal observation). Morgan and Heinrich [81] observed that the mimicry of many of the hoverflies they studied appeared most accurate in flight. They also showed that hoverflies (including *Eristalis*) were able to warm up using behavior such as basking or shivering, which they suggested might allow them to behave more like their endothermic models. There have been few studies, however, that have empirically measured or formatively studied these behaviors, even though the flights of hoverflies and bees have both been well studied.

The aerodynamics of both groups has been elegantly elucidated by Ellington [68]; the flight mechanism, wing design, and kinematics of hoverflies has been investigated by Ennos [82–84]; and other behavioral aspects, such as the mechanism by which hoverflies compute interception courses and manage to return to exactly the same spot, have been studied by Collett and Land [85,86]. Any flight mimicry between the two groups must be quite unlike the locomotor mimicry between butterflies. For a start, the warning coloration of Hymenoptera and their hoverfly mimics is displayed on the abdomen and thorax, not the wings. Second, the flight apparatus of hoverflies and Hymenoptera are quite different. Hoverflies have two wings and twist them on the upstroke in the manner described by Ennos [83]. In contrast, the Hymenoptera have four coupled wings, with positive camber at the base, which are twisted in the upstroke by the same mechanism as in butterflies [87] (personal observation). Therefore there is no possibility of convergence in their mechanics. Furthermore, convergence in wingbeat frequency cannot be involved in the mimesis because the wingbeat frequency of both groups is far too high at 150 to 250 Hz for predatory birds to detect or even for them to be able to see the wings in flight. These wingbeat frequencies also overlap extensively with each other and with those of other insects that use asynchronous muscles [68,84,88].

Hoverflies are generally regarded as having superior flight agility compared with hymenopterans because their center of body mass (CM_{body}) is closer to their wing base [68]; they use inclined stroke plane hovering; and they have the apparent ability to move the aerodynamic force vector independently of the stroke plane [84]. Therefore one would expect any flight mimicry to involve the body movements of the insects, not the wings, and that hoverflies would have to compromise their flight ability when foraging to appear more like a hymenopteran.

In the first quantitative study of flight mimicry in the group, Golding et al. [54] examined the flight of hoverflies of the genus *Eristalis*, which are known as droneflies and are considered to be Batesian mimics of honeybees (*Apis mellifera*). They are of similar overall shape and body mass, although female *Eristalis* spp. tend to be slightly larger than males. In appearance droneflies differ from honeybees in having shorter antennae, no discernable “waist,” one pair of wings, and often more orange or yellow markings on the abdomen. Filming from above a patch of flowers in the field, Golding measured the horizontal flight velocities and routes taken by insects free flying between individual flowers when foraging. She compared *E. tenax* with *A. mellifera* along with a control hoverfly (*Syrphus ribesii*) and a nonmimetic muscid fly. It was found that droneflies did indeed show more similar flight movements to the honeybee than to the other two insects [54]. The muscid flew faster than the other three species and took more direct routes between flowers. The control hoverfly flew at similar speeds to the dronefly and honeybee, but took longer to fly between the flowers because it took more convoluted routes and hovered more. The dronefly flew at similar speeds to the honeybee, took similarly convoluted routes, and hovered for similar amounts of time.

The dronefly also performed loops along the flight path similar to those performed by the honeybee. This behavior was not detectable to the human eye, but it may well be to a predatory bird because they can detect motion two to four times faster than humans [76]. The looped flight of honeybees may be connected with their ability to orientate their position in relation to the hive using the sun, as described by von Frisch [89]. It is surprising behavior for droneflies, however, as they are capable of sudden changes of direction without altering their body position; they can perform turns of over 90 degrees while traveling less than one body length [68,84]. The most likely explanation of the looping flight of foraging droneflies is that their flight behavior has been modified to be more similar to that of their hymenopteran model. This cannot be classed as locomotory mimicry as defined by Srygley [70] in his studies of more closely related butterflies because the organisms have such different flight apparatus, but it could accurately be described as *mimetic flight behavior*. Closer examination of these maneuvers using high speed cinematography might help determine whether the aerodynamic mechanisms used by the two species are the same. However, the droneflies have still retained the ability for fast accurate flight to escape predation and in males for patrolling territory and chasing females.

Golding et. al. (in preparation) are continuing with this work on other mimicry groups. The most recent results are from a study of flights between flowers made by social wasps (*Vespula vulgaris*) and four of their hoverfly mimics (*Sericomyia silentis*; *Myathropa florea*, *Helophilus pendulus*, and *Syrphus ribesii*). *Sericomyia silentis* is a large fly similar in size to *V. vulgaris*, and they occur together during late summer; *Sericomyia silentis* is a conspicuous, bright yellow and black species. The other three yellow and black species are smaller than wasps but are similar in size to each other although *M. florea* can be slightly larger. *H. pendulus* and *M. florea* have more elaborate markings, both on thorax and abdomen than *Syrphus ribesii*. To the human eye, *Sericomyia silentis* appears to be the best mimic and *Syrphus ribesii*, the poorest with *H. pendulus* and *M. florae* midway between the two. It might be expected therefore that *Sericomyia silentis* would show the most similar behavior to its hymenopteran model. In contrast, preliminary results from analysis of 115 flights performed by 53 individuals from the 5 species seem to be showing the opposite. *Syrphus ribesii* flies at similar speeds to wasps and has comparable flight trajectories; it takes similar, more convoluted routes as wasps and flies relatively slowly between flowers. *Sericomyia silentis, M. florae,* and *H. pendulus* have similar speeds and flight trajectories to each other; they fly straighter and faster (Figure 10.1).

An interpretation of these results is that *Syrphus ribesii* has to compensate for its poor morphological mimicry by showing better behavioral mimicry. It adopts

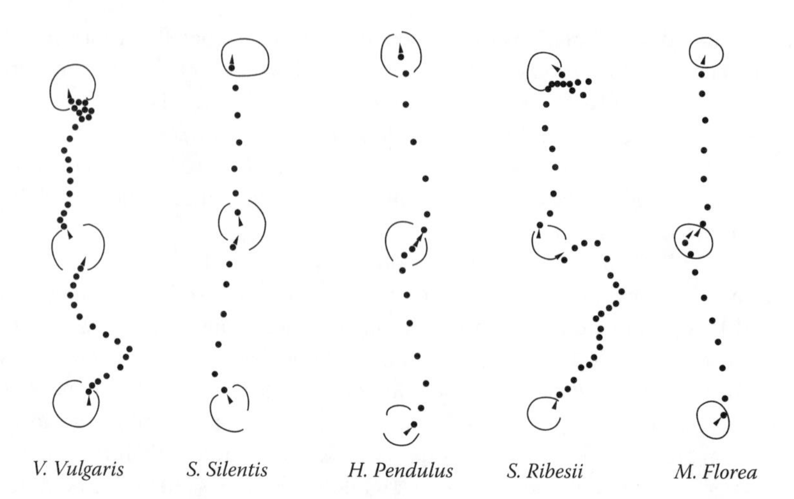

FIGURE 10.1 Typical flight trajectories between flowers of a wasp *Vespula vulgaris* and its potential mimics: *Serricomyia silentis*, *Helophilus pendulus, Syrphus ribesii*, and *Myathropa florea*. Flowers are approximately 10 cm apart, and the time between points is 0.04 sec.

slow, convoluted flight to mimic wasp flight even though it is capable of fast flight. The other species are more secure in their morphological mimicry and so do not attempt to behave like a wasp. There is some evidence for this hypothesis; if *Sericomyia silentis* is threatened, it adopts a characteristic wasplike zigzag flight pattern whereas *Syrphus ribesii* flies off rapidly (personal observation). These results support the hypothesis that poor visual mimics adapt their behavior to be more like their models but retain their ability for rapid escape flight. Better visual mimics may only use behavioral mimicry when they perceive danger. There is ongoing work that looks at another aspect of flight mimicry in hoverflies: the flight trajectories of bumblebees, which often have a particularly clumsy flight, and their hoverfly mimics, especially their movements in the vertical plane.

10.4 CONCLUSION

Like Driver and Humphries [37] in their seminal work on protean behavior, we make no apology for having delved back into nineteenth and early twentieth century descriptions of mimicry and animal behavior. The reasons are obvious: Mimicry has been known about for at least 150 years, and there were many early papers reporting much work, discussion, speculation, and argument about the subject. In the context of this article, there are many observations by early entomologists, notably Bates, Carpenter, Poulton, and Shelford, that emphasized that behavior is just as important as morphology for successful mimetic deception. However, as we have seen, many aspects of mimicry, particularly those relating to behavior, have only been recorded as anecdotal observations and have remained largely unstudied on a quantitative basis. Moreover, the interrelations between behavior, locomotion, and mimicry have barely been addressed.

Of course quantifying behavior and movement is not particularly easy, but it is surprising that almost no attempt has been made to quantify the running and walking movements and gaits of Hymenoptera such as ants and wasps and their Batesian mimics from other insect orders. Because land locomotion is by its nature carried out largely in two dimensions, filming and quantifying movement should present little problem. Of the quantitative research that has been carried out on the more complex subject of flight in butterflies and hoverflies, analysis has largely been confined to examining flight in two dimensions. Complex three-dimensional movement has not been tackled because of its technical difficulty. Neither has much effort been put into examining the aerodynamic basis of flight mimicry. Closer examination of the movements of the body and wings during maneuvers by mimetic and nonmimetic insects would help to determine whether mimics show convergence to their models in aerodynamics.

There are some fascinating examples of behavioral mimicry to study, so if anyone is looking for a new line of research, there is plenty of scope!

REFERENCES

1. Bates, H.W., Contributions to an insect fauna of the Amazon Valley, Lepidoptera: Heliconidae, *Trans. Linn. Soc. Lond.,* 23, 495, 1862.
2. Bates, H.W., Contributions to an insect fauna of the Amazon Valley, Coleoptera: Longicorns, *Ann. Mag. Nat. Hist.*, 9, 446, 1862.
3. Edmunds, M., *Defence in Animals*, Longman, Harlow, 1974.
4. Brower, J.V.Z., Experimental studies of mimicry IV. The reaction of starlings to different proportions of models and mimics, *Am. Nat.*, 94, 271, 1960.
5. Edmunds, M., Why are there good and poor mimics? *Biol. J. Linn. Soc.,* 70, 459, 2000.
6. Pough, F.H., Mimicry of vertebrates: are the rules different? *Am. Nat.*, 131, 67, 1988.
7. Rettenmeyer, C.W., Insect mimicry, *Ann. Rev. Ent.*, 15, 43, 1970.
8. Weins, D., Mimicry in plants, *Evol. Biol.*, 11, 365, 1978.
9. Müller, F., A remarkable case of mimicry in butterflies, *Proc. R. Ent. Soc. Lond.,* 1879, 20, 1879.
10. Brower, A.V.Z., Locomotor mimicry in butterflies? A critical review of the evidence, *Phil. Trans. R. Soc. Lond. B.*, 347, 413, 1995.
11. Ritland, D.B., Revising a classic butterfly mimicry scenario: demonstration of Müllerian mimicry between Florida viceroys (*Limenitis archippus floridensis*) and Florida queens (*Danaus gilippus berenice*), *Evolution,* 45, 918, 1991.
12. Fisher, R.A., *The Genetical Theory of Natural Selection,* Clarendon Press, Oxford, 1930.
13. Wickler, W., *Mimicry in Plants and Animals,* World University Library, London, 1968.
14. Pasteur, G., A classificatory review of mimicry systems, *Ann. Rev. Ecol. Syst.,* 13, 169, 1982.
15. Moulton, J.C., On some of the principal mimetic combinations of tropical American butterflies, *Trans. Ent. Soc. Lond.*, 1909, 585, 1909.
16. Shepherd, P.M., Genetics of mimicry, *Proceedings of the XVI International Congress of Zoology,* Natural History Press, Garden City, N.Y., 1963.

17. Turner, J.R.G., Evolution of complex polymorphism and mimicry in distasteful South American butterflies, *Proc. Int. Congr. Ent.,* 12, 267, 1965.
18. Owen, D., Mimetic polymorphism and the palatability spectrum, *Oikos,* 21, 333, 1970.
19. Smith, D.A.S. and Gordon, I.J., The genetics of the butterfly *Hypolimnas misippus* (L.) — The classification of phenotypes and the inheritance of forms *misippus* and *inaria*, *Heredity,* 59, 467, 1987.
20. Smith, D.A.S. et al., Genetics of the butterfly *Danaus chrysippus* (L.) in a broad hybrid zone, with special reference to sex ratio, polymorphism and intragenomic conflict, *Biol. J. Linn. Soc.*, 65, 1, 1998.
21. Carpenter, G.D.H., Observations and experiments in Africa by the late C.F.M. Swynnerton; non-wild birds eating butterflies and the preference shown, *Proc. Linn. Soc. Lond.*, 10, 1942.
22. Chai, P., Field observations and feeding experiments on the responses of rufous-tailed jacamars (*Galbula ruficauda*) to free-flying butterflies in a tropical rainforest, *Biol. J. Linn. Soc.*, 29, 161, 1986.
23. Jeffords, M.R., Sternburg, J.G., and Waldbauer, G.B., Batesian mimicry: field demonstration of the survival value of pipevine swallowtail and monarch colour patterns, *Evolution,* 33, 275, 1979.
24. Robinson, G.S. and Carter, D.J., The first vespiforme tineid moth (Lepidoptera: Tineidae), *Syst. Ent.*, 14, 259, 1989.
25. Heal, J.R., Colour patterns in Syrphidae II. *Eristalis intricarius*, *Heredity,* 43, 229, 1979.
26. Heal, J.R., Colour patterns in Syrphidae III. Sexual dimorphism in *Eristalis arbustorum*. *Ecol. Ent.* 6, 119, 1981.
27. Stubbs, A.E. and Falk, S.J., *British Hoverflies*, The British Entomological and Natural History Society, London, 1983.
28. Howarth, B., Clee, C., and Edmunds, M., The mimicry between British Syrphidae (Diptera) and aculeate Hymenoptera, *Br. J. Ent. Nat. Hist.*, 13, 1, 2000.
29. Walbauer, G.P., Asynchrony between Batesian mimics and their models, *Am. Nat.*, 131, 103, 1988.
30. Howarth, B., and Edmunds, M., The phenology of Syrphidae (Diptera): are they Batesian mimics of Hymenoptera? *Biol. J. Linn. Soc.*, 71, 437, 2000.
31. Howarth, B., Edmunds, M. and Gilbert, F., Does the abundance of hoverfly mimics (Syrphidae) depend on the numbers of their hymenopteran models? *Evolution,* 58, 367, 2004.
32. Mallet, J. and Gilbert, L.E., Why are there so many mimicry rings? Correlations between habitat, behaviour and mimicry in *Heliconius* butterflies, *Biol. J. Linn. Soc.*, 55, 59, 1995.
33. Beccaloni, G.W., Vertical stratification of ithomiine butterfly (Nymphalidae: Ithomiinae) mimicry complexes: the relationship between adult flight height and larval host-plant height, *Biol. J. Linn. Soc.*, 62, 3134, 1997.
34. Sherratt, T.N., Rashed, A., and Beatty, C.D., The evolution of locomotory behaviour in profitable and unprofitable simulated prey, *Oecologia,* 138, 143, 2004.
35. Carpenter, G.D.H. and Ford, E.B., *Mimicry,* Methuen, London, 1933.
36. Humphries, D.A. and Driver, P.M., Erratic display as a device against predator, *Science,* 156, 1767, 1967.
37. Driver, P.M. and Humphries, D.A., *Protean Behaviour: The Biology of Anarchization,* Clarendon Press, Oxford, 1988.

38. Roeder, K.D. and Treat, A.E., The acoustic detection of bats by moths, *Proc. 11th Cong. Ent.*, 3, 1, 1960.
39. Marden, J.H. and Chai, P., Aerial predation and butterfly design: how palatability, mimicry and the need for evasive flight constrain mass allocation, *Am. Nat.*, 138, 15, 1991.
40. Srygley, R.B., Incorporating motion into investigations of mimicry, *Evol. Ecol.*, 13, 691, 1999.
41. Hespenheide, H.A., Novel mimicry complex — beetles and flies, *J. Entomol. Ser. A*, 48, 49, 1973.
42. Gibson, D.O., Batesian mimicry without distastefulness, *Nature,* 250, 77, 1974.
43. Gibson, D.O., The role of escape in mimicry and polymorphism 1. The response of captive birds to artificial prey, *Biol. J. Linn. Soc.* 14, 201, 1980.
44. Oliveira, P.S., Ant-mimicry in some Brazilian salticid and clubionid spiders (Araneae: Salticidae; Clubionidae), *Biol. J. Linn. Soc.*, 33, 1, 1988.
45. Walbauer, G.P., Mimicry of hymenopteran antennae by Syrphidae, *Psyche* (Camb. Mass.), 77, 45, 1970.
46. Cott, H., *Adaptive Colouration in Animals*, Methuen, London, 1940.
47. Carpenter, G.D.H. and Poulton, E.B., *Mimicry*, Methuen, London, 1902.
48. Oliveira, P.S., On the mimetic association between nymphs of *Hyalymenus* spp. (Hemiptera, Alydidae) and ants, *Zool. J. Linn. Soc.*, 83, 371, 1985.
49. Reiskind, J., Ant-mimicry in Panamanian clubionid and salticid spiders (Araneae — Clubionidae, Salticidae), *Biotropica,* 9, 1, 1977.
50. Cushing, P.E., Myrmecomorphy and mymmecophily in spiders: a review, *Fla. Entomol.,* 80, 165, 1997.
51. McIver, J.D. and Stonedahl, G. Myrmecomorphy — morphological and behavioural mimicry of ants, *Annu. Rev. Entomol.,* 38, 351, 1993.
52. Linsley, E.G., Eisner, T., and Klots, A.B., Mimetic assemblages of sibling species of lycid beetles, *Evolution,* 15, 15, 1961.
53. Golding, Y.C. and Edmunds, M., Behavioural mimicry of honeybees (*Apis mellifera*) by droneflies (Diptera: Syrphidae: *Eristalis* spp.), *Proc. R. Soc. Lond. B.,* 267, 903, 2000.
54. Golding, Y.C., Ennos, A.R., and Edmunds, M., Similarity in flight behaviour between the honeybee *Apis mellifera* (Hymenoptera: Apidae) and its presumed mimic, the dronefly *Eristalis tenax* (Diptera: Syrphidae), *J. Exp. Biol.,* 204, 139, 2001.
55. Dudley, R., *The Biomechanics of Insect Flight*, Princeton University Press, Princeton, 2000.
56. Taylor, G.K., Mechanics and aerodynamics of insect flight control, *Biol. Rev.*, 76, 449, 2001.
57. Land, M.F., Motion and vision: why animals move their eyes, *J. Comp. Physiol. A.,* 185, 341, 1999.
58. Opler, P.A., Polymorphic mimicry of polistine wasps by a neotropical neuropteran, *Biotropica,* 13, 165, 1981.
59. Dressler, R.L., *Eulaema bombiformis, E. meriana,* and Müllerian mimicry in related species (Hymenoptera, Apidae), *Biotropica,* 11, 144, 1979.
60. Silbergleid, R.E. and Eisner, T.E., Mimicry of hymenoptera by beetles with unconventional flight, *Science,* 163, 486, 1969.
61. Kassarov, L., Are birds the primary selective force leading to evolution of mimicry and aposematism in butterflies? An opposing point of view, *Behaviour,* 140, 433, 2003.
62. Srygley, R.B. and Ellington, C.P., Discrimination of flying mimetic, passion-vine butterflies *Heliconius*, *Proc. R. Soc. Lond. B.*, 266, 2137, 1999.

63. Cuthill, I.C. and Bennett, A.T.D., Mimicry in the eye of the beholder, *Proc. R. Soc. Lond. B.*, 253, 203, 1993.
64. Chai, P. and Srygley, R., Predation and the flight, morphology, and temperature of neotropical rain-forest butterflies, *Am. Nat.*, 135, 748, 1990.
65. Srygley, R.B. and Chai, P., Flight morphology of neotropical butterflies: palatability and distribution of mass to the thorax and abdomen, *Oecologia,* 84, 491, 1990.
66. Srygley, R.B. and Dudley, R.T., Correlations of the position of center of body mass with butterfly escape tactics, *J. Exp. Biol.,* 174, 155, 1993.
67. Srygley, R.B., The aerodynamic costs of warning signals in palatable mimetic butterflies and their distasteful models, *Proc. R. Soc. Lond. B.*, 271, 589, 2004.
68. Ellington, C.P., The aerodynamics of hovering insect flight, *Phil. Trans. R. Soc. Lond. B.,* 305, 1, 1984.
69. Srygley, R.B. and Thomas A.L.R., Unconventional lift-generating mechanisms in free-flying butterflies, *Nature,* 420, 660, 2002.
70. Srygley, R.B., Locomotor mimicry in butterflies? The associations of positions of centre of mass among groups of mimetic, unprofitable prey, *Phil. Trans. R. Soc. Lond. B.*, 343, 145, 1994.
71. Wallace, A.R., On the phenomena of variation and geographical distribution as illustrated by the Papilionidae of the Malayan region, *Trans. Linn Soc. Lond.,* 24, 1, 1865.
72. Marden, J.H. and Chai, P., Aerial predation and butterfly design — how palatability, mimicry, and the need for evasive flight constrain mass allocation, *Am. Nat.* 138, 15, 1991.
73. Ohsaki, N., Preferential predation of female butterflies and the evolution of Batesian mimicry, *Nature,* 378, 173, 1995.
74. Srygley, R.B., Locomotor mimicry in *Heliconius* butterflies: contrast analyses of flight morphology and kinemetics, *Phil. Trans. R. Soc. Lond. B*, 354, 203, 1999.
75. Srygley, R.B. and Ellington, C.P., Estimating the relative fitness of local adaptive peaks: the aerodynamic costs of flight in mimetic passion-vine butterflies *Heliconius*, *Proc. R. Soc. Lond. B.*, 266, 2239, 1999.
76. Srygley, R.B. and Ellington, C.P., Discrimination of flying mimetic, passion-vine butterflies *Heliconius, R. Soc. Lond. B.*, 266, 2137, 1999.
77. Maier, C.T., Evolution of Batesian mimicry in the Syrphidae (Diptera) *J. N.Y. Entomol. S.* 86, 307, 1978.
78. Dlusski, G.M., Are dipteran insects protected by their similarity to stinging Hymenoptera? *Bull. Mosk. O-Va Ispytatelei Otd Biol.*, 89, 25, 1984.
79. Howarth, B., An ecological study of Batesian mimicry in the British Syrphidae (Diptera), Ph.D. Thesis, University of Central Lancashire, 1988.
80. Azmeh, S., Mimicry in the hoverflies, Ph.D. Thesis, University of Nottingham, 1999.
81. Morgan, K.R. and Heinrich, B., Temperature regulation in bee- and wasp-mimicking syrphid flies, *J. Exp. Biol.*, 133, 59, 1987.
82. Ennos, A.R., A comparative study of the flight mechanism of Diptera, *J. Exp. Biol.*, 127, 355, 1987.
83. Ennos, A.R., The importance of torsion in the design of insect wings, *J. Exp. Biol.*, 140, 137, 1988.
84. Ennos, A.R., The kinematics and aerodynamics of free flight of some Diptera, *J. Exp. Biol.*, 142, 49, 1989.
85. Collett, T.S. and Land, M.F., Visual spatial memory in a hoverfly, *J. Comp. Physiol.*, 100, 59, 1975.
86. Collett, T.S. and Land, M.F., How hoverflies compute interception courses, *J. Comp. Physiol.*, 125, 191, 1978.

87. Wootton, R.J., Leading edge section and asymmetric twisting in the wings of flying butterflies (Insecta, Papilionoidea), *J. Exp. Biol.,* 180, 105, 1993.
88. Unwin, D.M. and Corbet, S.A., Wingbeat frequency, temperature and body size in bees and flies, *Physiol. Ent.*, 9, 115, 1984.
89. Frisch, K. von, *The Language and Orientation of Bees,* Clarendon Press, Oxford, 1967.

11 Interindividual Variation in the Muscle Physiology of Vertebrate Ectotherms: Consequences for Behavioral and Ecological Performance

Carlos A. Navas, Rob S. James, and Robbie S. Wilson

CONTENTS

11.1 INTRODUCTION

The aim of this review is to promote the study of interindividual variation in the skeletal muscle of vertebrate ectotherms as a promising field of research integrating questions in evolution, ecology, behavior, and muscle function — levels of organization that are often addressed independently. To fulfill our goal, we offer an evolutionary background that illustrates the relevance of studies of individual variability in behavioral performance in the context of evolutionary physiology. Then, we focus on some of the best-studied examples of behavioral performance and relate this variability to underlying physiological factors, mainly morphology, muscle size, and mechanics, discussing in parallel the ecological consequences of such variation.

Most of our discussion addresses the measures of performance that are more frequently treated in the literature because of apparent ecological relevance, including sprint speed, power production, force production, efficiency, stamina (measured as either time before exhaustion or distance capacity), and maneuverability. For instance, many species of fish [1,2] and lizards [3,4] rely on fast bursts of activity to avoid predators, and faster lizards exhibit greater social dominance [5]. The efficiency of conversion of metabolic energy into locomotion has been considered fundamental in the evolution of the muscle traits of vertebrates [6] and may be an important factor modulating the evolution of locomotor physiology and behavior of vertebrate ectotherms. Force and power production are crucial to vertebrate ectotherms during a variety of behaviors, with power production being notably important for the jumping performance of anuran amphibians [7,8]. Force production in the forearms of male anurans is important for both holding females and avoiding takeovers by competitor males [9]. Maneuverability has also been considered a key aspect in antipredator behaviors of fish [1] and larval anurans [10,11], perhaps more important than sprint speed or endurance in at least some specific ecological settings.

Stamina is highly relevant for certain vertebrate ectotherms, such as juvenile *Bufo* during their dispersion phase [12] or male anurans where chorus tenure is a factor in mating success [13]. However, stamina might not be the primary factor modulating the evolution of exercise physiology in some vertebrate ectotherms such as small heliothermic lizards, in which intermittent locomotion further prolongs the ability of individuals to sustain locomotor activity [14]. Although some lizard taxa, such as the genera *Varanus* or *Cnemidophorous*, exhibit particularly high running endurance [15,16], phylogenetic analyses across lizard clades do not indicate unambiguous evolutionary shifts regarding running endurance [15]. Also, social dominance, an ecological factor that is clearly related to differences among individual lizards, does not relate to endurance in *Sceloporus occidentalis* [5].

11.2 EVOLUTIONARY IMPLICATIONS OF INDIVIDUAL VARIATION IN BEHAVIORAL PERFORMANCE AND MUSCLE PHYSIOLOGY

Whether the behavioral performance* of vertebrate ectotherms relates to underlying physiological characteristics and adaptive consequences has been a central question in evolutionary physiology. From the range of performance measures that may be investigated, studies of locomotor performance have offered particular insight into these questions. The evolution of specific traits underlying locomotor performance may occur during historical ecological transitions and are thus likely to have adaptive consequences. For example, phylogenetic comparative approaches were initially used to provide evidence for adaptive differences in locomotor performance among

* Behavioral performance admittedly involves more than active behaviors, but in this chapter we restrict the term to the quantifiable maximum performance of animals during active behaviors that involve the production of mechanical force by skeletal muscle.

lizard species [16,17]. These early patterns have been confirmed by more recent studies focusing simultaneously on sprint performance and muscle physiology. In the family Phrynosomatidae, which encompasses the genus *Sceloporus*, sand lizards (*Uta* and related lizards), and horned lizards (*Phrynosoma* and related lizards), sand lizards are the fastest in the clade and have comparatively faster glycolytic fibers [18]. Similarly, in the lizard genus *Tropidurus*, the sand dune endemic species *T. psamonastes* exhibits higher sprint speeds on sand and a higher proportion of glycolytic fibers than the sister species *T. itambere*, which is endemic to rocky outcrops [19]. Convergent evolution of performance traits suggesting adaptive trends have been reported in other taxa of ectothermic vertebrates, including the muscle design of unrelated fish taxa that are ecologically similar [20,21] or the peculiar ability of unrelated high-elevation tropical anurans to be active at low temperatures [22].

Given the observed and presumably adaptive variation in behavioral performance and muscle physiology among species, questions arise concerning the relationships between this higher-level variation and variation among individuals within a population. This is a central issue in evolutionary physiology because adaptive shifts in the behavioral performance traits of vertebrates are believed to begin within populations and have three requirements: (1) significant repeatable variation among individuals, (2) heritability of the trait, and (3) variation in the trait influencing the survival or reproductive success of the individual (i.e., fitness) [23,24]. Testing the set of requirements for the evolution of a physiological performance trait is a challenging enterprise, so most of the data available are fragmented (see next paragraph). Not all reported interindividual variation in performance is both repeatable and has a heritable basis. For example, nonheritable variation may result from health or reproductive state. Interindividual variation due to ontogenetic development might also be difficult to analyze in this context, particularly in amphibians, because it might involve heritable and environmentally induced variation. Metamorphosis involves dramatic morphological transitions, and behavioral performance ranks before and after metamorphosis may not be maintained [25,26]. Studies using adult individual vertebrate ectotherms in similar conditions of health, reproductive state, and season would be ideal to address the question of whether reproducible variability exists within populations.

Moving to the second postulate, the heritability of performance traits is assumed given interspecific adaptive trends, but direct evidence and attempts at quantification are scarce. Metabolic rates during intense exercise [24], sprint speed and endurance in lizards [27], as well as various other metabolic traits related to performance and exercise physiology, are at least partially heritable and respond to artificial selection in the laboratory [28,29]. Finally, and regarding the relationship between behavioral performance and fitness, faster garter snakes have more chance of survival than slower counterparts [30], and lizard hatchling survival is positively correlated with body size, sprint speed, and stride length, but negatively correlated with growth rate [4,31]. In amphibians, phenotypic plasticity in larval forms leads to deeper tails in environments with odonate larvae, and these morphs have increased chances of survival [32], partially due to increased swimming speed [33] but perhaps more importantly due to the deflection of attacks away from the vulnerable head region to the tail [32].

The occurrence of reproducible individual variation in the behavioral performance of vertebrates has been well documented in lizards, fish, and amphibians [26,34–39], although this literature encompasses two different approaches to the question. On the one hand, some researchers focus on what animals do under natural conditions, testing the reproducibility of behavioral performances in undisturbed animals. However, behavioral performance in nature might vary among individuals independently of their capacity to produce force of work. A cleaner link between behavior and physiology would be expected when attempts are made to elicit the maximal performance in an individual. Maximal performance is more likely to be affected by body size (see Section 11.3), as clearly illustrated in the jumping performance of anuran amphibians [40–43] either via effects on whole body mechanics or on muscle performance [41,44]. Interindividual differences in maximum performance may also be related to body shape and morphometric traits such as relative hind limb length and relative tail length, as shown for the locomotor performance of hatchling lizards [27]. In the iguanid lizard *Ctenosaura similes*, running endurance is mass dependent and related to relative differences in organ mass, particularly thigh muscle mass [45], and compatible findings have been reported for salamanders [46]. In the tree-frog *Hyla multilineata*, a significant amount of variance in maximum jump distance among individuals is related to a combination of factors, including the metabolic profile and mass of the plantaris muscle and total hind limb muscle mass [47].

Most of the information available regarding interindividual variation in the natural behaviors of vertebrate ectotherms comes from studies with amphibians. Both discrete and continuous behavioral diversity have been reported within populations of salamanders [48] and frogs [49], showing, for example, consistent variation among individuals in antipredator behaviors that might depend only partially or not at all on body size. For most anurans, females select males as reproductive partners based on call structure and intensity [50,51]. However, individuals of some species are silent "satellite males" that do not call but attempt to grasp females on their way to a calling male [51]. The relationship between this sort of interindividual variation in behavioral strategies and variation in physiological traits, particularly muscle physiology, has been poorly studied, and the data available do not point to a simple pattern. For the North American spring peepers (*Pseudacris crucifer*), persistent calling is considered fundamental for the ability of males to obtain copulations [52]; however, both satellite and calling males coexist within populations. Analyses of the muscle aerobic capacity of satellite males indicate that they are not physiologically inferior to calling males, suggesting these alternative mating strategies are not necessarily related to physiological potential [53]. Similarly, individual *Bufo* toads differ in their number of attempts to clasp females, but this does not appear to be related to the metabolic physiology of individuals [54]. Interindividual differences in behavior have been reported in various species of male anurans, for example regarding the extent to which calling activity decreases through the night or the time of night at which animals retreat [55], but differences of this sort seem unrelated to an eventual depletion of muscle glycogen reserves [56,57]. These papers, although focused on specific physiological traits, suggest that natural behavioral variation among individuals might occur in the absence of concomitant physiological differentiation. In contrast, clear consistency between behavior and physiology exist in *Scinax hiemalis*, a tree-frog that when threatened in the field chooses between feigning death, remaining immobile, or jumping away. In the laboratory, behavioral responses when faced with sudden disturbance are associated with both body size and absolute ability of individuals to jump, a trait probably associated with the muscle physiology of individuals [49].

11.3 SCALING EFFECTS ON VERTEBRATE ECTOTHERM MUSCLE AND WHOLE BODY PERFORMANCE

As discussed in Section 11.2, body size has a large effect on the physiology and locomotor performance of animals; therefore, it is important to elucidate the extent to which body size accounts for interindividual differences in behavioral and underlying muscle performance. However, in many such scaling relationships, there is a large amount of scatter, suggesting that mass-independent differences in behavioral performance are also important in evolutionary physiology.

Generally, as animal size increases, skeletal muscle shortening velocity and contraction kinetics become slower, a pattern confirmed in both interspecific and interindividual studies. Interindividual studies within anuran [41,44,46], lizard [58,59], and fish [2,60] species have highlighted that as body size increases, there are usually decreases in temporal traits of muscle functions, including relative muscle shortening velocity and twitch activation and relaxation rates, i.e., muscle in smaller individuals of a species will tend to produce force quicker, shorten relatively faster, and relax quicker. In contrast, maximum isometric muscle stress and normalized power output are generally unaffected by body size [41,44,46,58,60]. Body size–dependent changes in the mechanical properties of skeletal muscle have in some cases been linked to underlying muscle structure and/or fiber type (see Section 11.4) and are paralleled by compatible alterations in locomotor performance, with smaller individuals often being relatively as fast or even faster (i.e., when speed is expressed relative to body size). Body size is likely to be a confounding factor when comparing the performance of ectothermic individuals because when analogous muscles are compared in endotherms, those from larger animals have an increased proportion of slow myosin and lower maximal shortening velocity (V_{max}) within the same fiber type [61]. Indeed 78% of the interspecific variability in V_{max} of muscles taken from a range of endotherms and ectotherms can be explained by variation in body size [62]. Among species of terrestrial animals (including endotherms and ectotherms), there is a significant correlation between maximal shortening velocity and maximal stress (P_0) in skeletal muscle [62] with muscles that have relatively high V_{max} also having, although to a lesser extent, relatively higher P_0.

In general, larger individuals within fish, salamander, and lizard species are constrained to comparatively lower stride and tail-beat frequencies, thus, achieving higher absolute locomotory speeds (because of their greater body size) but lower relative (length-specific) locomotory speeds [46,58,63,64]. Also, smaller individuals have higher rates of acceleration during jumping [42,65] and swimming [66]. However, it is apparent in aquatic vertebrates that as body size increases, maneuverability decreases [63,66], with larger individuals turning more slowly and through greater angles. Therefore, larger animal species and individuals can sometimes be at a disadvantage when trying to catch smaller but more maneuverable species and individuals. Some species of large aquatic vertebrates (e.g., killer whales) have countered this problem by developing means of prey capture that maximize acceleration of part of the predator's body (e.g., tail slaps) rather than relying on whole body acceleration [66]. Body size also affects maneuverability during running in terrestrial vertebrates, which apparently affects predator prey choice and tactics used during prey capture [67].

A number of integrative studies have attempted to determine body size effects on both muscle contractile properties and locomotory performance. Muscle twitch duration has been used to predict minimum tail-beat frequency in fish, which in turn was used to calculate maximum swimming speed [2]. Larger fish had slower muscle mechanical properties that led to slower (i.e., constrained) tail-beat frequency, which in turn limited swimming speed. During sprinting in the lizard *Dipsosaurus dorsalis*, the decrease in stride frequency with body size was matched by an increase in twitch times in fast glycolytic muscle fibers, whereas increases in stride length were related to increased hindlimb length [59] (Figure 11.1). The slope of scaling relationships was identical in *D. dorsalis* for both limb cycle frequency during sprinting and optimal cycle frequency for maximal muscle power output, further supporting the notion that muscle fiber type is matched to locomotor performance in this species [68]. At lower temperatures in this lizard species, the cycle frequency for maximal muscle power matches the stride frequency used [69], i.e., the muscle in smaller individuals tends to cycle at a faster rate (frequency) with faster activation, shortening, and relaxation rates tending to enable higher stride frequencies by minimizing the ground contact time [70,71]. The higher stride frequencies in smaller individuals are vital to enable them to achieve the same, or similar, sprint velocity as larger individuals of the same species [59]. In contrast, larger individuals within a species have slower, more economical muscle fibers optimized to work at lower limb cycle frequency during sprinting, yet still achieve the same sprint speed because of their longer limb length (but see Ref. [70] for a full discussion on lower stride frequency in larger lizards).

Optimal cycle frequency for maximal muscle work loop-power output decreases with increasing body size [44,58,72]. Altringham and co-workers [44] have shown that interindividual variation in the optimal cycle frequency for work loop-power output in *Xenopus* muscle fiber bundles is more dependent on body size in slower muscle fibers (from adductor magnus) than in faster muscle fibers (from sartorius), with scaling exponents of $M_b^{0.23}$ and $M_b^{0.07}$, respectively. It has been argued that in faster muscles and muscle fiber bundles, contractile speed decreases less dramatically with body size as animals attempt to maximize speed and power available for escape responses [44]. In contrast, in slower muscle and muscle fibers, rapid muscle shortening velocity or deactivation rate is not as crucial to organismal performance; therefore, large energetic savings can be made by decreasing these variables, hence, the relatively high scaling exponent [71].

Despite the general patterns of body size–related changes in locomotory performance, the effect of body size can vary between species. Within some species of

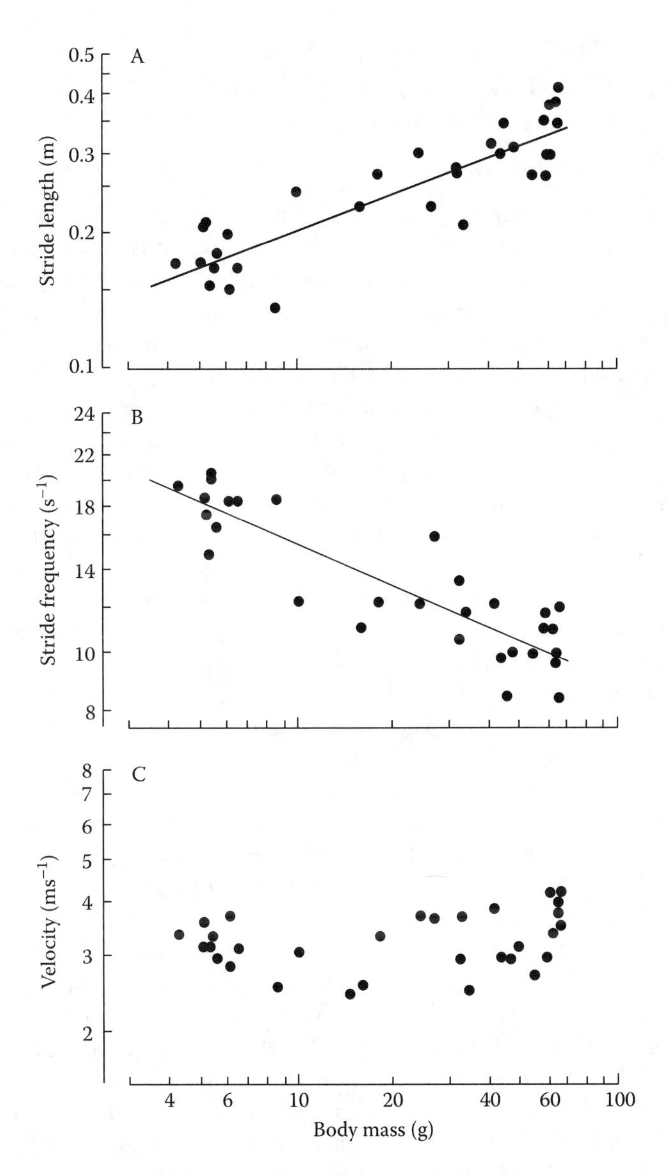

FIGURE 11.1 (A) Stride length, (B) stride frequency, and (C) running velocity as functions of body mass during sprint runs with a body temperature of 35°C in the lizard *Dipsosaurus dorsalis*. Data are plotted on log–log coordinates. Solid lines represent regression lines. (From Marsh, R.L., *J. Exp. Biol.*, 137, 119, 1988, with permission.)

Anolis lizard, maximum acceleration and distance achieved during jumping increased with hindlimb length, yet within other species the relationship between performance and hindlimb length was negative [73].

11.4 RELATIONSHIPS BETWEEN MUSCLE SPECIALIZATION AND THE INDIVIDUAL BEHAVIORAL PERFORMANCE OF VERTEBRATE ECTOTHERMS

As mentioned in the introduction to the present chapter, within the context of vertebrate ectotherms, sprint speed has been more frequently studied and seems of particular ecological relevance in lizards and fish. Speed during burst running and swimming is dependent on stride length and stride frequency. Stride length in lizards is dependent on limb length and gait whereas stride frequency is dependent on skeletal muscle having rapid activation, high shortening velocity, and fast relaxation. It has been demonstrated that Lacertid species living in more open environments tend to have longer limbs and actually modulate their sprint speed by altering their stride length, and consequently the range of motion of their muscles [74]. In contrast, those species that live in more closed environments (where more vegetation is present to restrict locomotion) have shorter limbs and alter their speed by changing their stride frequency (therefore, altering muscle cycle frequency) [74].

The relationship between sprint speed and muscle activation rate, relaxation rate, and shortening velocity suggests a connection between individual differences in performance and muscle molecular traits. This is because any muscle with rapid activation and relaxation must rely on rapid release and reuptake of calcium, respectively. Rapid relaxation is correlated with greater parvalbumin content in skeletal muscle [75] but comes at a high energetic cost because it requires rapid calcium pumping and fast cycling cross-bridges, two processes that require ATP consumption [76]. Rapid activation and relaxation can also require an increased proportion of the muscle to be devoted to the sarcoplasmic reticulum (SR) t-tubule system to decrease diffusion distances between this system and the thin filaments [77]. Therefore, any muscle that cycles at very high frequencies has a disproportionately large volume of SR. Increased SR volume leads to a proportional decrease in the space available for the contractile filaments, causing muscles that cycle at very high frequency, such as rattlesnake tailshaker and toadfish swimbladder muscles, to produce relatively low force per unit cross-sectional area [76,77]. Higher levels of SR calcium ATPase activity and SR calcium uptake (i.e., two other SR indicators of contractile speed) in sprint-trained horses have been linked with improved sprint performance [78]. Therefore, as has previously been suggested for lizards running at suboptimal temperatures [69], the rate of muscle activation and relaxation can limit sprint performance, because muscles used during sprinting need high power output. Given that both muscle power production and activation and relaxation rates can in some cases affect sprint speed, and that morphological tradeoffs might constrain the maximization of both (see Section 11.5), it is possible that various muscle configurations might, for different reasons, lead to similar sprint speeds among individuals.

Maximal shortening velocity (V_{max}) of muscle within a species is primarily dependent on the myosin heavy chain isoforms expressed within the muscle, with significant differences in V_{max} existing between fiber types [79]. Other sources of variation in the maximum shortening velocity of muscle may include variation in myosin light chains, troponin, tropomyosin, and various calcium regulatory proteins

[80,81]. In *Xenopus laevis*, maximal stress, maximal shortening velocity, and maximal power output vary in the same manner between muscle fiber types decreasing from type 1 to type 2 to type 3 to type 4 to tonic [79]. Higher muscle force in both ectotherms and endotherms is thought to be primarily due to relatively higher myofibrillar volume [77].

Muscle architecture can be varied to match the task performed. Evidence from human limb muscles suggests that muscles with a higher pennation angle should produce relatively higher forces (because of a larger physiological cross-sectional area) and may produce greater force at any muscle-shortening velocity (because of lower fiber-shortening velocity) when compared with more parallel fibered muscles [82]. However, muscles that are less pennate are better suited to larger length changes, with the result that antigravity extensor muscles are more pennate and flexor muscles are more parallel fibered. It is possible that interindividual variability in muscle architecture may occur leading to changes in muscle size, muscle pennation, and the position of the origin and insertion of muscles such as have been found between species. For example, projectile theory suggests that jumping performance could be altered by changing: (1) hindlimb length to affect the distance or time period over which the propulsive force acts; (2) hindlimb jumping muscle size to alter the power available for propulsion; and (3) moving the position of the origins or insertions of the jumping muscles to alter the angular velocity of the limbs [83]. Within a given clade, jumping specialists usually have longer hindlimbs [40,43,65,84] and more enlarged hindlimb musculature, across or within species [65], or across species [40]. One notable example of muscle specialization is evidenced in the longissimus dorsi muscle of amphisbaenians, which provides the high forces required for digging. In the amphisbaenid species *Leposternon microcephalum*, this muscle is highly pennate (to increase the physiological cross-sectional area and hence force) and overall has fast moderately oxidative muscle fibers. Subtle biochemical changes along the longitudinal axes suggest higher glycolytic capacity in the distal fibers that seems specialized to match the function of this section of the muscle and perhaps enhance force toward the end of the digging strokes, when greater forces are required [85].

Skeletal muscle can perform many different tasks during animal behavior, acting to rapidly produce high power output, produce power efficiently over long periods, produce high force for long periods, produce force isometrically to stabilize a limb, transmit force, or absorb power to act as a brake [86]. Essentially the mechanical properties of muscle are determined by their composition, which in turn reflects tradeoffs between the differing tasks performed by the muscle [77]. In mammalian muscles, the fiber type of a muscle has been correlated to its daily activity patterns [87].

In many animals, including fish, frogs, and lizards, different muscles may contain different proportions of fiber type, and within muscles, fiber types are separated into distinct regions. The mechanical properties of the different muscles and fiber types and the proportions of different fiber types within muscles match the force and power output required by their different functions [79,88,89]. For example, in *Rana pipiens*, the hindlimb muscles primarily used to power jumping (89%) consist of the fastest, most powerful, type-1 muscle fibers, which should maximize burst jumping performance to enable rapid escape from predators [90].

In contrast, hindlimb muscles not used to power jumping are composed of far fewer (29%) type-1 muscle fibers.

Skeletal muscle is also involved in sound production, which is a highly specialized activity. The trunk muscles that support calling in males of many species of frog and some other vertebrate ectotherms are highly aerobic and cycle continuously over many hours a day during the mating season [13]. This behavior relates to the mating success of males, which in many species is highly dependent on energetically demanding vocal displays involved in female attraction and male-to-male interactions [51]. Males able to produce the most energetic calls are often more attractive to females [91]. The structure of trunk muscles reflects both the aerobic nature and high cycle frequency requirements of the behavior they support, including 100% fast oxidative glycolytic fibers [92], high citrate synthase activity [93], and high densities of mitochondria and capillaries [94]. Differences in calling rate between species have been linked to differences in mechanical properties of calling muscles [92].

The amplexus behavior of male anurans requires the forearms of the male to be maintained in a relatively fixed position for long periods to hold the female during breeding. The forearm muscles involved in this behavior can produce high, sustained isometric force (Navas and James, unpublished results) [9]. These sustained forces can even be maintained via only short periods of electrical activation because of very slow muscle relaxation rates. These mechanical requirements are only needed by the males, and indeed, there is marked sexual dimorphism as evidenced by muscular hypertrophy [9] and increased proportion of tonic fibers in males [95]. Reproductively successful toads in the species *Bufo marinus* (those found in amplexus) had significantly larger forelimb muscle mass (when corrected for body size) than those calling males not found in amplexus [96]. Therefore, these muscles seem well adapted and vital for their role in producing high forces to successfully grip the female during amplexus but would, for instance, be of limited use during cyclic locomotion due to the prolonged relaxation times.

11.5 TRADEOFFS IN WHOLE-MUSCLE FUNCTION AND ITS ECOLOGICAL IMPORTANCE

Performance tradeoffs represent constraints on the direction of adaptive physiological evolution and therefore are fundamental for understanding the widespread occurrence of individual variation in physiological traits. Identification of the physiological mechanisms underlying constraints on vertebrate physical performance are an important research objective for evolutionary-minded functional biologists. Functional constraints can occur from negative interactions between pairs of ecologically relevant performance traits. For example, one of the most intuitive tradeoffs in muscle performance is between sprint (high power output) and endurance (high fatigue resistance) activities, which arise via differential expression of fast and slow muscle fiber types that maximize these differing activities [97]. As a consequence, increases in the performance of one task can lead to decreases in the performance of another, with the net result a compromised phenotype that is optimal for neither task. The existence and mechanistic basis of functional tradeoffs can be studied in a variety of ways, including optimality models, quantitative genetic analyses, interindividual analyses, and interspecific comparisons. The vertebrate locomotor system must often perform a variety of activities that are often antagonistic in their requirements (see Section 11.4), and as a consequence, it has been used as a model system for examining performance tradeoffs.

In parallel with the previous discussion on muscle specialization (see Section 11.4), the study of functional performance tradeoffs was first addressed within the context of interspecific comparisons, which have been very successful at identifying the range of possible functional tradeoffs that can limit the evolution of vertebrate physical performance. These studies have also recently been complemented with several interspecific comparative analyses within a phylogenetic context [74,98,99] and comparative studies of different breeds of domesticated animals [100]. For example, by comparing two dog breeds (greyhound and pit bulls) that have experienced intense artificial selection for certain physical attributes, Pasi and Carrier [100] investigated the functional tradeoffs that prevent the simultaneous optimization of both running performance and physical fighting. The same authors found that pit bulls possessed greater muscle mass distally in their limbs, larger muscles in their forelimbs than their hindlimbs, and a lower capacity for elastic energy storage in their ankle joint tendons than greyhounds [100]. Thus, physical attributes that were associated with fighting were antagonistic to running performance and thus could not have evolved simultaneously.

Regarding interindividual variation, most analyses of locomotor tradeoffs have been conducted using analyses of whole-animal performance. Somewhat surprisingly, most of these analyses have not detected the existence of functional tradeoffs in locomotor performance, despite the logical underlying basis of these physiological tradeoffs. For example, although a tradeoff between speed and endurance capacity is predicted based on individual differences in the proportions of fast, powerful fibers and slow, more fatigue-resistant fibers, when tested at the whole-animal level, most studies have found no evidence for this performance tradeoff [27,35,101–105]. However, recent studies on human decathletes have argued that such interindividual comparisons are complicated by the masking effects of variation in general physical quality [106]. This often prevents the detection of functional tradeoffs in performance despite their probable importance within each individual and their influence on the directions of adaptive evolutionary change.

Although previous inter- and intraspecific comparative studies have been successful at identifying the possible functional constraints operating on the vertebrate locomotor system, they have not allowed detailed analyses of the mechanistic bases of these hypothesized tradeoffs. Studies of performance tradeoffs at the level of the whole muscle allow direct investigations of the importance of functional tradeoffs at the level at which they are presumed to operate, and they help in understanding how some of these constraints might be overridden by processes occurring at higher levels of organization, so as to be obscured in the context of interindividual studies. Few studies have tested the extent of performance tradeoffs in whole-muscle performance. Interindividual analyses of muscle physiology can determine both whether functional tradeoffs can be detected in whole muscle and the magnitude of any possible performance tradeoffs. Based on comparisons of different isolated muscle fiber types, we know that variation in the specific force- and velocity-dependent properties as well as the mechanical power production capabilities of a muscle are related to the specific muscle fiber type and the proportion of certain myosin isoforms expressed within the fibers (see Section 11.4). A recent comparative analysis of 11 different species of Phrynosomatid lizards [18] also showed that the percentage of fast oxidative glycolytic muscle fibers was negatively correlated with the percentage of fast glycolytic fibers. Several recent studies have also directly investigated the extent of functional tradeoffs in whole-muscle performance by using the work-loop technique to analyze the contractile properties of isolated skeletal muscle [105,107,108].

Using interindividual analyses of whole-muscle performance, tradeoffs between maximum power output and fatigue resistance were detected in both isolated mouse and frog muscle [105,107,108]. Despite the lack of a detectable tradeoff between speed and endurance in whole-animal performance of the frog *Xenopus laevis*, a clear tradeoff between power and fatigue resistance was still found in peroneus muscle [105]. Interindividual analyses of whole gastrocnemius muscle performance of the toad *Bufo viridus* demonstrated tradeoffs between fatigue resistance and both maximum power output and stress of the muscle [108] (Figures 11.2A and 11.2B). In addition, this correlative analysis showed a significant positive relationship between power output and maximum stress [108]. Thus, increases in maximum stress of an individual muscle leads to simultaneous increases in maximum power output (Figure 11.2C) and decreases in fatigue resistance [108]. Surprisingly, no significant correlations were detected between whole muscle performance of *B. viridus* and muscle fiber type composition. Future studies should investigate the relationship between interindividual variation in whole-muscle performance and the predicted underlying relationship with the muscle's morphological structure.

11.6 CONCLUSIONS

Throughout this chapter, we have argued that interindividual differences are of great importance in evolutionary physiology because they are likely to reflect intrinsic variability upon which natural selection can potentially act [109]. Such variability might have an underlying basis in the physiological traits of individuals, among which those related to skeletal muscle traits, including size, morphology and physiology, are particularly relevant for being closely related to behavioral performance and, likely, to the ecological success of individuals [39,46,108,110]. The assumption of a logical relationship between muscle performance and ecological success permeates the literature, but studies directly investigating such relationships are scarce. Similarly, attempts to investigate the strength of phenotypic selection on biomechanical quantitative traits are scant (see, e.g., Ref. [4]), an observation that is true of physiological traits in general [111]. The study of the nature, reproducibility, and ecological consequences of interindividual variation of muscle physiology, therefore, is an integrative and contemporary field of research.

Although the theoretical links relating muscle mechanics, behavioral performance, and evolution seem rather well-established, empirical evidence is missing. Such empirical support is not just a formality, but also an essential task to understanding how

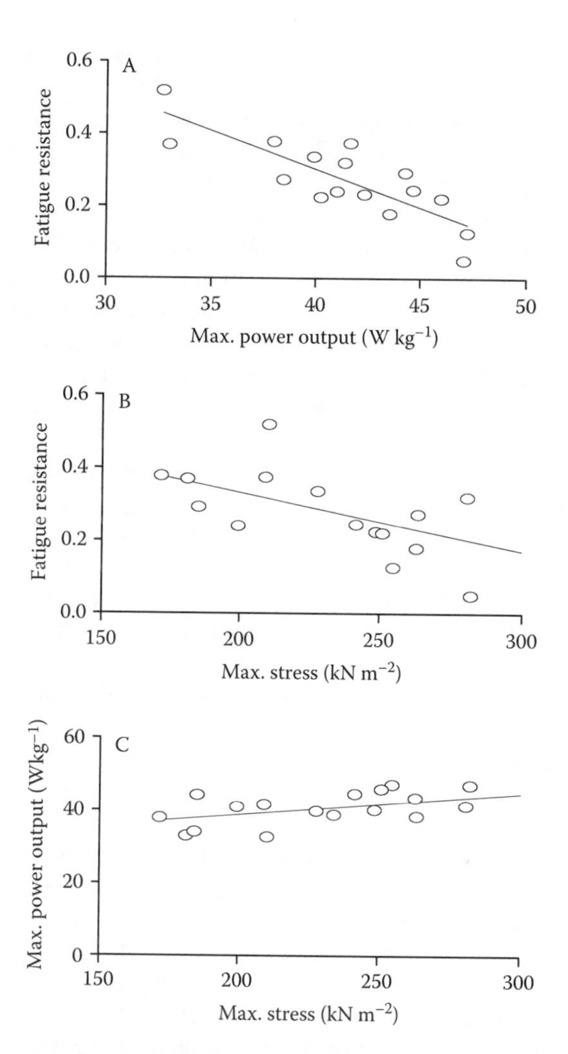

FIGURE 11.2 Correlations between several measures of *in vitro* muscle performance for the gastrocnemius muscle of the toad *Bufo viridus*. Measures of isolated muscle performance reported are (A) maximum net power output, (B) maximum isometric stress, and (C) fatigue resistance of power output during a fatigue run at 5 Hz. All correlations were statistically significant at the level of $P < 0.05$. (Modified after Wilson, R.S. et al., *J. Comp. Physiol.*, 174B, 453, 2004.)

muscle physiology evolves. For example, studies showing short-term reproducibility of behavioral performance are becoming abundant, but studies across years [34], temperatures [39], or ecological settings [38] are still uncommon. If behavioral performance ranks are context dependent, the relationship between behavioral performance and fitness would be of a more complex nature than usually stated in the literature. Understanding how the morphophysiology of skeletal muscle changes along the lifetime of individuals in response to relevant ecophysiological variables and how these changes affect the behavioral performance of individuals, would be an important contribution of muscle physiologists to the evolution of behavioral performance, a main issue in evolutionary physiology. Interesting contributions from this standpoint include how skeletal muscle responds to changes in ecophysiological variables such as ontogeny, level of activity, temperature, availability of resources, and duration of hypometabolic states, for example, estivation or hibernation. Additionally, very few studies have used training as a tool to explore the scope for changes in behavioral performance in vertebrate ectotherms (see, e.g., Ref. [112]).

The study of functional tradeoffs in skeletal muscle also emerges as a field that deserves further attention. There is enough evidence to state clearly that certain muscle traits cannot be simultaneously maximized, and research along these lines still has much to offer. Phenotypic tradeoffs, however, might relate muscle properties to other important traits of natural history. As an example, hatchling survival and sprint speed seem inversely correlated in some lizards [31]. Such findings pose questions regarding how an investment in the production of skeletal muscle might vary among individuals in ways that might be consequential for fitness. If such patterns abound, it would be quite incorrect to assume simplistically that the fastest lizards are the fittest. Studies relating muscle function with life-history traits such as time for maturity, growth rate, hatchling survival, and other traits are very uncommon but are needed. The relationship between fitness and behavioral performance is further complicated by the observation that differences in muscle performance among individuals may not necessarily be coupled with differences in undisturbed (e.g., field) behavioral performance (see Section 11.2). Related questions emerge, however, given that differences among individuals in their activity patterns could, at least in theory, cause differentiation in various aspects of exercise physiology through training effects.

Many questions remain to be answered regarding the role of body size as a factor explaining interindividual variation in muscle performance. It is clear that no one scaling law is universal in its prediction of changes in performance or function with body size. For example, the scaling of maximal acceleration during jumping within and across *Anolis* lizard species demonstrates positive allometry in smaller lizards but no relationship in larger lizards [73]. Wilson et al. [42] have also demonstrated a discontinuous scaling relationship with positive allometry between maximum jump distance and body mass in frogs of smaller size and then body size independence in larger frogs. The nature of such discontinuities is fundamental to evolutionary physiology because scaling patterns might differ at contrasting body masses due to different selective pressures associated with ecological performance. Perhaps in some species of frogs, small individuals must jump a minimum absolute distance to be ecologically functional, whereas larger individuals exhibit mass independence of jump distance because they have exceeded a critical level of ecologically relevant performance [42].

Although the scaling relationships of locomotion in running animals have received a large amount of attention, they have not been considered in such a clear ecological context as in the jumping studies mentioned above. For instance, various predictions have been made that limb dimensions should scale according to geometric [113] or elastic similarity [114], and that the energetic cost of locomotion is dependent on the rate at which force is produced and stride frequency [115,116].

The generally assumed extension of these hypotheses is that such physical constraints of body size dictate the stride frequency used at any animal size, which in turn dictates the maximal shortening velocity of muscle (for a review, see [117]). However, as discussed in Section 11.3 (see also [62]), the power produced at any stride frequency is dependent on a range of muscle mechanical properties, not just maximum shortening velocity. Pellegrino and collaborators [61] have also postulated that the higher energy cost of locomotion in small animals is due to their higher relative muscle power output, i.e., they have faster muscles that produce more power, incurring a greater energetic cost. Other important constraints may act, however, because viable individuals within a species are likely to be those attaining locomotory performances that grant appropriate ecological performance. Among lizards, for example, larger individuals with longer legs and stride lengths could reach a threshold speed by using a lower stride frequency. They could, therefore, exhibit slower maximal muscle shortening velocity and contraction kinetics, and hence relatively lower energetic cost.

Multidisciplinary studies involving muscle physiology and biomechanics, ecology, and evolution are important to fully understand the relationships between organismal performance and fitness. They are needed also to understand how the possible mechanistic bases of interindividual behavioral variation may indicate the role physiology plays in constraining the behavioral tactics used by ectotherms. The relationships between behavioral and physiological performance are evident only when experiments involve animals forced to perform near their maximal possibilities, and yet studies with vertebrate ectotherms in the field do not show a clear relationship between behavior and exercise physiology. Little is known about the nature and significance of interindividual variation in the skeletal muscle of vertebrate ectotherms: Is interindividual diversity in muscle performance similar among populations of related species? If not, why not? To what extent is interindividual variation merely noise, or does it reflect behavioral strategies coexisting within populations? Why do relationships between individual field behavior and physiology seem absent is some species? Perhaps many such taxa most frequently perform well below their maximal potential, but if so, what would be the real impact of behavioral performance on fitness? This relationship is assumed to exist, but it is mainly a sensible theoretical premise for which direct empirical support is limited to very few studies [118,119]. Further efforts favoring an integrative perspective of the mechanistic causes and ecological consequences of interindividual variation are, therefore, very much needed.

REFERENCES

1. Domenici, P. and Blake, R., The kinematics and performance of fish fast-start swimming, *J. Exp. Biol.,* 200, 1165, 1997.
2. Wardle, C.S., Limit of fish swimming speed, *Nature,* 255, 725, 1975.
3. Hertz, P.E., Huey, R.B., and Nevo, E., Homage to Santa Anita: thermal sensitivity of sprint speed in agamid lizards, *Evolution,* 37, 1075, 1983.

4. Miles, D.B., The race goes to the swift: fitness consequences of variation in sprint performance in juvenile lizards, *Evol. Ecol. Res.,* 6, 63, 2004.
5. Garland, T., Hankins, E., and Huey, R.B., Locomotory capacity and social dominance in male lizards, *Funct. Ecol.,* 4, 243, 1990.
6. Alexander, R.M., Energy-saving mechanisms in walking and running, *J. Exp. Biol.,* 160, 55, 1991.
7. Lutz, G.J. and Rome, L.C., Built for jumping: the design of the frog muscular system, *Science,* 263, 370, 1994.
8. Roberts, T.J. and Marsh, R.L., Probing the limits to muscle-powered accelerations: lessons from jumping bullfrogs, *J. Exp. Biol.,* 206, 2567, 2003.
9. Peters, S.E. and Aulner, D.A., Sexual dimorphism in forelimb muscles of the bullfrog, *Rana catesbeiana*: a functional analysis of isometric contractile properties, *J. Exp. Biol.,* 203, 3639, 2000.
10. Brown, R.M. and Taylor, D.H., Compensatory escape mode trade-offs between swimming performance and maneuvering behavior through larval ontogeny of the wood frog, *Rana sylvatica*, *Copeia,* 1995, 1, 1995.
11. D'Aout, K. and Aerts, P., Kinematics and efficiency of steady swimming in adult axolotls (*Ambystoma mexicanum*), *J. Exp. Biol.,* 200, 1863, 1997.
12. Goater, C.P., Semlitsch, R.D., and Bernasconi, M.V., Effects of body size and parasite infection on the locomotory performance of juvenile toads, *Bufo bufo*, *Oikos,* 66, 129, 1993.
13. Bevier, C.R., Utilization of energy substrates during calling activity in tropical frogs, *Beh. Ecol. Sociobiol.,* 41, 343, 1997.
14. Weinstein, R.B. and Full, R.J., Intermittent locomotion increases endurance in a gecko, *Physiol. Biochem. Zool.,* 72, 732, 1999.
15. Garland, T., Phylogenetic analyses of lizard endurance in relation to body size and body temperature, in *Lizard Ecology: Historical and Experimental Perspectives*, Vitt, L.J. and Pianka, E.R., Eds., Princeton University Press, Princeton, 1994, p. 237.
16. Garland, T. and Losos, J.B., Ecological morphology of locomotor performance in squamate reptiles, in *Ecological Morphology: Integrative Organismal Biology*, Wainwright, P.C. and Reilly, S.M., Eds., University of Chicago Press, Chicago, 1994, p. 240.
17. Bauwens, D., Garland, T., Castilla, A.M., and Van Damme, R., Evolution of sprint speed in lacertid lizards: morphological, physiological, and behavioral covariation, *Evolution,* 49, 848, 1995.
18. Bonine, K.E., Gleeson, T.T., and Garland, T., Jr., Comparative analysis of fiber-type composition in the iliofibularis muscle of phrynosomatid lizards (Squamata), *J. Morphol.,* 250, 265, 2001.
19. Kohlsdorf, T., James, R.S., Carvalho, J.E., Wilson, R.S., Dal Pai-Silva, M., and Navas, C.A., Locomotor performance of closely related *Tropidurus* species: relationships with physiological parameters and ecological divergence, *J. Exp. Biol.,* 207, 1183, 2004.
20. Donley, J.M., Sepulveda, C.A., Konstantinidis, P., Gemballa, S., and Shadwick, R.E., Convergent evolution in mechanical design of lamnid sharks and tunas, *Nature,* 429, 61, 2004.
21. Franklin, C.E., Studies of evolutionary temperature adaptation: muscle function and locomotor performance in Antarctic fish, *Clin. Exp. Pharmacol. Physiol.,* 25, 753, 1998.
22. Navas, C.A., Metabolic physiology, locomotor performance, and thermal niche breadth in neotropical anurans, *Physiol. Zool.,* 69, 1481, 1996.

23. Bennett, A.F., Interindividual variability: an underutilized resource, in *New Directions in Ecological Physiology*, Feder, M.E., Bennett, A.F., Burggren, W., and Huey, R.B., Eds., Cambridge University Press, Cambridge, 1987, p. 147.
24. Pough, H.F. and Andrews, R.M., Individual and sibling-group variation in metabolism of lizards: the aerobic capacity model for the origin of endothermy, *Comp. Biochem. Physiol.,* 79A, 415, 1984.
25. Shaffer, H.B., Austin, C.C., and Huey, R.B., The consequences of metamorphosis on salamander (*Ambystoma*) locomotor performance, *Physiol. Zool.,* 64, 212, 1991.
26. Watkins, T.B., The effect of metamorphosis on the repeatability of maximal locomotor performance in the Pacific tree frog *Hyla regilla, J. Exp. Biol.,* 200, 2663, 1997.
27. Tsuji, J.S., Huey, R., Berkum, F.V., Garland, T., and Shaw, R.G., Locomotor performance of hatching fence lizards (*Sceloporus occidentalis*): quantitative genetics and morphometric correlates, *Evol. Ecol.,* 3, 240, 1989.
28. Bennett, A.F. and Lenski, R.E., Experimental evolution and its role in evolutionary physiology, *Am. Zool.,* 39, 346, 1999.
29. Dohm, M.R., Richardson, C.S., and Garland, T., Exercise physiology of wild and random bred laboratory house mice and their reciprocal hybrids, *Am. J. Physiol.,* 36, 1098, 1994.
30. Jayne, B.C., Bennett, A.F., and Lauder, G.V., Muscle recruitment during terrestrial locomotion: how speed and temperature affect fiber type use in a lizard, *J. Exp. Biol.,* 152, 101, 1990.
31. Warner, D.A. and Andrews, R.M., Laboratory and field experiments identify sources of variation in phenotypes and survival of hatchling lizards, *Biol. J. Linn. Soc.,* 76, 105, 2002.
32. Van Buskirk, J., Phenotypic lability and the evolution of predator-induced plasticity in tadpoles, *Evolution,* 56, 361, 2002.
33. Van Buskirk, J. and McCollum, S.A., Influence of tail shape on tadpole swimming performance, *J. Exp. Biol.,* 203, 2149, 2000.
34. Huey, R. and Dunhan, A., Repeatability of locomotor performance in natural populations of the lizard *Sceloporus merriami, Evolution,* 41, 1116, 1987.
35. Huey, R.B., Dunham, A.E., Overall, K.L., and Newman, R.A., Variation in locomotor performance in demographically known populations of the lizard *Sceloporus merriami, Physiol. Zool.,* 63, 845, 1990.
36. Kolok, A.S., Plaisance, E.P., and Abdelghani, A., Individual variation in the swimming performance of fishes: an overlooked source of variation in toxicity studies, *Env. Toxicol. Chem.,* 17, 282, 1998.
37. Marker, G.M. and Gatten, R.E., Individual variability in sprint performance, lactate production, and enzyme activity in frogs (*Rana pipiens*), *J. Herpetol.,* 27, 294, 1993.
38. Martinez, M., Guderley, H., Nelson, J.A., Webber, D., and Dutil, J.D., Once a fast cod, always a fast cod: maintenance of performance hierarchies despite changing food availability in cod (*Gadus morhua*), *Physiol. Biochem. Zool.,* 75, 90, 2002.
39. Navas, C.A., James, R.S., Wakeling, J.M., Kemp, K.M., and Johnston, I.J., An integrative study of the temperature dependence of whole animal and muscle performance during jumping and swimming in the frog *Rana temporaria, J. Comp. Physiol.,* 169, 588, 1999.
40. Choi, I.H., Shim, J.H., and Ricklefs, R.E., Morphometric relationships of take-off speed in anuran amphibians, *J. Exp. Zool.,* 299A, 99, 2003.
41. Marsh, R.L., Jumping ability of anuran amphibians, *Adv. Vet. Sci. Comp. Med.,* 38B, 51, 1994.

42. Wilson, R.S., Franklin, C.E., and James, R.S., Allometric scaling relationships of jumping performance in the striped marsh frog *Limnodynastes peronii*, *J. Exp. Biol.*, 203, 1937, 2000.
43. Zug, G.R., Anuran locomotion — structure and function, 2: Jumping performance of semiaquatic, terrestrial, and arboreal frogs, *Smith. Cont. Zool.*, 276, 1, 1978.
44. Altringham, J.D., Morris, T., James, R.S., and Smith, C.I., Scaling effects on muscle function in fast and slow muscle of *Xenopus laevis*, *J. Exp. Biol. Online*, 1, 1, 1996.
45. Garland, T., Physiological correlates of locomotory performance in a lizard: an allometric approach, *Am. J. Physiol.*, 247, R806, 1984.
46. Bennett, A.F., Garland, T., and Else, P., Individual correlation of morphology, muscle mechanics, and locomotion in a salamander, *Am. J. Physiol.*, 256, R1200, 1989.
47. James, R.S., Wilson, R.S., Carvalho, J.E., Kohlsdorf, T., Gomes, F.R., and Navas, C.A., Physiological and morphological basis of individual variation in jumping performance of *Hyla multilineata*, *Comp. Biochem. Physiol.*, 137A, S87, 2004.
48. Dowdey, T.G. and Brodie, E.D., Antipredator strategies of salamanders: individual and geographical variation in responses of *Eurycea bislineata* to snakes, *Anim. Beh.*, 38, 707, 1989.
49. Gomes, F.R., Bevier, C.C., and Navas, C.A., Environmental and physiological factors influence antipredator behavior in *Scinax hiemalis* (Anura: Hylidae), *Copeia*, 2002, 994, 2002.
50. Bourne, G.R., Lekking behavior in the neotropical frog *Ololygon rubra*, *Beh. Ecol. Sociobiol.*, 31, 173, 1992.
51. Wells, K.D., The social behaviour of anuran amphibians, *Anim. Behav.*, 25, 666, 1977.
52. Forester, D.C., Lykens, D.V., and Harrison, W.K., The significance of persistent vocalisation by the spring peeper, *Pseudacris crucifer* (Anura, Hylidae), *Behaviour*, 108, 197, 1989.
53. Lance, S.L. and Wells, K.D., Are spring peeper satellite males physiologically inferior to calling males? *Copeia*, 4, 1162, 1993.
54. Wells, K.D. and Taigen, T.L., Reproductive behavior and aerobic capacities of male American toads (*Bufo americanus*): is behavior constrained by physiology? *Herpetologica*, 40, 292, 1984.
55. Runkle, L.S., Wells, K.D., Robb, C.C., and Lance, S.L., Individual, nightly, and seasonal variation in calling behavior of the gray treefrog, *Hyla versicolor*: implications for energy expenditure, *Beh. Ecol. Sociobiol.*, 5, 318, 1994.
56. Schwartz, J.J., Ressel, S.J., and Bevier, C.R., Carbohydrate and calling: depletion of muscle glycogen and the chorusing dynamics of the neotropical treefrog *Hyla microcephala*, *Beh. Ecol. Sociobiol.*, 37, 125, 1995.
57. Wells, K.D., Taigen, T.L., Rusch, S.W., and Robb, C.C., Seasonal and nightly variation in glycogen reserves of calling gray treefrogs (*Hyla versicolor*), *Herpetologica*, 5, 359, 1995.
58. Johnson, T.P., Swoap, S.J., Bennett, A.F., and Josephson, R.K., Body size, muscle power output and limitations on burst locomotor performance in the lizard *Dipsosaurus dorsalis*, *J. Exp. Biol.*, 174, 199, 1993.
59. Marsh, R.L., Ontogenesis of contractile properties of skeletal muscle and sprint performance in the lizard *Dipsosaurus dorsalis*, *J. Exp. Biol.*, 137, 119, 1988.
60. James, R., Cole, N., Davies, M., and Johnston, I.A., Scaling of intrinsic contractile properties and myofibrillar protein composition of fast muscle in the fish *Myoxocephalus scorpius* L, *J. Exp. Biol.*, 201, 901, 1998.

61. Pellegrino, M.A., Canepari, M., Rossi, R., D'Antona, G., Reggiani, C., and Bottinelli, R., Orthologous myosin isoforms and scaling of shortening velocity with body size in mouse, rat, rabbit and human muscles, *J. Physiol.,* 546, 677, 2003.
62. Medler, S., Comparative trends in shortening velocity and force production in skeletal muscles, *Am. J. Physiol.,* 283, R368, 2002.
63. James, R. and Johnston, I.A., Scaling of muscle performance during escape responses in the fish *Myoxocephalus scorpius* L., *J. Exp. Biol.,* 201, 913, 1998.
64. Videler, J.J. and Wardle, C.S., Fish swimming stride by stride: speed limits and endurance, *Rev. Fish. Biol. Fisher.,* 1, 23, 1991.
65. Emerson, S.B., Allometry and jumping in frogs: helping the twain to meet, *Evolution,* 32, 551, 1978.
66. Domenici, P., The scaling of locomotor performance in predator-prey encounters: from fish to killer whales, *Comp. Biochem. Physiol.,* 131A, 169, 2001.
67. Iriarte-Diaz, J., Differential scaling of locomotor performance in small and large terrestrial mammals, *J. Exp. Biol.,* 205, 2897, 2002.
68. Johnson, D.C., Burt, C.T., Perng, W.C., and Hitzig, B.M., Effects of temperature on muscle pH-i and phosphate metabolites in newts and lungless salamanders, *Am. J. Physiol.,* 265, R1162, 1993.
69. Swoap, S.J., Johnson, T.P., Josephson, R.K., and Bennett, A.B., Temperature, muscle power output and limitations on burst locomotor performance in the lizard *Dipsosaurus dorsalis*, *J. Exp. Biol.,* 174, 185, 1993.
70. Farley, C.T., Maximum speed and mechanical power output in lizards, *J. Exp. Biol.,* 200, 2189, 1997.
71. Marsh, R.L., Deactivation rate and shortening velocity as determinants of contractile frequency, *Am. J. Physiol.,* 259, R223, 1990.
72. Altringham, J.D. and Johnston, I.A., Scaling effects on muscle function: power output of isolated fish muscle fibres performing oscillatory work, *J. Exp. Biol.,* 151, 1990.
73. Toro, E., Herrel, A., Vanhooydonck, B., and Irschick, D.J., A biomechanical analysis of intra- and interspecific scaling of jumping and morphology in Caribbean *Anolis* lizards, *J. Exp. Biol.,* 206, 2641, 2003.
74. Vanhooydonck, B., Van Damme, R., and Aerts, P., Variation in speed, gait characteristics and microhabitat use in lacertid lizards, *J. Exp. Biol.,* 205, 1037, 2002.
75. Heizmann, C.W., Berchtold, M.W., and Rowlerson, A.M., Correlation of parvalbumin concentration with relaxation speed in mammalian muscles, *Proc. Natl. Acad. Sci. USA,* 79, 7243, 1982.
76. Rome, L.C., Some advances in integrative muscle physiology, *Comp. Biochem. Physiol.,* 120B, 51, 1998.
77. Lindstedt, S.L., McGlothlin, T., Percy, E., and Pifer, J., Task-specific design of skeletal muscle: balancing muscle structural composition, *Comp. Biochem. Physiol.,* 120B, 35, 1998.
78. Wilson, J.A., Kronfeld, D.S., Gay, L.S., Williams, J.H., Wilson, T.M., and Lindinger, M.I., Sarcoplasmic reticulum responses to repeated sprints are affected by conditioning of horses, *J. Anim. Sci.,* 76, 3065, 1998.
79. Lutz, G.J. and Lieber, R.L., Myosin isoforms in anuran skeletal muscle: their influence on contractile properties and *in vivo* muscle function, *Microsc. Res. Tech.,* 50, 443, 2000.
80. Bottinelli, R. and Reggiani, C., Human skeletal muscle fibres: molecular and functional diversity, *Prog. Biophys. Mol. Biol.,* 73, 195, 2000.
81. Pette, D. and Staron, R.S., Myosin isoforms, muscle fiber types, and transitions, *Microsc. Res. Tech.,* 50, 500, 2000.

82. Lieber, R.L. and Friden, J., Functional and clinical significance of skeletal muscle architecture, *Muscle Nerve,* 23, 1647, 2000.
83. Emerson, S.B., Jumping and leaping, in *Functional Vertebrate Morphology*, Hildebrand, M., Bramble, D., Liem, K., and Wake, D., Eds., Harvard University Press, Cambridge, 1985, p. 58.
84. Toro, E., Herrel, A., and Irschick, D., The evolution of jumping performance in Caribbean *Anolis* lizards: solutions to biomechanical trade-offs, *Am. Nat.,* 163, 844, 2004.
85. Navas, C.A., Antoniazzi, M.M., Carvalho, J.E., Chaui-Berlink, J.G., James, R.S., Jared, C., Kohlsdorf, T., Pai-Silva, M.D., and Wilson, R.S., Morphological and physiological specialization for digging in amphisbaenians, an ancient lineage of fossorial vertebrates, *J. Exp. Biol.,* 207, 2433, 2004.
86. Dickinson, M., Farley, C., Full, R., Koehl, M., Kram, R., and Lehman, S., How animals move: an integrative review, *Science,* 288, 100, 2000.
87. Kernell, D., Hensbergen, E., Lind, A., and Eerbeek, O., Relation between fibre composition and daily duration of spontaneous activity in ankle muscles of the cat, *Arch. Ital. Biol.,* 136, 191, 1998.
88. Putnam, R.W., Gleeson, T.T., and Bennett, A.F., Histochemical determination of the fiber composition of locomotory muscles in a lizard, *Dipsosaurus dorsalis*, *J. Exp. Zool.,* 214, 303, 1980.
89. Rome, L.C., The design of the vertebrate muscular systems: comparative and integrative approaches, *Clinic. Orthop. Relat. Res.,* 403S, 58, 2002.
90. Lutz, G.J., Bremner, S., Lajevardi, N., Lieber, R.L., and Rome, L.C., Quantitative analysis of muscle fibre type and myosin heavy chain distribution in the frog hindlimb: implications for locomotory design, *J. Muscle. Res. Cell. Motil.,* 19, 717, 1998.
91. Gerhardt, H.C., The evolution of vocalization in frogs and toads, *Ann. Rev. Ecol. Syst.,* 25, 293, 1994.
92. Marsh, R.L., Contractile properties of muscles used in sound production and locomotion in two species of gray tree frog, *J. Exp. Biol.,* 202, 3215, 1999.
93. Bevier, C., Biochemical correlates of calling activity in neotropical frogs, *Physiol. Zool.,* 68, 1118, 1995.
94. Ressel, S.J., Ultrastructural properties of muscles used for call production in neotropical frogs, *Physiol. Zool.,* 69, 952, 1996.
95. Rubinstein, N.A., Erulkar, S.D., and Schneider, G.T., Sexual dimorphism in the fibers of a "clasp" muscle of *Xenopus laevis*, *Exp. Neurol.,* 82, 424, 1983.
96. Lee, J.C., Evolution of a secondary sexual dimorphism in the toad, *Bufo marinus*, *Copeia,* 2001, 928, 2001.
97. Goldspink, G., Selective gene expression during adaptation of muscle in response to different physiological demands, *Comp. Biochem. Physiol.,* 120, 5, 1998.
98. Van Damme, R. and Vanhooydonck, B., Origins of interspecific variation in lizard sprint capacity, *Funct. Ecol.,* 15, 186, 2001.
99. Vanhooydonck, B., Van Damme, R., and Aerts, P., Speed and stamina trade-off in lacertid lizards, *Int. J. Org. Evol.,* 55, 1040, 2001.
100. Pasi, B.M. and Carrier, D.R., Functional trade-offs in the limb muscles of dogs selected for running vs. fighting, *J. Evol. Biol.,* 16, 324, 2003.
101. Dohm, M.R., Hayes, J.P., and Garland, T., Quantitative genetics of sprint running speed and swimming endurance in laboratory house mice (*Mus domesticus*), *Evolution,* 50, 1688, 1996.
102. Garland, T., Jr., Else, P.L., Hulbert, A.J., and Tap, P., Effects of endurance training and captivity on activity metabolism of lizards, *Am. J. Physiol.,* 252, R450, 1987.

103. Jayne, B.C. and Bennett, A.F., Selection of locomotor performance capacity in a natural population of garter snakes, *Evolution,* 44, 1204, 1990.
104. Sorci, G., Swallow, J.G., Garland, T., Jr., and Clobert, J., Quantitative genetics of locomotor speed and endurance in the lizard *Lacerta vivipara*, *Physiol. Zool.,* 68, 698, 1995.
105. Wilson, R.S., James, R.S., and Van Damme, R., Trade-offs between speed and endurance in the frog *Xenopus laevis*: a multi-level approach, *J. Exp. Biol.,* 205, 1145, 2002.
106. Van Damme, R., Wilson, R., Vanhooydonck, B., and Aerts, P., Performance constraints in decathlon athletes, *Nature,* 415, 755, 2002.
107. Wilson, R.S. and James, R.S., Constraints on muscular performance: trade-offs between power output and fatigue resistance, *Proc. R. Soc. Lond. B.,* 271, 222, 2004.
108. Wilson, R.S., James, R.S., Kohlsdorf, T., and Cox, V.M., Interindividual variation of isolated muscle performance and fibre-type composition in the toad *Bufo viridus*, *J. Comp. Physiol.,* 174B, 453, 2004.
109. Spicer, J.I. and Gaston, K.J., *Physiological Diversity and Its Ecological Consequences,* Blackwell Science, London, 1999.
110. Gleeson, T.T. and Harrison, J.M., Muscle composition and its relation to sprint running in the lizard *Dipsosaurus dorsalis*, *Am. J. Physiol.,* 255, R470, 1988.
111. Kingsolver, J.G., Hoekstra, H.E., Hoekstra, J.M., Berrigan, D., Vignieri, S.N., Hill, C.E., Hoang, A., Gibert, P., and Beerli, P., The strength of phenotypic selection in natural populations, *Am. Nat.,* 157, 245, 2001.
112. Miller, K. and Camilliere, J.J., Physical training improves swimming performance of the African clawed frog, *Xenopus laevis*, *Herpetologica,* 37, 1, 1981.
113. Hill, A.V., The dimensions of animals and their muscular dynamics, *Science Progress (London),* 38, 209, 1950.
114. McMahon, T.A., Using body size to understand the structural design of animals: quadrupedal locomotion, *J. Appl. Physiol.,* 39, 619, 1975.
115. Rome, L.C., Scaling of muscle fibres and locomotion, *J. Exp. Biol.,* 168, 243, 1992.
116. Taylor, C.R., Heglund, N.C., McMahon, T.A., and Looney, T.R., Energetic cost of generating muscular force during running: a comparison of large and small animals, *J. Exp. Biol.,* 86, 9, 1980.
117. Lindstedt, S.L., Hoppeler, H., Bard, K.M., and Thronson, H.A., Estimate of muscle-shortening rate during locomotion, *Am. J. Physiol.,* 248, R699, 1985.
118. Angilletta, M.J., Niewiarowski, P.E., and Navas, C.A., The evolution of thermal physiology in ectotherms, *J. Therm. Biol.,* 27, 249, 2002.
119. Johnston, I.A. and Temple, G.K., Thermal plasticity of skeletal muscle phenotype in ectothermic vertebrates and its significance for locomotory behaviour, *J. Exp. Biol.,* 205, 2305, 2002.

12 Power Generation during Locomotion in *Anolis* Lizards: An Ecomorphological Approach

Bieke Vanhooydonck, Peter Aerts, Duncan J. Irschick, and Anthony Herrel

CONTENTS

12.1 INTRODUCTION

Movement requires the generation of muscular power. How much mechanical power a muscle can produce is determined by its physiological properties, such as force–velocity relationship, pattern of stimulation, and strain waveform and amplitude [1–5]. High power outputs are typically associated with explosive movements such as fast starts, jumping, or take off in flying organisms [6–8]. Among species or higher taxa, mechanical power output and muscle physiology can vary dramatically. For instance, Wakeling and Johnston [9] showed that among six teleost fish species, total (hydrodynamic) power output during fast starts varied 10-fold, i.e., from around 16 W kg^{-1} to over 160 W kg^{-1}, and that this interspecific variation was associated with differences in body shape and natural temperature regime. Moreover, muscle power output limits the fast-start performance of these animals [9]. Similarly, large differences in muscle power output have been measured in other studies comparing species, such as birds [10] and lizards [11], or life-history stages within one species, such as eels [12] and frogs [13].

Muscle power output is determined by the interaction between the active muscles and the external environment through which the organism is moving [2]. This becomes clear when comparing muscle power output in individuals moving through different media, such as water and air [14,15]. Ducks, for instance, produce less power when they are swimming than when they are walking, and their work, forces, and muscle strains differ between the two locomotor modes [15]. Additionally, different locomotor modes in the same medium, such as walking vs. running or running vs. jumping, might impose different power requirements on the locomotor musculature of an organism. Similarly, in legged animals, the configuration or movement pattern of a limb during different locomotor behaviors can affect muscle power output. *In vivo* changes in configuration or posture and power output can be studied by measuring muscle performance, such as power output, in concert with morphological and/or kinematical analyses [16,17].

At an even larger scale, different (micro) habitats might impose demands on the locomotor muscle system with regard to power output. For instance, arboreal habitats are extremely complex and often require organisms to jump from branch to branch. Because jumping requires high power outputs [8,18], arboreal organisms are expected to be able to generate high muscular powers. Similarly, lizards such as geckoes that often move on vertical substrates against gravity while escaping from predators [19] are expected to have high muscle mass–specific power outputs [20].

Given the potential for differences in limb configuration during different locomotor modes and its potential effect on power output, we compare mechanical power output in *Anolis* lizards while accelerating from a standstill and while jumping. We examined these two locomotor modes because *Anolis* lizards typically use burst running and jumping to escape from predators or when searching for food [21], suggesting that the selective forces acting on the generation of high mechanical power output should be strong. For two of the *Anolis* species, we furthermore quantify joint angles during running and jumping to investigate whether differences in limb configuration might explain differences in power output.

Additionally, we investigate whether muscle mass–specific power output has evolved within *Anolis* lizards in response to variation in microhabitat use and behavior. Previous ecomorphological studies have shown that *Anolis* lizards have evolved into ecologically and phenotypically distinct forms termed *ecomorphs*, including trunk-ground, trunk-crown, crown-giant, and twig ecomorphs [22–25]. Thus, although most species are classified as "arboreal," some are more ground dwelling, while others occur almost exclusively in vertical microhabitats. Consequently, we expect higher mechanical power outputs to evolve in species frequently occupying arboreal habitats because they often have to move against gravity. Similarly, we expect high power outputs to evolve in species that escape upward toward the canopy when threatened by a predator. Therefore, we expect to find correlations between muscle power output during locomotion and the habitat occupied by these animals. Specifically, we address the following questions: (1) Have muscle mass–specific power output during running and jumping coevolved? (2) Are interspecific differences in muscle mass–specific power output correlated with differences in ecology, i.e., microhabitat use? (3) Do differences in limb configuration explain differences in power output?

12.2 MATERIAL AND METHODS

12.2.1 Animals

We captured male individuals of 10 species *Anolis* by hand or noose at different field sites between November 2001 and June 2002. *A. carolinensis*, *A. sagrei*, *A. distichus*, and *A. garmani* were captured at mainland U.S. sites (*A. carolinensis* at New Orleans, LA; all the others at Miami, FL); *A. grahami*, *A. lineatopus*, and *A. valencienni* were caught in Jamaica (Discovery Bay), and *A. evermanni*, *A. gundlachi*, and *A. pulchellus* were caught in Puerto Rico (El Verde). We transported these species back to the laboratory at Tulane University. Upon arrival in the laboratory, we kept the lizards in pairs in 40-L terraria lined with leaf litter. We fed them live crickets dusted with calcium three times a week and sprayed them with water daily.

12.2.2 Morphology

We used digital calipers (Mitutoyo CD-15DC; ± 0.01 mm) to measure snout–vent length (measured from the tip of the snout to the cloaca) on all individuals used in the running trials. Between 1 and 3 individuals of the 10 species included in our analysis were sacrificed for morphological analysis. Specimens were preserved in 10% aqueous formalin and stored in 70% aquous ethanol solution. Before dissection, we weighed the intact specimens to the nearest 0.01 g on an electronic balance (A & D Instruments, FX 3200; ± 0.01 g). We subsequently dissected out all the hind limb muscles of one hind limb. We then weighed all femur retractor muscles and all knee and ankle extensor muscles separately on a Mettler MT 5 electronic balance (± 0.00001 g). We used the mass of these muscles in further analyses because these are the only muscles potentially responsible for generating propulsion. Muscle and body masses were averaged per species.

12.2.3 Running Trials

We induced the lizards to run up a 2-m long dowel (diameter of 8 cm) by clapping our hands and/or by tapping the base of their tails. The plastic dowel, covered with a mesh to provide sufficient traction, was placed against the wall at an angle of 45°. Individual lizards were filmed laterally using a high-speed video camera (Redlake

Motionscope PCI camera) at 250 Hz. We conducted between 5 to 10 trials per individual over several nonconsecutive days. Prior to experimentation and in between trials, we placed the lizards in an incubator set at 32°C for at least 1 h to attain body temperatures similar to their preferred field body temperatures [26].

The running trials for the Puerto Rican species (i.e., *A. pulchellus*, *A. gundlachi*, and *A. evermanni*) were conducted at the field station in Puerto Rico within 24 hr of capture. The experimental setup differed from the laboratory setup in the following ways: lizards were filmed at 240 Hz using a JVC high-speed video camera. Prior to filming and in between trials, the lizards were placed in individual bags in the shade to attain temperatures equal to or near the environmental temperature (30°C).

For analysis, we selected sequences in which the lizard (1) started from a complete standstill, (2) ran over a total distance of at least 25 cm, and (3) ran on top of the dowel (i.e., in a straight line). For these sequences, we digitized the tip of the lizard's snout for every frame using MOTUS software by Peak Performance. A sequence began 20 frames before the first movement and lasted until the lizard ran out of view or stopped. We subsequently smoothed the data using the Quintic Spline Processor implemented in MOTUS. Only data from individuals that performed at least three "good" trials were used in further analyses.

Based on the smoothed displacement data, instantaneous speed and acceleration (i.e., per frame) were calculated in MOTUS. Instantaneous body mass–specific power output for level locomotion can be calculated as the product of instantaneous velocity and instantaneous acceleration, and is derived by using the following formulas:

$$\text{Power output} = \text{rate of doing work} = \text{force} \times \text{distance/time} = \text{force} \times \text{velocity}$$

and

$$\text{Force} = \text{mass} \times \text{acceleration}$$

Then,

$$\text{Power output} = \text{mass} \times \text{acceleration} \times \text{velocity}$$

and

$$\text{Body mass–specific power output} = \text{acceleration} \times \text{velocity}$$

Because the animals were running at an angle of 45°, we had to take into account gravitational forces. Specifically, we used the following formula to calculate body mass–specific power output:

$$\text{Power output (W Kg}^{-1}\text{ body mass)} = [(\text{instantaneous acceleration}) + (\text{acceleration due to gravity} \times \cos 45°)] \times \text{instantaneous velocity}$$

The maximal value of body mass–specific power output was selected from each sequence.

To estimate peak muscle mass–specific power output, we took into account which step in each running sequence was associated with peak body mass–specific power output. If peak power output occurred within the first step, then both hind limbs were considered to be generating the propulsion because lizards typically started running from standstill with both hind limbs pushing off the substrate simultaneously. Muscle mass–specific power output was then calculated as body mass–specific power output divided by the ratio of the propulsive muscle mass of both hind limbs to total body mass. However, if peak power output occurred in later steps, then only one hind limb at a time was responsible for generating propulsion, and muscle mass–specific power output was calculated by dividing body mass–specific power output by the ratio of the propulsive muscle mass of only one hind limb to total body mass. The front limbs were not considered as contributing to the propulsion because they do not contribute to acceleration [27].

As an estimate of an individual's maximal muscle mass–specific power output, we selected the single highest peak muscle mass–specific power output from all the sequences for that individual, i.e., one value per individual. We subsequently averaged these values per species. We refer to this variable as "muscle mass–specific power output during running."

12.2.4 Jumping Trials

Jumping trials were performed on the same individuals and under similar laboratory settings as the running trials. As opposed to running, jumping is a single, discrete event making it possible to use a force platform to record the forces and ultimately power output (see further). We recorded forces of individual lizards jumping from a custom-made force platform to a branch positioned just outside the presumed reach of the individual (see Ref. [26] for detailed description). The lizards were induced to jump by clapping our hands or by tapping slightly on the tail. Prior to experimentation, lizards were placed in an incubator set at 28°C for *A. gundlachi* and 32°C for all the other species for at least 1 h [26,28]. Each animal was subjected to five separate trials, each on a different and nonconsecutive day. In each trial, we induced the lizard to perform as many separate jumps as possible until it was exhausted. In most cases, the animals performed three or more good jumps per trial. Body mass–specific peak (i.e., instantaneous) power output was calculated using an algorithm written in Superscope 11 (see Ref. [26,28] for details of the calculations). Because lizards always used both hind limbs to push off the substrate during jumping, we calculated muscle mass–specific power output by taking into account the propulsive force of two hind limbs as described above. As an estimate of an individual's maximal muscle mass–specific power output, we selected the single highest instantaneous muscle mass–specific power output from all jumping trials for that individual, i.e., one value per individual. We subsequently averaged these values per species. We refer to this variable as "muscle mass–specific power output during jumping."

For two of the species (*A. valencienni* and *A. carolinensis*), jumping trials were filmed in lateral view at 250 frames s^{-1} using the Redlake camera.

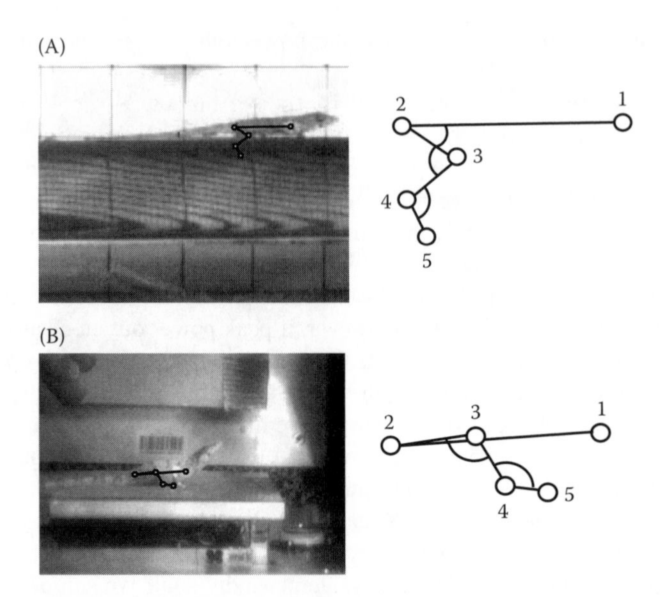

FIGURE 12.1 Lateral view and stick figures of *A. valencienni*. (A) Footfall prior to peak power (running). (B) Takeoff (jumping). In the right panel, the stick figure (enlarged) is shown with the three two-dimensional angles that were measured. Numbers refer to anatomical landmarks on the body: (1) shoulder, (2) pelvis, (3) knee, (4) ankle, and (5) base of the second toe.

12.2.5 Configuration of Hind Limbs

We quantified three joint angles for the two species (*A. valencienni* and *A. carolinensis*) for which movies of both running and jumping were available. In both cases, we only used those sequences in which the individual was producing the highest power output. For the running trials, joint angles were quantified on the frame of footfall of the step in which peak power was reached; in those sequences in which peak power was reached in the first step, joint angles were quantified on the frame prior to the start of any movement. For jumping trials, we used the last frame prior to the start of the jump.

The frames were subsequently imported into CorelDraw (version 10; Corel Corporation 2000), and four lines were drawn connecting (1) the hip to the shoulder, (2) the knee to the hip, (3) the ankle to the knee, and (4) the base of the second toe to the ankle. We then calculated the three angles between the different lines, i.e., the hip, knee, and ankle angles (Figure 12.1).

12.2.6 Ecology

We quantified the time each species spent off the ground, i.e., its degree of arboreality, by observing individual lizards in their natural habitat. Observations were done on the same populations as those sampled for the locomotor trials. Whenever we observed an animal, we noted the substrate type (ground, leaf litter, grass, trunk, branch, or boulder or stone wall) and dimensions (height and width) of the perch it was on at first sight. An individual was subsequently scored as being "on the ground" if the substrate type was ground, leaf litter, or grass; perch height was less than 10 cm; and perch width was greater than 400 cm. We subsequently approached the lizard and noted in which direction it fled (up, down, or horizontally).

As an estimate of "degree of arboreality," we used the ratio of number of observations the individual lizards were not sitting on the ground against the total number of observations for that species. As an estimate of "proportion of observations fleeing upward," we used the ratio of number of observations fleeing upward against the total number of observations for that species.

Each individual was only observed once. Except for *A. garmani*, only observations on males were included in the analyses. Since we were unable to observe any undisturbed *A. garmani* male in the field, we used observations on females instead.

12.2.7 Statistical Analysis

Because the values for the variables under study were not normally distributed, muscle mass–specific power output during running and jumping and all morphological variables were logarithmically ($\log_{10}$) transformed and the ecological variables were transformed by taking their arcsine [29] before statistical analyses.

To compare muscle mass–specific power output during running to muscle mass–specific power output during jumping for all 10 species, we performed a two-way ANOVA with the locomotor mode (i.e., running or jumping) and species as the factors. Two species, *A. carolinensis* and *A. valencienni*, were further analyzed in detail for two reasons: (1) They showed marked differences in muscle mass–specific power output during running and jumping (see Section 12.3), and (2) movies of both jumping and running were available, thus allowing the quantification of the two-dimensional joint angles. For just *A. carolinensis* and *A. valencienni*, we repeated the two-way ANOVA as described above. Since the analysis showed a significant locomotor mode–species interaction effect and species effect (see Section 12.3 for details), we performed one-way ANOVA (with species as factor) on the three joint angles for each locomotor mode separately.

We followed the procedure described below to test whether muscle mass–specific power output during running and/or jumping were intercorrelated and whether the variation in muscle mass–specific power output was explained by the variation in ecology (i.e., degree of arboreality and proportion of time escaping upward). In the latter case, we only used the maximal muscle mass–specific power output for each species, regardless of whether it was attained running or jumping. We refer to this variable as "maximal muscle mass–specific power output."

Because species share parts of their evolutionary history, they cannot be regarded as independent data points in statistical analyses [30–32]. Various methods and computer programs have been developed over the years, however, in which phylogenetic relationships among different species are taken into account in statistical analyses [30–33]. In this study, we used the independent contrast approach [30,31].

We calculated the standardized independent contrasts using the PDTREE program [34] on the transformed means per species of muscle mass–specific power output during running and jumping, maximal muscle mass–specific power output

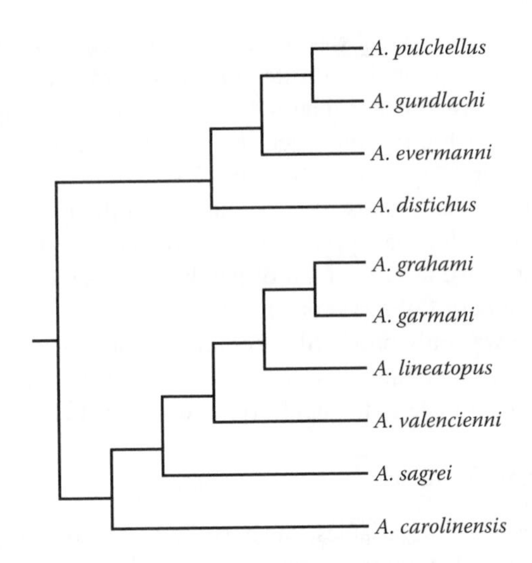

FIGURE 12.2 Phylogenetic tree of the 10 *Anolis* species used in this study. Relationships are based on mitochondrial DNA data by Harmon L.J. et al., *Science*, 301, 961, 2003. Branch lengths are not to scale.

snout-vent length (SVL), degree of arboreality, and proportion of time escaping upward. We subsequently performed two multiple regression analyses (backward method). In the first regression, we used the contrasts in muscle mass–specific power output during running as the dependent variable and the contrasts in muscle mass–specific power output during jumping as the independent variable. In the second regression, we used the contrasts in maximal muscle mass–specific power output as the dependent variable, and the contrasts in SVL, degree of arboreality, and proportion of time escaping upward as the independent variables. All the regressions were forced through the origin [35].

The independent contrast method requires information on the topology and branch lengths of the phylogenetic tree. The phylogeny of the 10 *Anolis* species under study here is based on a phylogenetic analysis of a much larger number of anole species by Harmon et al. [36], using mitochondrial DNA sequences (Figure 12.2). Branch lengths are available upon request from L. Harmon. Moreover, checks of branch lengths, using the diagnostics in the PDTREE program, did not show any significant correlation between the absolute values of the standardized contrasts and their standard deviations [35]. Recently, the phylogenetic relationships among *Anolis* lizards have been reexamined, resulting in minor changes [37]. Because, however, branch lengths have not been made available for this phylogeny, we preferred to use the phylogeny data of Harmon et al. [35].

12.3 RESULTS

The 10 species varied 10-fold in body mass, 15-fold in hind limb muscle mass, and about 2-fold in muscle mass–specific power output both during running and jumping (Table 12.1). As for interspecific variation in field behavior, the degree of arboreality varied between 45 and 100%, while the incidence of escaping upward varied between 27 and 100% (Table 12.1).

The 10 species of *Anolis* lizards differed significantly in muscle mass–specific power output (two-way ANOVA, species effect, $F_{9,178} = 9.76$, $p < 0.0001$) (Figure 12.3). Muscle mass–specific power output, however, did not differ between jumping and running (two-way ANOVA, locomotor mode effect, $F_{1,178} = 0.06$, $p = 0.80$), and the difference among species was similar in both locomotor modes (two-way ANOVA, locomotor mode–species effect, $F_{9,178} = 1.56$, $p = 0.13$). Additionally, the contrasts in muscle mass–specific power output during running were positively correlated with the contrasts in muscle mass–specific power output during jumping ($r = 0.86$, $F_{1,8} = 6.94$, $p = 0.030$) (Figure 12.4).

A multiple regression model (backward method) with the independent contrasts of maximal muscle mass–specific power output as the dependent variable and the independent contrasts of SVL, degree of arboreality, and proportion of time escaping upward as independent variables was performed; only an analysis of the latter independent variable was found to be significant ($r = 0.68$, $F_{1,8} = 7.05$, $p = 0.029$). Thus, only the variation in (contrasts in) the relative numbers escaping upward explained a significant proportion of the variation in (contrasts in) maximal muscle mass–specific power (Figure 12.5). Separate bivariate regressions showed the same results, i.e., only the contrasts in proportion of time escaping upward show a significant correlation with the contrasts in maximal power output.

In the two-species comparison, muscle-mass specific power output differed significantly between *A. carolinensis* and *A. valencienni* (two-way ANOVA, species effect, $F_{1,31} = 4.71$, $p = 0.038$), and the interspecific difference in muscle mass–specific power output varied significantly according to locomotor mode (two-way ANOVA, locomotor mode–species effect, $F_{1,31} = 4.88$, $p = 0.035$). Muscle mass–specific power output during running, however, did not differ significantly from the muscle mass–specific power output during jumping (two-way ANOVA, locomotor mode effect, $F_{1,31} = 0.12$, $p = 0.73$). Because there were no significant differences in power output between locomotor modes, only between species, we further compared muscle mass–specific power output for each locomotor mode separately between the two species using one-way ANOVA. During running, muscle mass–specific power output did not differ between *A. carolinensis* and *A. valencienni* (one-way ANOVA, $F_{1,26} = 0.94$, $p = 0.34$). Muscle mass–specific power output during jumping, on the other hand, was significantly greater in *A. valencienni* than in *A. carolinensis* (one-way ANOVA, $F_{1,21} = 5.31$, $p = 0.031$).

Because of the significant species–locomotor mode interaction effect and species effect (two-way ANOVA, see above), we compared the three joint angles for the two species for each locomotor mode seperately using one-way ANOVA. For running, both the hip and knee angle were not significantly different between *A. valencienni* and *A. carolinensis* (one-way ANOVA for both, $F_{1,16} < 0.27$, $p > 0.61$). The ankle angle in running trials, however, was significantly smaller in *A. valencienni* ($F_{1,16} = 5.76$, $p = 0.029$). For jumping, all three joint angles were significantly different between the two species (hip: $F_{1,18} = 11.84$, $p = 0.003$; knee: $F_{1,18} = 38.11$, $p < 0.0001$; ankle: $F_{1,18} = 15.55$, $p = 0.001$). Whereas the femur was more protracted

TABLE 12.1
Descriptive Statistics for Muscle Mass–Specific Power during Running and Jumping Trials, Propulsive Muscle Mass (of One Hind Limb), Body Mass, Degree of Arboreality, and Proportion of Time Escaping Upward per Species

	Muscle Mass–Specific Power (W kg^{-1})				Mass (g)			Field Behavior		
Species	N	Running	N	Jumping	N	Muscle	Body	N	Arboreality	Escaping up
A. pulchellus	16	574.32 ± 26.96	5	472.26 ± 36.49	2	0.076 ± 0.003	2.37 ± 0.07	23	0.78	0.43
A. gundlachi	18	524.84 ± 44.98	6	516.42 ± 65.83	3	0.317 ± 0.030	6.68 ± 0.21	35	0.97	0.69
A. evermanni	15	671.71 ± 46.25	3	767.66 ± 17.85	3	0.201 ± 0.006	5.42 ± 0.35	23	0.83	0.48
A. distichus	12	735.46 ± 61.73	5	647.43 ± 13.32	3	0.075 ± 0.025	2.97 ± 0.15	25	1	0.72
A. grahami	13	438.26 ± 40.47	4	601.50 ± 64.28	3	0.241 ± 0.040	5.54 ± 0.13	23	0.86	0.78
A. garmani[a]	8	570.55 ± 57.59	7	598.01 ± 58.51	1	1.054	26.31	6	1	0.83
A. lineatopus	14	540.73 ± 37.91	10	543.98 ± 31.94	1	0.247	5.19	14	0.93	0.42
A. valencienni	8	725.21 ± 70.36	8	848.71 ± 57.82	2	0.159 ± 0.021	6.93 ± 0.96	5	1	1
A. sagrei	19	419.61 ± 24.49	9	373.68 ± 39.69	2	0.226 ± 0.057	4.19 ± 0.45	22	0.45	0.27
A. carolinensis	12	714.90 ± 37.32	7	595.91 ± 89.60	3	0.222 ± 0.012	7.33 ± 0.33	30	0.97	0.57

Note: Given are the number of individuals (*N*) and – where applicable – means and standard errors (mean ± SE) per species.

[a] Observations of behavior are of females only. We were unable to gather data on males for this species.

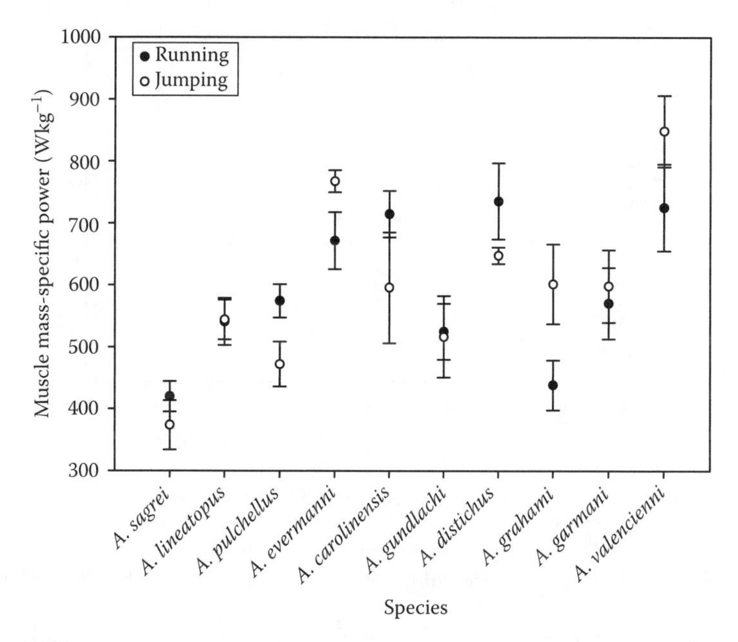

FIGURE 12.3 Mean and standard error of muscle mass–specific power output during running (closed symbols) and jumping (open symbols) per species. Species are ordered from low to high incidence of escaping upward when confronted with a (human) predator.

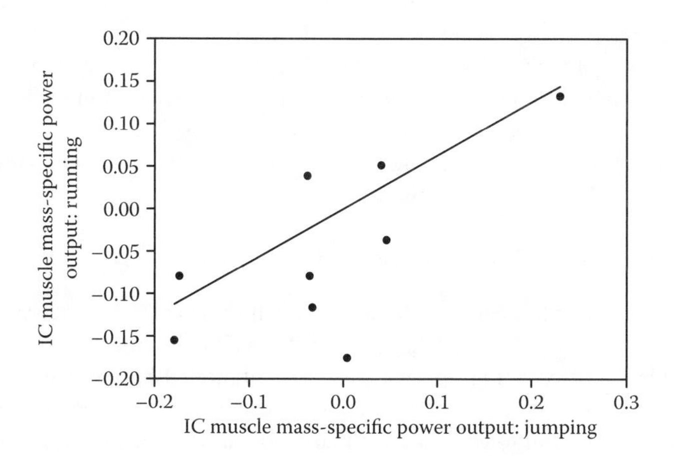

FIGURE 12.4 Regression through the origin of the independent contrasts (IC) of muscle mass–specific power output during running against the independent contrasts of muscle mass–specific power output during jumping ($r = 0.86$, $p = 0.030$).

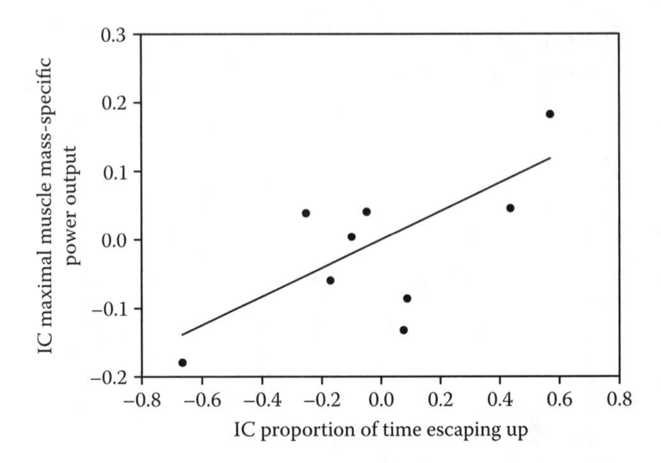

FIGURE 12.5 Regression through the origin of independent contrasts of maximal muscle mass–specific power output against the independent contrasts of the proportion of time lizards escape upward ($r = 0.68$, $p = 0.029$). Evolution toward higher incidence of escaping upward has been parallelled by evolution toward higher power outputs among the 10 *Anolis* species.

in *A. valencienni*, the knee and ankle angle in this species were significantly smaller than in *A. carolinensis* (Figure 12.6).

12.4 DISCUSSION

The aim of this study was to investigate whether muscle mass–specific power output has evolved within a closely related group of organisms in response to variation in microhabitat use and behavior. To our knowledge, ours is the first study that quantitatively links the variation in muscle power output to variation in ecology using a comparative approach. Our data show that muscle mass–specific power output varies considerably among species (more than twofold) and is indeed linked to the ecology and behavior of the species investigated.

12.4.1 Ecological Correlates of Power Output

Our results show a significant correlation between maximal power output and ecology across 10 species of *Anolis* lizards. More specifically, the evolution toward a higher incidence of escaping upward has been parallelled by the evolution toward higher muscle mass–specific power output. Muscle mass–specific power output, however, does not seem to correlate with an arboreal lifestyle *per se*. *Anolis* lizards are typically "active" lizards that run and jump around their natural habitat to search for food and/or partners, to defend their territories, or to escape predators [21]. However, it is clear that, on top of the large ecological variation, such as the degree of arboreality, there is also a large behavioral variation, such as antipredator behavior. Moreover, perching off the ground does not necessarily mean that the animals move up and down a lot, i.e., in a vertical direction. While some species typically jump down to other trees or to the ground, others "squirrel" to the opposite side of the

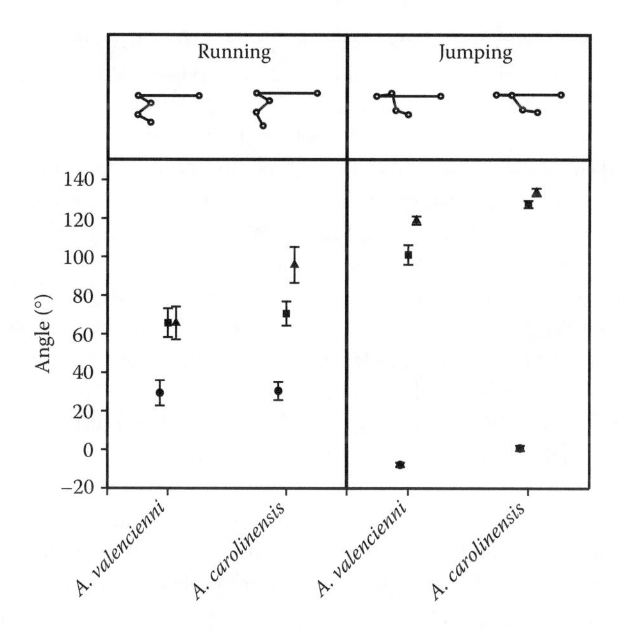

FIGURE 12.6 Plot of two-dimensional angles at footfall in running trials (left) and takeoff in jumping trials (right) for *A. valencienni* and *A. carolinensis*. Shown are means and standard error bars. For running, only the ankle angle differs between the two species ($F = 5.76$, $p = 0.029$), whereas all three angles differed significantly for jumping (all $F > 11.84$, all $p < 0.003$). Stick figures representing the mean configuration of the limb segments during running and jumping, per species are shown in the top panel. ● = hip; ■ = knee; ▲ = ankle.

trunk, i.e., stay at the same height, and/or run up [38]. In our sample, for instance, *A. lineatopus* typically does not escape upward when confronted with a (human) predator while *A. valencienni* does, although both show similar degrees of arboreality. Moreover, maximal power output is greater in *A. valencienni* (see Table 12.1 for raw values). Thus, it seems that higher power outputs are selected for in arboreal microhabitats, specifically in those species that frequently need to move rapidly by running or jumping against gravity.

12.4.2 Interspecific Variation

When comparing all 10 species, we found no overall difference in muscle mass–specific power output between jumping and running. Moreover, both power outputs seem to have coevolved. This might be explained by the fact that the same muscles are used in running and jumping.

Still, it is striking that even within a very closely related group of lizards such as *Anolis* lizards, large variation in muscle mass–specific power output exists. In our sample of 10 species, muscle mass–specific power output varies between 374 and 849 W kg^{-1}, i.e., more than a twofold difference (Table 12.1). In general, differences in maximal power output, when executing the same task, are attributed to differences in the physiological properties of the muscle. At the moment, we do not have data on the physiology, for example muscle fiber–type of the different muscles in the different species. However, it does seem likely that differences in muscle fiber–type composition are present because similar results have been found in other lizard species [11,39]. On the contrary, while (theoretically) also being of great importance to muscular power output, the configuration of the musculoskeletal system, such as the origin and insertion of the muscles, is generally considered to be more conservative, i.e., less variation is found among species. We have some evidence that this might not be the case among *Anolis* lizards. When comparing the hind limb musculature of two extreme *Anolis* species — a ground–trunk anole, *A. sagrei*, and a twig anole, *A. valencienni* — we found distinct differences in the insertion of the hip retractor and knee extensor muscles between these two species [40]. Thus, our limited data set on *Anolis* lizards at least suggests that muscular morphology might be more variable than previously thought.

Surprisingly, the values for muscle mass–specific power output we find for some *Anolis* species, such as jumping in *A. valencienni*, are within the range of so-called extremely high power outputs in jumping frogs [41,42] and flying or running birds [5,10,43]. In the past, it has been generally assumed that power amplifiers are necessary to produce such high powers at the whole animal level [18,41,42]. Recently, however, it has been shown that power outputs of around 1000 W kg^{-1} are possible at the muscular level [5,44]. Although likely, it remains to be tested whether in *Anolis* lizards high power outputs at the whole animal level reflect power output at the muscular level. However, our data do show that high power outputs, i.e., in the range of 500 to 800 W kg^{-1}, might not be as exceptional as previously claimed [5,10,41–43].

12.4.3 Power Output during Running and Jumping: A Two-Species Comparison

The comparison between *A. carolinensis* and *A. valencienni* shows that while muscle mass–specific power output is similar during running, it differs significantly during jumping. We suggest that biomechanical differences lie at the basis of this discrepancy. While the configuration of the hind limb segments is similar in both species in running trials, it differs greatly in jumping events. In the species producing the highest muscle mass–specific power during jumping, i.e., *A. valencienni*, the femur is significantly more protracted, i.e., the hip angle is negative (Figure 12.6), while the knee and ankle are more flexed, i.e., the knee and ankle angles are smaller (Figure 12.6). Could these differences be responsible for the differences in muscle mass–specific power output between the two species? The different configuration of the hind limbs may indeed result in changes in the instantaneous moment arms of the hip retractors and knee and ankle extensors that power locomotion. For example, inspection of Figure 12.6 shows that in, e.g., *A. valencienni*, the femur is maximally protracted. In this configuration, the dominant femur retractor (m. caudofemoralis) has a negligible moment arm and initially cannot contribute to femur retraction. This opens up the possibility for elastic energy storage in the broad tendon of the *M. caudofemoralis* inserting on the femur. However, for this to be the case, the muscle has to be active prior to femur retraction such that energy can effectively be stored. In *A. carolinensis*, the femur is significantly less protracted prior to the onset of the jump and may thus have less potential for energy storage. Although these differences may explain the observed differences in power output (and also the magnitude of power output in these animals), this hypothesis needs to be investigated further by electromyographic and sonometric recordings of the *M. caudofemoralis* during jumping.

ACKNOWLEDGMENTS

We would like to thank Daniel Ballow, Esteban Toro, and Margarita Ramos for their tremendous help with hours and hours of digitization, Kiisa Nishikawa (Northern Arizona University) for use of the high-speed JVC camera, and the organizers of the symposium "Ecology and Biomechanics" at the Annual Meeting of the Society of Experimental Biology (2004) for the invitation to contribute to their symposium and book. Luke Harmon provided the information on the topology and branch lengths of the phylogenetic tree. Esteban Toro and Jay Meyers kindly let us use their high-speed movies of jumping *A. valencienni* and *A. carolinensis*. Bieke Vanhooydonck and Anthony Herrel are postdoctoral fellows of the Fund for Scientific Research Flanders (FWO-Vl). This work was supported by NSF grants to Duncan Irschick (IBN 9983003 and 0421917).

REFERENCES

1. James, R.S., Atringham, J.D., and Goldspink, D.F., The mechanical properties of fast and slow skeletal muscles of the mouse in relation to their locomotory function, *J. Exp. Biol.*, 198, 491, 1995.
2. Marsh, R.L., How muscles deal with real-world loads: the influence of length trajectory on muscle performance, *J. Exp. Biol.*, 202, 3377, 1999.
3. Askew, G.N. and Marsh, R.L., The effects of length trajectory changes on the mechanical power output of mouse skeletal muscles, *J. Exp. Biol.*, 200, 3119, 1997.
4. Askew, G.N. and Marsh, R.L., Optimal shortening velocity (V/V_{max}) of skeletal muscle during cyclical contractions: length-force effects and velocity-dependent acitvation and deactivation, *J. Exp. Biol.*, 201, 1527, 1998.
5. Askew, G.N. and Marsh, R.L., The mechanical power output of the pectoralis muscle of blue-breasted quail (*Coturnix chinensis*): the *in vivo* length cycle and its implications for muscle performance, *J. Exp. Biol.*, 204, 3587, 2001.
6. Lutz, G.J. and Rome, L.C., Built for jumping: the design of the frog muscular system, *Science*, 263, 370, 1994.
7. Askew, G.N. and Marsh, R.L., Muscle designed for maximum short-term power output: quail flight muscle, *J. Exp. Biol.*, 205, 2153, 2002.
8. Biewener, A.A., Future directions for the analysis of musculoskeletal design and locomotor performance, *J. Morphol.*, 252, 38, 2002.
9. Wakeling, J.M. and Johnston, I.A., Muscle power output limits fast-start performance in fish, *J. Exp. Biol.*, 201, 1505, 1998.
10. Askew, G.N., Marsh, R.L., and Ellington, C.P., The mechanical power of the flight muscles of blue-breasted quail (*Coturnix chinensis*) during take-off, *J. Exp. Biol.*, 201, 3601, 2001.

11. Kohlsdorf, T. et al., Locomotor performance of closely related *Tropidurus* species: relationships with physiological parameters and ecological divergence, *J. Exp. Biol.*, 207, 1183, 2004.
12. Ellerby, D.J., Spierts, I.L.Y., and Altringham, J.D., Slow muscle power output of yellow- and silver-phase European eels (*Anguilla anguilla* L.): changes in muscle performance prior to migration, *J. Exp. Biol.*, 204, 1369, 2001.
13. Wilson, R.S., Franklin, C.E., and James, R.S., Allometric scaling relationships of jumping performance in the striped marsh frog *Limnodynastes peronii*, *J. Exp. Biol.*, 203, 1937, 2000.
14. Gillis, G.B. and Biewener, A.A., Hindlimb extensor muscle function during jumping and swimming in the toad (*Bufo marinus*), *J. Exp. Biol.*, 203, 3547, 2000.
15. Biewener, A.A. and Corning, W.R., Dynamics of mallard (*Anas platyrhynchos*) gastrocnemius function during swimming versus terrestrial locomotion, *J. Exp. Biol.*, 204, 1745, 2001.
16. Hedrick, T.L., Tobalske, B.W., and Biewener, A.A., How cockatiels (*Nymphicus hollandicus*) modulate pectoralis power output across flight speeds, *J. Exp. Biol.*, 206, 1363, 2003.
17. Hedrick, T.L., Usherwood, J.R., and Biewener, A.A., Wing inertia and whole-body acceleration: an analysis of instantaneous aerodynamic force production in cockatiels (*Nymphicus hollandicus*) flying across a range of speeds, *J. Exp. Biol.*, 207, 1689, 2004.
18. Aerts, P., Vertical jumping in *Galago senegalensis*: the quest for a hidden power amplifier, *Phil. Trans. R. Soc. B.*, 353, 1607, 1998.
19. Vanhooydonck, B., et al., Effects of substrate structure on speed and acceleration capacity in climbing geckos, *Biol. J. Linn. Soc.*, 85, 385, 2005.
20. Irschick, D.J. et al., Effects of loading and size on maximum power output and gait characteristics in geckos, *J. Exp. Biol.*, 206, 3923, 2003.
21. Irschick, D.J. and Losos, J.B., A comparative analysis of the ecological significance of maximal locomotor performance in Caribbean *Anolis* lizards, *Evolution*, 52, 219, 1998.
22. Rand, A.S., The ecological distribution of the anoline lizards around Kingston, Jamaica, *Breviora*, 272, 1, 1967.
23. Schoener, T.W., The *Anolis* lizards of Bimini: resource partitioning in a complex fauna, *Ecology*, 49, 704, 1968.
24. Williams, E.E., Ecomorphs, faunas, island size, and diverse end points in island radiations of *Anolis*, in *Lizard Ecology: Studies of a Model Organism*, Huey, R.B. et al., Eds., Harvard University Press, Cambridge, 1983, p. 326.
25. Losos, J.B., Ecomorphology, performance capability, and scaling of West Indian *Anolis* lizards: an evolutionary analysis, *Ecol. Mon.*, 60, 1990.
26. Toro, E. et al., A biomechanical analysis of intra- and interspecific scaling of jumping biomechanics and morphology in Caribbean *Anolis* lizards, *J. Exp. Biol.*, 206, 2641, 2003.
27. Aerts, P. et al., Bipedalism in lizards: whole-body modelling reveals a possible sprandrel, *Phil. Trans. R. Soc. B.*, 358, 1525, 2003.
28. Toro, E., Herrel, A., and Irschick, D.J., The evolution of jumping performance in Caribbean *Anolis* lizards: solutions to biomechanical trade-offs, *Am. Nat.*, 163, 844, 2004.
29. Sokal, R.R. and Rohlf, F.J., *Biometry: The Principals and Practice of Statistics in Biological Research*, 3rd ed., W.H. Freeman, New York, 1995, p. 859.
30. Felsenstein, J., Phylogenies and the comparative method, *Am. Nat.*, 125, 1, 1985.

31. Felsenstein, J., Phylogenies and quantitative characters, *Ann. Rev. Ecol. Syst.*, 19, 445, 1988.
32. Harvey, P.H. and Pagel, M.D., *The Comparative Method in Evolutionary Biology*, Oxford University Press, Oxford, 1991, p. 248.
33. Losos, J.B. and Miles, D.B., Adaptation, constraint, and the comparative method: phylogenetic issues and methods, in *Ecological Morphology: Integrative Organismal Biology*, Wainwright, P.C. and Reilly, S.M., Eds., University of Chicago Press, Chicago, 1994, p. 60.
34. Garland, T., Jr., Midford, P.E., and Ives, A.R., An introduction to phylogenetically based statistical methods, with a new method for confidence intervals on ancestral states, *Am. Zool.*, 39, 374, 1999.
35. Garland, T., Jr., Harvey, P.H., and Ives, A.R., Procedures for the analysis of comparative data using phylogenetically independent contrasts, *Syst. Biol.*, 41, 18, 1992.
36. Harmon, L.J. et al., Tempo and mode of evolutionary radiation in Iguanian lizards, *Science*, 301, 961, 2003.
37. Nicholson, K.E. et al., Mainland colonization by island lizards, *J. Biogeogr.*, 32, 1, 2005.
38. Regalado, R., Approach distance and escape behavior of three species of Cuban *Anolis* (Squamata, Polychrotidae), *Carib. J. Sci.*, 34, 211, 1998.
39. Bonine, K.E., Gleeson, T.T., and Garland, T. Jr., Comparative analysis of fiber-type composition in the iliofibularis muscle of phrynosomatid lizards (Squamata), *J. Morphol.*, 250, 265, 2001.
40. Herrel, A., unpublished data, 2005.
41. Marsh, R.L. and John-Alder, H.B., Jumping performance of hylid frogs measured with high-speed cine film, *J. Exp. Biol.*, 188, 131, 1994.
42. Peplowski, M.M. and Marsh, R.L., Work and power output in the hindlimb muscles of Cuban tree frogs *Osteopilus septentrionalis* during jumping, *J. Exp. Biol.*, 200, 2861, 1997.
43. Roberts, T.J. and Scales, J.A., Mechanical power output during running accelerations in wild turkeys, *J. Exp. Biol.*, 205, 1485, 2002.
44. Curtin, N., Woledge, R., and Aerts, P., Muscle directly meets the vast power demands in agile lizards, *Proc. Roy. Soc. Lond. B.*, 272, 581, 2005.

13 Implications of Microbial Motility on Water Column Ecosystems

Karen K. Christensen-Dalsgaard

CONTENTS

13.1 INTRODUCTION

All aquatic bodies in the world, from the smallest forest ponds to the open ocean, house complex and diverse microbial ecosystems. When it comes to things such as number of species and carbon and nutrient turnover, unicellular organisms completely dominate many aquatic ecosystems, and organisms on the size scale of fish are only minor players, contributing little to the overall balance. In environments dominated by open water, such as marine systems and those of large lakes, most of the photosynthetic activity is carried out by microscopic phytoplankton cells. In many ways, the microbial biota is as fascinating and complex as the apparently more flashy systems of tropical forests or coastal marine macroscopic ecosystems.

Pelagic microbial biota differ from systems of larger organisms in many respects. At the microbial level, aquatic systems are highly heterogeneous, and microorganisms live in and are adapted to a world of alternating feasts and famines [1]. This requires an ability to survive the famines as well as an ability to move in response to chemical gradients, which the microorganisms cannot perceive over the length of their bodies, in order to utilize the patchy resources. Microorganisms function at very low Reynolds numbers and thus in an environment entirely dominated by viscous forces; typically, microorganisms are too small to have sensory organs with which they can perceive prey unless they make contact with them. This leaves them only a few possible means of locomotion and feeding. Nevertheless, a large and diverse range of morphologies has developed (Figure 13.1), and numerous species

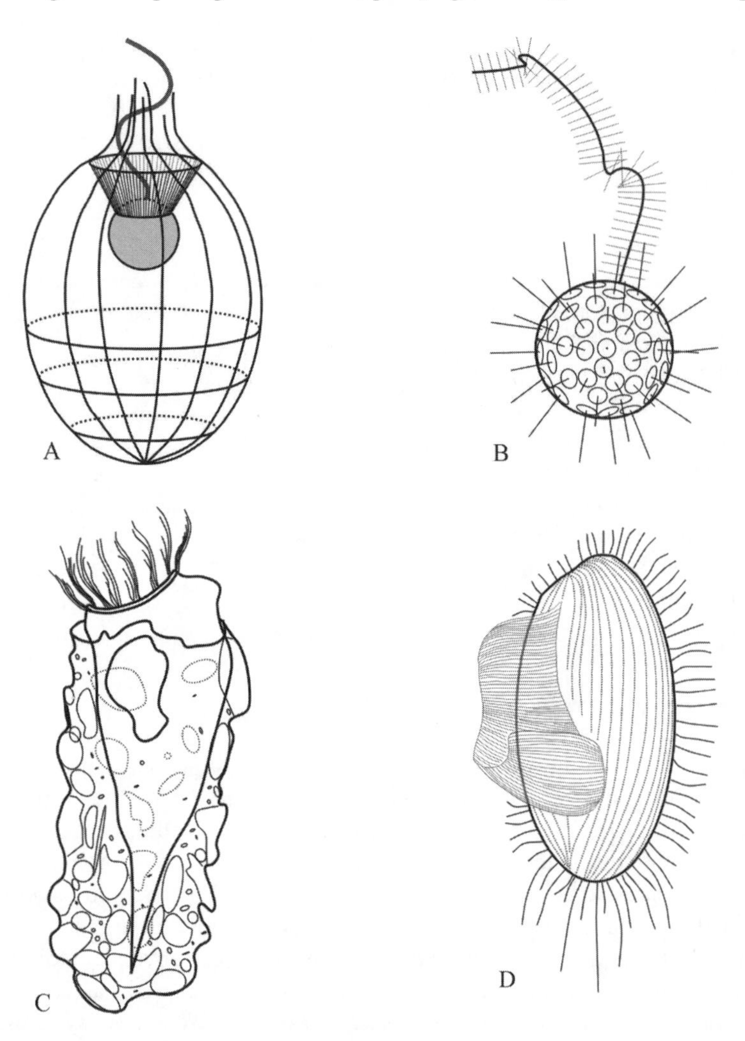

FIGURE 13.1 The diversity of the microbes represented by four genera, not drawn to scale. (A) *Diaphanoeca*, a flagellate with a smooth flagellum and a "cage" around the cell that increases the drag on the organism and thus the filtration efficiency (see Section 13.4.2). (B) *Paraphysomonas*, a flagellate with one short, smooth flagellum, one hispid flagellum (see Section 13.2), and a cell body covered by silica spikes. (C) The ciliate *Tintinnidium* incorporates agglutinated material into its lorica and has most of its cilia confined to its anterior end. (D) *Pleuronema* has a cell body entirely covered by cilia.

that occupy apparently similar niches coexist. As in other ecosystems, the peculiar characteristics of microbial ecosystems are nothing but a sum of the characteristics of each of the individual organisms of which it is composed and their response to the prevailing environment. Unlike systems of larger organisms, however, the microbial biota benefits from an absence of complex behavioral responses. Because of this, the characteristics of these ecosystems can in principle be understood directly from the mechanics and physiology of the individual organisms and their (to a large extent) predictable responses to stimuli.

In this chapter, I review some aspects of the existing knowledge on this topic and present some new calculations. I focus on pelagic systems and organisms that swim by cilia and flagella, which is the case for almost all motile pelagic microbes. I do not consider factors affecting the photosynthetic rate of autotrophic protozoa but deal only with the heterotrophic aspects of the microbial ecosystem. Unlike larger organisms, the distinction between autotrophic and heterotrophic in the microbial world is not clear. Many heterotrophic flagellates also contain chloroplasts [2]; even within the same species, there may be individuals with or without chloroplasts. However, because autotrophic flagellates are capable of ingesting particles at rates similar to those of apochlorotic flagellates [2,3], in this review they are grouped together with the rest of the heterotrophic flagellates.

13.1.1 Microbial Ecology in a Larger Context

It is well known that the classical textbook food chain of phytoplankton being eaten by copepods being eaten by fish and so forth is a huge oversimplification and describes at best only a minor part of the aquatic food chain [4]. Depending on conditions, 1 to 60% of the phytoplankton primary production is lost immediately as dissolved organic matter (DOM), probably mainly through lysing of phytoplankton cells [5–7]. Because of their small size and thus high surface to volume ratio, bacteria are highly efficient in the uptake of DOM [8], and the bacterial production based on phytoplankton exudates can be as much as 18 to 45% of the primary production.

Heterotrophic protozoa such as flagellates and ciliates are efficient grazers on bacteria and other small particles. Through their large numbers and high volume specific grazing rates, they are capable of clearing 3 to 100% of the entire water column for small particles per day. The average values lie between 7 and 90%, depending on the area studied [9–12]. Much of the grazing seems to be carried out by minute eukaryotic organisms not much larger than bacteria [13]; this, however, varies. Flagellates are generally shown to be the most important grazers on bacteria; ciliates, like flagellates, mainly graze on larger particles, but ciliates can also be important bacteriovors [12]. The importance of heterotrophic microorganisms in the ocean seems to vary with the season; they may mainly be important after the spring bloom under summer stratification when the phytoplankton is dominated by small forms [11]. Bacterial numbers remain fairly constant over time in marine systems, being typically around 0.5 to 3 × 10^6 ml^{1}; fluctuations in nutrient and DOM availability are apparent instead in fluctuations in the numbers of bacterivorous protozoa.

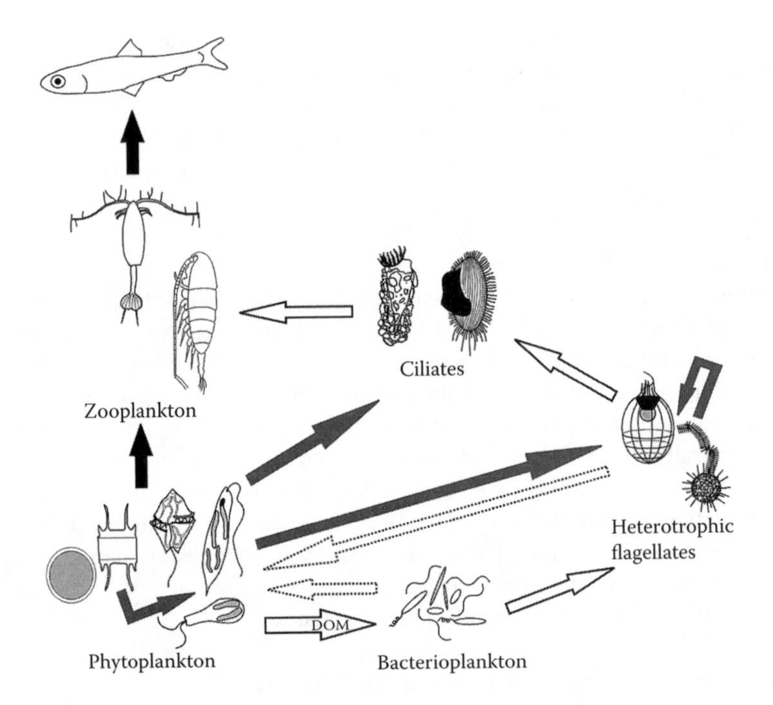

FIGURE 13.2 A much simplified version of the pelagic ecosystem. Black solid arrows indicate the classical food chain. Open arrows with a solid line represent the microbial loop functioning as a link that returns DOM to the higher trophic levels. Gray arrows show how the microbial loop can function also as a sink, which results in a higher number of trophic levels and thus higher respiratory costs. Open arrows with dotted lines show ingestion by heterotrophic phytoplankton.

Thus, the bacterial communities seem top-down controlled by grazing rather than limited by nutrients [4,9,10,13].

Flagellates may, for example, be consumed by ciliates that are in turn consumed by larger organisms such as copepods. Thus, the carbon originally lost as DOM is returned rather inefficiently to the traditional food chain through what has been denoted the microbial loop (Figure 13.2) [4]. In this way, protozoa form an important link from the DOM to higher organisms. They do not, however, specifically prey on bacteria but instead ingest particles in the right size class, and many flagellates are capable of ingesting particles their own size or in a few cases even larger, (e.g., see Refs. [14,15]). Small phytoplankton cells as well as other protozoa may be ingested as efficiently as, or even more efficiently, than bacteria [14,16–18]. Thus, protozoa may also function as sinks removing carbon from the system through increasing the number of trophic levels and so respiratory costs (Figure 13.2). Whether protozoa function mainly as sinks or links depends on the relative abundance of bacteria and small phytoplankton.

13.2 GENERATING MOTION WITH CILIA OR FLAGELLA

Microorganisms operate at Reynolds numbers (Re) << 1, and so in a world where all fluid motions are reversible, they are excluded from any form of propulsion that makes use of the inertia of the water. Octopuses or jellyfish shrunk to the size of protozoa and trying to move would simply be moving back and forth on one spot. In order to move, microorganisms instead make use of the difference in drag of a cylinder moving perpendicular compared to parallel to the flow; the resistance to normal motions of a cylinder is somewhat higher than the resistance to tangential motions. This is the principle behind motion by smooth and hispid flagella as well as cilia. Because of the insignificance of inertial effects at low Re, the motion of an object is only possible as long as a force acts upon it. If one attempted at low Re to throw a ball, it would never leave the hand (see, e.g., Ref. [19]).

13.2.1 Smooth Flagella

Because of the differences in drag, a moving cylinder tilted toward the direction of motion will exert a force on the fluid normal to the length of the cylinder (Figure 13.3). This is the principle behind flagellar propulsion first noted by Taylor [20,21]. Thus, contrary to appearance, the mechanics of flagellar motion is more closely related to that of a snake moving through sand than that of eels or water snakes swimming. Motion is generated by the propagation of planar or three-dimensional helical waves along the length of the flagellum. This generates a force normal to the segments of the flagellum that are tilted to the direction of the wave propagation (Figure 13.3). The propulsive effect depends on this force exceeding the retarding components of tangential forces acting along the body [22–24].

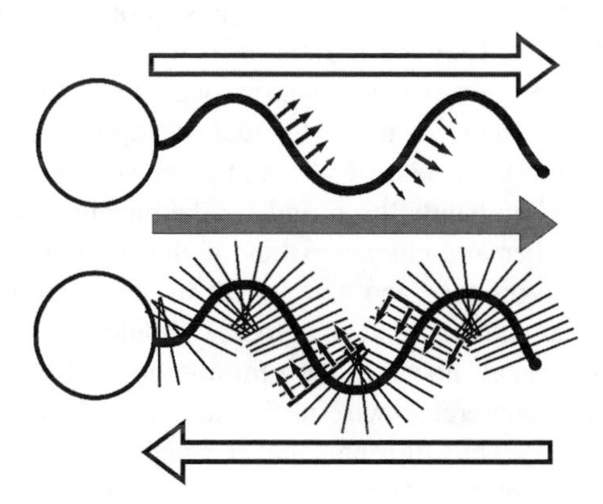

FIGURE 13.3 The generation of motion with smooth (upper) or hispid (lower) flagella. The black arrows represent propulsive force or thrust, the gray arrow shows the direction of flagellar wave propagation, and the white arrows indicate the overall direction of the resulting fluid motion with respect to the cell.

The fluid dynamics of bacterial and eukaryotic flagella are similar, but they differ in all other respects. Eukaryote cilia and flagella are around 0.2 μm in diameter and composed of the well-known 9 + 2 structure of inflexible microtubules that slide relative to each other. The bacterial flagellum is about 0.02 μm in diameter and is in itself completely immobile. It is composed of molecules of the protein flagellin that form a hollow tube. Perhaps unique in the biological world, the bacterial flagellum rotates continuously around its own axis [25] because of two rings that rotate relative to each other [26–28]. In this way, helical waves are propagated along the flagellum. Bacteria often have many flagella, which tend to form a bundle or bundles because beating filaments in the vicinity of each other tend to be synchronized through viscous coupling [20,29]. The flagella of this bundle are only separated during tumbles (see Section 13.5).

Both helical and planer waveforms have energetic disadvantages. In planar waves, the segments of the flagellum nearly parallel to the wave direction produce only drag and no thrust [30]. All segments of the helical waves produce thrust but also generate a torque on the organism that must be counterbalanced by the counter-rotation of the cell body. This reduces the swimming speed proportionally to the effective rotation rate [31]. It has been proposed that the body movements of micro-organisms rotated around their axis could contribute to the thrust in the manner of a rotating inclined plane [32,33], but this contribution would in most cases be negligible [34]. One interesting exception is the bacterium *Spirillum*, which has a spiral-shaped body. It uses the flagella to generate a rotation of the cell body, which then moves through the fluid much in the manner of a corkscrew through a cork [35–37].

13.2.2 Hispid Flagella

Hispid flagella have rigid hairs, or mastigonemes, protruding from the flagellum (Figures 13.1B and 13.3). They are curious in that they pull the cell body in the same direction as that of the wave propagation, opposite to that of smooth flagella. This is because the movement of the individual mastigonemes produces thrust in the direction of the cell body (Figure 13.3), and given sufficiently large numbers of mastigonemes, it is predominantly these, and not the flagellum itself, that moves the fluid [38,39]. The number and characteristics of the mastigonemes required may depend on the relative amplitude and wavelength of the flagellum [38,40,41]. Theoretically, only relatively inflexible mastigomenes should be capable of moving the fluid [39,42]. Dinoflagellates, however, have mastigonemes that appear flexible, and yet they always rotate counterclockwise in the direction of the flagellar beat of the transverse flagellum [43]. Thus, it seems that our understanding of the functioning of mastigonemes is not yet complete.

Flagellates with hispid flagella are common in all aquatic habitats, and this type of flagellum is present in a number of unrelated families. Because the presence of mastigonemes is not a primary character, it must provide important competitive advantages in microbial ecosystems. The importance of mastigonemes in evolution and ecology is, however, as yet poorly understood. They could improve the swimming efficiency of the cells, because the previously described energetic disadvantages of smooth flagella are not present in hispid flagella. The flagellum itself, however, will work against the motion of the fluid, so the picture is not clear. Another possibility is that the mastigonemes can function as mechanoreceptors or in feeding because the anterior position of the hispid flagellum would make it well placed to work also as a sensory or food-intercepting organelle. The mastigonemes also make it possible for hispid flagella to move fluid across or even perpendicular to the flagellar axis, something that is not possible for smooth flagella [44]. It has already been shown that *Paraphysomonas* uses its flagellum for intercepting prey and increases its effective feeding area by utilizing this possibility [44]. This, however, may be a special case.

13.2.3 Cilia

Cilia make direct use of the differences in drag between cylinders moving normal compared to tangential to the fluid in generating motion. The movement of the cilia consists of a power stroke in which the cilia forms a cylinder with a motion normal to the fluid, and a recovery stroke where most of the cilia moves tangentially with respect to the fluid (Figure 13.4). Whereas flagellates typically have only one or few flagella, ciliates must have thousands of cilia to produce motion. They are typically arranged in rows in which the cilia beat in metachronical waves, i.e., waves formed by a slight phase lag between adjacent cilia. These waves seem to be fluid dynamical in origin because they can be explained largely through the viscous coupling of adjacent cilia [45,46].

Another important aspect of the functioning of the cilia is the proximity of the cell wall. During the power stroke, when the cilium is extended away from the cell wall, it is capable of carrying along with it a large envelope of fluid. During the recovery stroke, the cilium moves close to the surface of the cell, and because of the viscous interactions with the wall, the cilium cannot carry as much fluid with it. Hence, there is a net movement of fluid down the surface of the cell, which contributes to the motion of the organism [47]. The fact that the ciliates move by moving fluid over the cell surface results in a much steeper velocity gradient over the surface than that found in inert bodies being pulled through the fluid by an external force, such as sedimenting organisms. Bodies pulled in this way carry more fluid along with them, and so they disturb the fluid much more than swimming cells [47,48]. Because predators may perceive prey through fluid dynamic signals such as shear [49,50], this reduces the visibility of the ciliates.

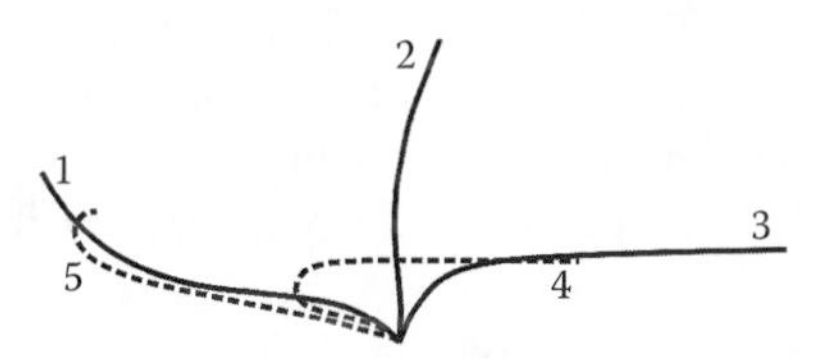

FIGURE 13.4 Movement of a cilium. The movement from position 1 to 3 constitutes the power stroke and from 3 over to 4 and 5 and back to 1 constitutes the recovery stroke. Movement purely related to the recovery stroke is drawn with dashed lines.

Flagellates, which pull an inert cell body through the fluid with a flagellum, do not have this advantage, but they are also typically smaller and slower. Hence, it is not clear under which conditions flagellates are hydrodynamically more visible than ciliates. Ciliates that have the cell body only partially, as compared to fully, covered by cilia could also generate a larger scale flow field around the cell body. Thus, it seems that the exact mechanism by which protozoa generate motion influences their relative visibility toward different types of predators and so influences their relative predation rates. Though this has important ecological implications, there have to my knowledge not been any thorough investigations of this phenomenon on protozoa. Only work on how foraging behaviors influence hydrodynamic visibility in copepods [51] and on how size and velocity of an assumed nonciliated particle affects its visibility [49] have been carried out so far.

13.3 THE ENERGETICS OF MOTION

One of the curious aspects of microbial ecosystems compared to those of larger motile organisms is the apparent lack of an optimal swimming speed. It has often been stated that the energy spent by microorganisms on swimming constitutes only a minute part of their metabolism (e.g., Refs. [52,53]). In light of this, it should always be advantageous for microorganisms to swim as fast as possible because increasing their swimming speed will increase the contact rate with prey and thus the feeding rate. In spite of this, there are large differences in swimming speed between different species within the same size class and feeding on the same prey. The extent of this cannot be explained by differences in drag of the feeding apparatuses, as even within species that do not have retractable feeding structures, large variations in swimming velocity occur [54]. This variation should provide a firm basis for natural selection toward higher swimming speeds. So why do many microorganisms still swim relatively slowly?

One answer could be to reduce the probability of being preyed upon themselves. Contact rates between predator and prey are dependent not only on the swimming speed of the predator, but also on the swimming speed of the prey [55]. However, flagellates often seem bottom–up rather than top–down regulated (e.g., for instance Refs. [10,56]). Furthermore motility, while increasing the contact rate, may decrease the interception rate, hence in reality potentially providing protection from predation [57]. Thus, this does not seem to be satisfactory as the only answer.

In most previous studies, only the drag on the cell body itself was used in the energy budget calculations. The drag on the cell body, however, constitutes only a minute part of the overall energy expenditure of motion; most of the energy is used in overcoming the tangential drag on the flagellum [22,30,31,39,58]. The total power consumption by a flagellum propelling a microorganism by helical motions is given by [31]:

$$P = \eta^{-1} 6r\pi\mu U^2 \tag{13.1}$$

where P (J s^{-1}) is the average power consumption, r (m) is the radius of the organism, μ (N s m^{-2}) is the viscosity of water, and U (m s^{-1}) is the swimming speed of the organism. η^{-1} is a nondimensional parameter defining the swimming efficiency of the organism. The optimal (smallest possible) value of η^{-1} that is achievable for a given organism depends on factors such as relative size of the cell body and relative length and width of the flagellum. The larger the cell body or the thicker the flagellum compared to its length (L), the less the possible efficiency. Whereas an organism with $L/r = 10$ and $r = 50 \times$ the radius of the flagellum can, in principle, achieve an optimal η^{-1} of 125, an organism with the same radius of the flagellum but $L/r = 5$ will not be able to do better than a η^{-1} of 210. The actual value of η^{-1} for a given organism can in principle range from this optimum to infinity, depending on flagellar parameters such as amplitude and frequency [31]. The value will only approach infinity if, for instance, the amplitude of the flagellum is going toward zero, resulting in very inefficient swimming.

I will assume that the flagellates are swimming at constant speed until they encounter a food particle, then stop their motion for the time it takes to ingest the food particle, and then immediately reassume constant forward motion, as is seen for, e.g., *Paraphysomonas vestita*. The ingestion rate over time can then be calculated to be:

$$I = \frac{1}{(CV)^{-1} + i} \tag{13.2}$$

where I (particles s^{-1}) is the ingestion rate, C (particles m^{-3}) is the concentration of food particles, V (m^3 s^{-1}) is the volume of liquid that passes through the area swept for particles over time ($V = UA$, where A is the area swept for particles), and i (s $particle^{-1}$) is the time it takes for the protozoon to ingest one food particle.

I have assumed the following: The bacteria have a radius of 0.3 μm and are similar to *E. coli* with 26% of their volume composed of organic compounds, of which 8% is lipids and 92% is other organic compounds [59]. The energetic value of lipids is taken to be 37 KJ g^1, and the energetic value of other compounds is taken to be 17 KJ g^1. The bacteria do not swim. The protozoa have a radius of 3 μm, and the flagellum has a length of 30 μm and a radius of 0.1 μm. At these values, the energetic optimum of the organism lies at approximately $\eta^{-1} = 150$. The protozoa are assumed not to spend any energy on motion while in the process of ingesting a particle, and when swimming, do so close to their energetic optimum. They can ingest 60% of their volume per hour in accordance with the data in, e.g., Ref. [52], giving an average ingestion time of 6 sec. Ingested carbon not used in respiration for the purpose of motion is used in growth, with a growth efficiency of 40%. In this case, A is assumed to be equal to the transectional area of the body of the protozoa, as is seen in *P. vestita* [54]. From these assumptions, I have calculated the growth rate of the protozoa as a function of swimming velocity at different bacterial concentrations (Figure 13.5A and 13.5B).

How the growth rate varies with swimming speed depends greatly on the concentration of food particles. At concentrations of 10^{11} particles m^{-3}, the growth rate

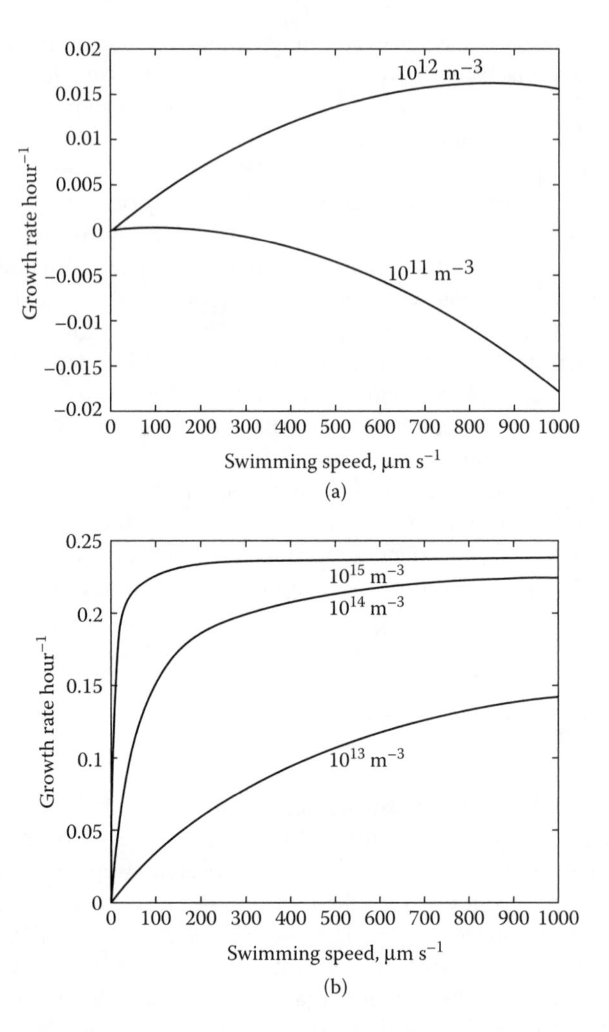

FIGURE 13.5 Theoretical growth rate of protozoa as function of their swimming speed at bacterial concentrations of (A) 10^{11} and 10^{12} bacteria per cubic meter and (B) 10^{13} to 10^{15} bacteria per cubic meter.

is effectively zero until the velocity is 180 μm sec^{-1}, with a small optimum of 1.85×10^{-4} hr^{-1} at 91 μm sec^{-1}. At velocities above 180 μm sec^{-1}, the growth rate becomes increasingly negative. At 10^{12} particles m^{-3}, the growth rate increases to a maximum of 1.61×10^{-2} at a velocity of 850 μm sec^{-1}, then decreases again. At higher particle concentrations than this, the growth rate increases continuously with velocity for realistic values of potential swimming speed. The value however reaches an asymptotic value toward the maximum growth rate and so, after the asymptotic level is reached, a further increase in swimming speed does not significantly increase growth rates. The growth rates found are well in accordance with previous values obtained experimentally of growth rates for the given particle concentration (e.g., Refs. [52,60].

As mentioned earlier, in aquatic systems, bacterial concentrations are typically around 0.5 to 3 × 10^{12} bacteria m^{-3}. Based on this, one would expect protozoa to swim at higher velocities than the 150 to 300 μm sec^{-1} commonly observed for protozoa not hindered significantly by feeding structures. However, as is discussed in Section 13.5.2, bacteria in the ocean are not uniformly distributed but accumulate in patches that can have concentrations orders of magnitude higher than background concentrations [61–63]. An individual protozoon in an environment with an average bacterial concentration of 10^{12} bacteria m^{-3} will rarely experience this bacterial concentration at any given time.

More realistically, we may imagine that the protozoa experience a lower background concentration of say 10^{11} bacteria m^{-3} with small patches of highly elevated concentrations, say, 10^{14} bacteria m^{-3}. At the low background concentrations, swimming speeds above 180 μm sec^{-1} will give the protozoa increasingly negative growth rates, thus, not allowing it to survive as long and lengthening the recovery time when it once again encounters higher concentrations of food. It should be noted, however, that the magnitude of the negative growth rate is exaggerated because protozoa undergo physiological changes that enable them to survive when starved [64]. At the higher patch concentration, swimming speeds above 300 μm sec^{-1} will only benefit it little. Hence, under these conditions, one would expect the optimal swimming speed of the organism to be below 300 μm sec^{-1}, in accordance with what is actually observed in nature. Because of the heterogeneous nature of the food resources, however, no single swimming speed is always optimal but will vary continuously with conditions. Every possible concentration between 10^{11} and 10^{14} bacteria m^{-3} will be encountered by the protozoa at some stage. This explains the high variation seen within these organisms when it comes to swimming speed; there is no unidirectional selective pressure. This is enhanced because swimming serves not only to directly enhance encounter rates with food particles, but also to move to and locate areas of higher food concentration (see Section 13.5). Whether this increases, decreases, or leaves unaltered the optimal swimming speed for a given organism depends on its abilities to orient itself to stimuli and its mechanisms for doing so.

The proportion of the energy budget used for motion obviously depends on bacterial concentration as well as swimming speed (Figures 13.6A and 13.6B). The higher the bacterial concentration, the higher the gain per distance the cell swims. For concentrations of above 10^{13} bacteria m^{-3}, less than 1% of the ingested carbon is utilized for energy production for swimming almost irrespective of velocity. Because this is a concentration at which experiments are frequently conducted, it is not surprising that it has often been concluded that protozoa spend only a very small fraction of their energy budget on swimming. However, at 200 μm sec^{-1} and 10^{12} bacteria m^{-3}, the figure is 10%, and at 10^{11}, the figure is around 100%, explaining why the threshold concentration above which protozoa are capable of balanced growth is typically above 10^{11} bacteria m^{-3} [52,65]. The rate with which flagellates are ingested by ciliates, and hence with which the organic carbon is channelled up to higher trophic levels in the food chain, depends among other things on the swimming speed of the flagellates. At the same time, the respiration required for the energy production results in fixed carbon being lost from the aquatic ecosystems

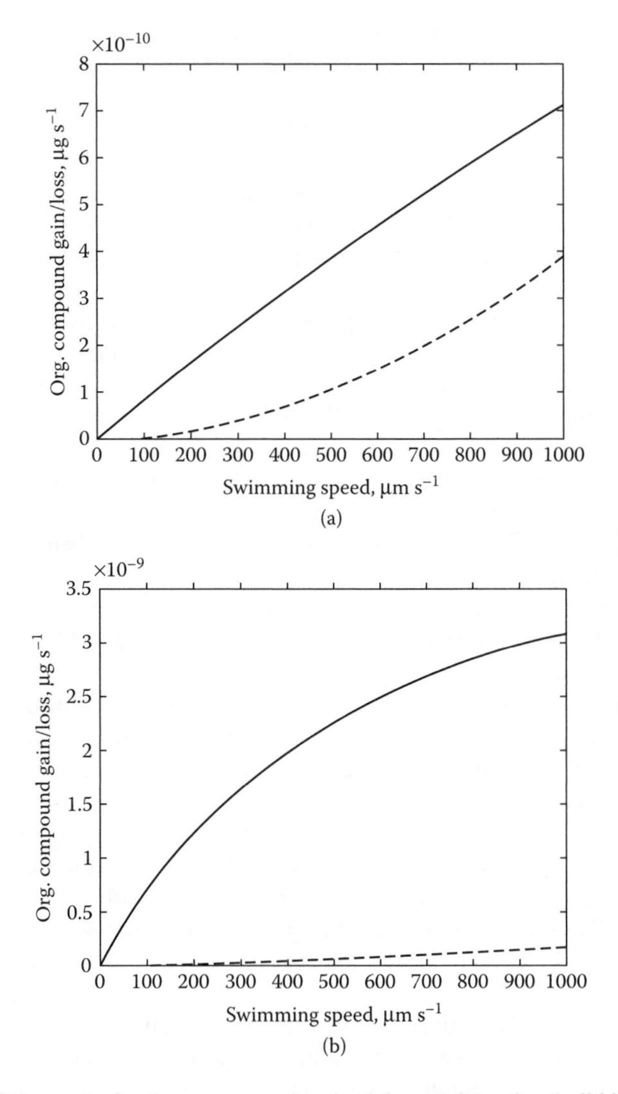

FIGURE 13.6 Amount of carbon compounds gained through ingestion (solid lines) and lost through respiration (dashed lines) as a function of swimming speed. Bacterial concentrations: (A) 10^{12} bacteria per cubic meter and (B) 10^{13} bacteria per cubic meter.

as CO_2. It seems that the potential importance of the energetic balance involved in motion of the individual protozoa so far has been underestimated, and it may be an important shaping factor in microbial ecosystems.

13.4 FEEDING MECHANISMS

Because of the low Reynolds number at which microorganisms live, they are excluded from any type of feeding that makes use of the inertia of the prey or their own bodies. At the same time, they are too simple to have elaborate organs such as complex eyes, which would enable them to perceive potential prey at a distance, and too small to react swiftly to chemical gradients. Thus, they are dependent on passive contact of their feeding structures with potential food organisms. They can improve the chances of this happening by swimming or generating feeding currents. The feeding mechanisms employed by heterotrophic microorganisms can broadly be divided in three categories: diffusion feeding, interception feeding, and filter feeding.

Diffusion feeding is the simplest form of feeding and involves passively waiting for motile prey to make contact with the feeding structure. This is not a widespread feeding mechanism but is used, e.g., in the heliozoan *Ciliophrys*, which is immobile with extended pseudopodia when feeding; *Ciliophrys*, however, is capable of contracting its pseudopodia and swimming when threatened [66].

In interception feeding, organisms swim or generate feeding currents and intercept particles that make contact with a feeding structure, such as the flagellum in *Paraphysomonas* [44]. An interesting and unusual example of an interception feeder is *Noctiluca scintillans*, which is not in itself motile but is capable of moving by being positively buoyant. In this way, it encounters food particles, often using a long thread of mucus to which the particles attach. It can control its velocity to some extent by altering its buoyancy by swelling; it moves approximately three times faster if starved [67]. Hence, the feeding state of the individual cells can alter the probability of blooms of *N. scintillans* occurring at the surface.

The term *interception feeder* is often used to imply that only particles in streamlines no more than one particle radius away from the feeding structure are intercepted [68,69]. In reality, electrodynamic, electrostatic, solvation, or steric forces may result in the particle being attracted or repelled, and so may cause it to cross streamlines, which theoretically may greatly affect particle retention rates [70]. Experimentally, it has been shown that the hydrophobicity of the bacteria may indeed affect feeding rates [71]. For small particles, diffusion encounter may also be of importance [72], even for swimming cells.

In filter feeding, the organism makes use of a filtering structure through which water is sieved by the motion of the organism or the generation of feeding currents. All particles larger than the distance between filtering elements are removed from the flow and transported to the "mouth." Particles smaller than this may also be removed if they make contact with and stick to the feeding structures in the same way as described for interception feeding. Thus, the feeding mechanisms may be overlapping. Filter feeding is used by a large number of species, including many ciliates as well as numerous flagellates such as choanoflagellates and pedinellids. The filter element can be composed of cilia, as is the case in ciliates, or pseudopodia in the case of flagellates.

Turbulence may in theory increase the contact rate, and thus the feeding rate, of aquatic organisms by increasing the relative velocity of predator and prey [73]. It is not clear, however, whether turbulence in reality has a positive effect on feeding rate, even among larger organisms such as copepods. This is, among other things, because turbulence may disturb the feeding currents and alter the behavior of the organisms (e.g., Refs. [74,75]).

For protozoa and bacteria, which are well below the Kolmogorov length scale and so live in a world of laminar flow, the picture is even more obscure. There have been reports of increased feeding rates of some flagellates in the presence of turbulence [76,77], but they are not entirely convincing. To some extent, the increased feeding rate could be related to the flagellates' increase in numbers but decrease in size in the presence of turbulence [78]. Thus, although it seems that turbulence does affect the microbial ecosystems, I will largely ignore it in this review because the nature of this phenomenon is as yet very poorly understood.

13.4.1 The Coexistence of Filter Feeders

A majority of the heterotrophic species of most pelagic microbial ecosystems are filter feeders. Many overlap in the particles that they are capable of ingesting, and yet they coexist. In systems of larger animals, this would not be possible because one species would tend to outcompete the others. The reason why this is possible can be found in the nature of filter feeding at low Reynolds numbers, which is illustrated here by a simplified model.

Protozoa seem to be capable of upholding only relatively small hydrostatic pressures in the range of 0.8 to 1.5 N/m^2 [79,80]. The unrestricted velocity through the area occupied by the filter is directly proportional to the pressure drop over the filter and to the distance between adjacent filtering elements compared to their diameter. Assuming that the filtering elements are parallel cylinders, the unrestricted velocity of the fluid can be given by [79]

$$U_0 = \frac{\Delta p(1 - 2\ln\tau + \tau^2/6 - \tau^4/144)b}{8\pi\mu} \tag{13.3}$$

where U_0 is the velocity of the fluid when unrestricted by the filter, Δp is the pressure drop over the filter, τ equals $\pi d/b$ (where d is the diameter of the individual filtering elements and b is the distance between the centers of adjacent filtering elements) (Figure 13.7), and μ is the viscosity of the fluid.

Assuming that the filtering elements are rigid, filter feeders can retain any particle with a diameter larger than b. Particles smaller than this can only be retained if they make contact with and are intercepted by the filtering element. In reality, it seems that only very few particles smaller than b are intercepted by filter feeders [81,82]. Thus, there is a trade-off: the greater the distance between filtering elements, the larger the velocity and hence volume of fluid filtered for particles, but the smaller the proportion of the particles that are retained. The power loss in driving fluid through a filter can be found by using Equation 13.3.

$$P = \frac{8\pi\mu\alpha U_0^2}{b(1 - 2\ln\tau + \tau^2/6 - \tau^4/144)} \tag{13.4}$$

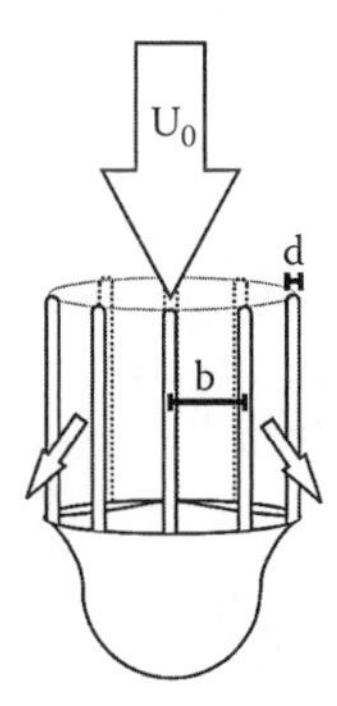

FIGURE 13.7 An idealized filter feeder as used in the model. Water enters the opening of the filter at velocity U_0 and passes out through the sides at a lower velocity as a result of the energy loss due to the pressure drop over the filtering elements. Particles in the water are intercepted by the filtering elements. The distance between the centers of adjacent filtering elements is given by b, and the diameter of the cylindrical filtering elements is given by d.

where P is the power loss and α is the area of the surface of the cylinder spanned by the feeding elements.

The growth rate of filter-feeding protozoa with various distances between adjacent filtering elements as a function of the size of the bacterium is presented in Figure 13.8. The calculations were based on the following assumptions. The pressure drop over the filtering structure is in all cases assumed constant at 1 N m^{-2}. The distances between adjacent filtering elements varies from 0.2 to 1.2 μm and are evenly distributed in this interval. The radii of the bacteria vary from 0.1 to 1.4 μm. As the size of the bacterium changes, the total bacterial volume is held constant at 0.11×10^{13} μm^3 m^{-3}, equivalent to 10^{13} bacteria per cubic meter at a radius of the bacteria of 0.3 μm. This number was chosen because it has been shown to be well above that required for balanced growth in protozoa (e.g., Refs. [52,60]). Because the total volume is held constant, the number changes from 2.7×10^{14} to 1×10^{11} m^{-3} as the radius of the bacteria changes from 0.1 to 1.4. This was done for comparative purposes to avoid an excessive increase in the growth rate of all the protozoa with increasing bacterial volume.

The protozoa are assumed to draw water through a filtering apparatus as shown in Figure 13.7. The average radius is 3 μm and the area of the opening of the filter structure (A) is 4.5 μm. The velocity of the fluid through the opening is assumed to be not yet influenced by the presence of the filtering elements and thus be equivalent to U_0, which seems valid from video microscopy analysis [54]. As my model only serves to illustrate a principle and does not attempt to be accurate, forces on the particle other than the viscous ones are ignored. Thus, for smaller particles to be intercepted, they should be on a streamline that passes the filtering element at a distance of no more than that of the radius of the particle [69]. For particles smaller than the size of the filtering element, 20% of the particles that made contact were assumed to "stick" to the feeding apparatus and thus be retained. Otherwise, the assumptions are similar to those of the power calculations used in Section 13.3. Particles are ingested with a rate as described by Equation 13.2, but with a maximum

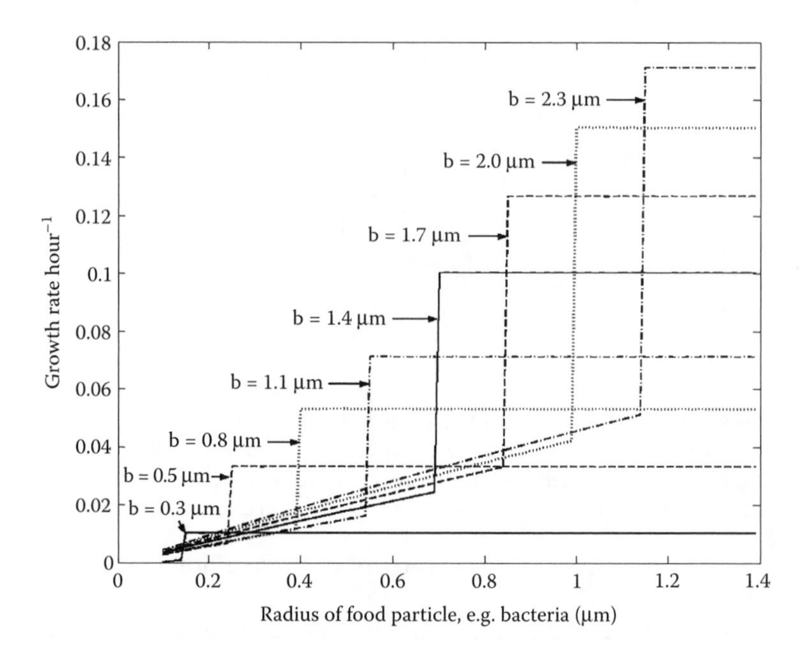

FIGURE 13.8 Growth rate of protozoan filter feeders as a function of the size of the food particles. The growth rate is calculated for different filter feeders with distances b between the centers of adjacent filtering elements ranging from 0.3 to 2.3 μm.

ingestion of 100% of the cell volume of the protozoa per hour since this seems an amount that is more realistic for filter feeders (e.g., Refs. [52,82]).

As can be seen in Figure 13.8, all of the protozoa have a positive growth rate at any of the sizes of bacteria. For the individual protozoon, the growth rate increases gradually with bacterial size until this exceeds the distance between filter elements. Here, there is a jump in growth rate marking the difference between being an inefficient interception feeder and being capable of retaining all the particles by means of the filter. After this there is no further increase in growth rate because the volume of bacteria per cubic meter is constant. Though all of the protozoa have positive growth rates with every size of bacteria, they each have a unique size range in which they have a maximum growth rate. Analogously, it is possible to calculate the number of bacteria ingested by each of the protozoa, assuming an equal number of bacteria in each bacterial size class. Though all of the protozoa are ingesting bacteria of every size, each is the main predator for a size range of bacteria and for this size range ingests the largest number of bacteria.

This provides some understanding of how filter-feeding protozoa can coexist in nature. In a natural setting, a range of sizes of bacteria will be present at any given time. Protozoa with different distances between filtering elements will be the main predators of different size ranges of bacteria. However, because the protozoa can survive on classes of bacteria for which they are not the main predators, they can be present in low numbers also in systems where they do not have the competitive advantage. They are then present to take advantage of possible changing conditions.

The system of filter feeders is dynamic because a large number of any given type of protozoa will tend to result in the "grazing down" of the size range of bacteria on which it is the dominant predator. This is especially true for protozoa specializing in larger food particles because of the much higher clearance rates. This can be illustrated by looking at the development over time in a system starting off with 10^{12} bacteria per cubic meter of each of the nine size groups described above, and 10^8 protozoa per cubic meter with each corresponding distance between filtering elements (Figures 13.9A and 9B). Bacteria are assumed to have a growth rate of 0.05 h^{-1}. The modeling has been done computationally with time steps of 120 sec. Larger bacteria are grazed down much more quickly than smaller ones, with only the smallest having positive growth rates under these conditions. In addition, the relative growth rates of the different types of protozoa change over time as the relative size distribution of the bacteria changes.

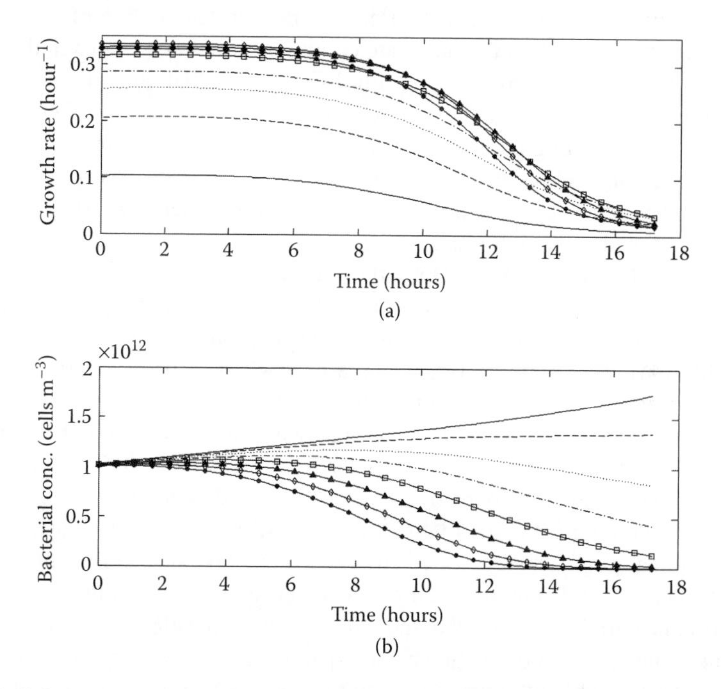

FIGURE 13.9 Development over time in (a) the growth rate of protozoa and (b) bacterial concentration. The distance between the surface of adjacent filtering elements (function of b and d) as well as the size of the bacteria range from 0.2 μm (solid line with no markers) to 2.3 μm (lines with solid circles) with steps of 0.3 μm between size groups. The larger bacteria are grazed down faster, and so the relative growth rate of filter feeders with smaller distances between filtering elements compared to those with larger distances increases.

13.4.2 Attaching to Particles while Feeding

Much organic matter in pelagic ecosystems is found in the form of aggregates, or particulate organic matter (POM). As these sediment out of the pelagic zone, they constitute a major pathway for loss of organic carbon from pelagic ecosystems. Many protozoa as well as bacteria attach to the surfaces of POM using mucus threads, flagella, or cilia. Bacteria associated with aggregates are to a large extent responsible for the degradation of the aggregates that occur while sinking to the seafloor [83] and are therefore important in recycling organic matter and nutrients to this system. The colonization of aggregates by bacteria is facilitated by the chemotactic behavior of some bacteria [84,85]. The colonization rate may be either positively [86] or negatively [87] affected by the presence of other bacteria on the aggregates.Much of the reason for the attachment of microbes to aggregates can undoubtedly be found in their wealth of organic matter. Bacteria feed directly on the organic matter of the aggregate, and flagellates feed on attached or suspended bacteria associated with the aggregates (e.g., Refs. [88,89]). Because the flagellates seem capable of controlling the bacterial communities, they affect the rate of degradation of the particles. Ciliates may in turn feed on the flagellate community, and aggregates in the pelagic environment are typically associated with diverse microbial communities [90–92]. Though microorganisms in this way may attach to remain in an environment that is advantageous for feeding purposes, protozoa not only attach to organic particles or in the presence of food particles, but also attach to glass or other artificial surfaces; they may even attach to inorganic surfaces in sterile water [54,93]. Hence, there must be additional reasons for attachment.

Attaching during feeding has important fluid dynamical implications, and they may in some cases be the main reason for attachment. Swimming microorganisms produce thrust only to overcome the drag of the cell body and so produce only local flow fields [94]. When motile organisms are attached to particles, the force they produce against the fluid will result in thrust balancing the drag of the particle to which it is attached, thus generating a large-scale flow field around the attached organism. With the same application of force, the velocity of the fluid past the feeding structure of the organism may be higher if attached and generating feeding currents than if free swimming (Figure 13.10). In this way, it may be advantageous from a fluid dynamical point of view for microorganisms to be attached, as measured by an increase in the feeding rate compared to the energy expenditure, even if the bacterial concentration around the aggregate is no higher than ambient concentrations [54]. This has also been shown for copepods [51] as well as ciliates [95].

The extent to which the organism can increase its feeding efficiency by attachment depends on the relative size of the feeding area as well as on flagellar parameters and characteristics of the stalk, and will thus vary between organisms (Figure 13.10). In this way, the particle composition of an aquatic environment may influence clearance rates of the protozoa as well as the relative competitive abilities of species. The process may be self-enhancing because the feeding currents of the protozoa may contribute to building up the aggregate [96]. Sessile organisms with a contractile stalk may further increase the filtration rate by altering the length of the stalk periodically, generating chaotic filtration currents that also sample the nearby liquid

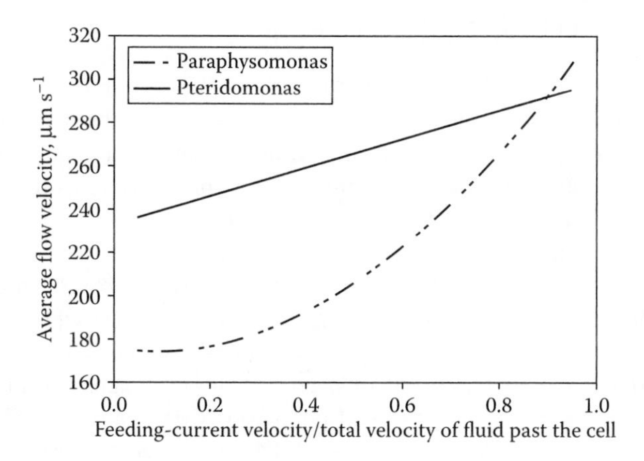

FIGURE 13.10 Increase in the filtration efficiency of flagellates when attaching to particles. The average velocity of the fluid past the feeding appendages, and hence the potential feeding efficiency, is shown as a function of how large a fraction of the velocity results from feeding currents as opposed to forward velocity of the cell. This ranges from 0 in the case of completely free-swimming cells to 1 for cells attached to particles that are too large for them to pull. The curves are calculated from data presented in Christensen-Dalsgaard, K.K. and Fenchel, T., *Aquat. Microb. Ecol.*, 33, 77, 2003.

[97]. Even in species without contractile stalks, chaotic flow patterns might increase feeding rates [98].

The large-scale flow fields of attached organisms make fluid dynamical interactions between feeding currents of adjacent cells possible and so the generation of communal currents. Though intuitively this would seem an advantage, it may not necessarily be so. For bryozoans, which obviously are not protozoans but also generate flow at low Reynolds numbers, it has been shown theoretically that the hydrodynamic interactions between zooids have negative effects on the ability of the individual zooids to generate feeding currents. This is due to a high pressure in the interior part of the colony caused by the viscous resistance to excurrant flow [99]. It may be possible to reduce this effect by forming communal holes for the outflow of water [100] in a manner similar to that of chimney-forming bryozoans [101]. Further research is needed to understand the effect of fluid dynamical interactions between organisms on aggregates, which presents a very different system to the dense and planar colonies of bryozoans.

Even less clear is the net effect of attachment on the predation on protozoa. Attaching reduces or eliminates the forward velocity of the cell and, in this way, reduces the probability of encountering a predator. But at the same time, the large scale flow field of the attached protozoa will make them more visible to predators that perceive their prey through hydrodynamical signals. Also, because copepods feed on organic aggregates, which they find though the plume of dissolved substances tailing behind the aggregate [102], attached protozoa may be eaten as a by-product. Attached bacteria, which enhance the degradation of the aggregates, increase the "visibility" of these aggregates to copepods and the chance that the aggregates may be eaten. So the overall effect of attachment on predation is not clear.

13.5 ORIENTATION TO STIMULI

As with all organisms, an ability to orient with respect to stimuli greatly increases the competitive ability of the microbial species. It allows them to increase their feeding and photosynthetic rate, and remain in chemically favorable environments as well as avoid potentially damaging situations. As previously mentioned, however, microorganisms are too small to be able to perceive chemical gradients over the length of their bodies and too simple to develop complex eyes. Hence, they have been forced to develop other means of orienting to stimuli based on a short term "memory" that allows them to detect changes in stimuli intensity [103]. The degree to which they have succeeded in doing this is remarkable and is an example of how complex and interesting the world of such seemingly simple organisms can be. However, as the topic is generally well described (see Refs. [104,105] as well as the comprehensive introductions in Refs. [106,107]), I restrict myself to a brief overview here.

Bacteria are so small and swim at such low velocities that swimming in a straight path is made impossible by Brownian motions. Instead they swim in what has been named the "random walk," which involves alternating "runs" and "tumbles." In the run, the bacterium swims in an almost straight path with the flagella rotating in a bundle. In tumbles, the flagella change the direction of the rotation, for example, from counterclockwise to clockwise, and the individual flagella separate, which causes the bacterium to stop and change direction randomly (e.g., [108]). Though runs and tumbles always occur, the presence of attractant or repellents can change the frequency with which they occur (e.g., Refs. [103,109]). When swimming up a concentration gradient of a positive attractant or down a gradient of a negative repellant, the tumbles become less frequent and the runs consequently become longer. When swimming in the opposite direction, tumbles become more frequent. This occurs through a direct reaction of the bacterium to the attractants or repellents via chemoreceptors in the membrane [110]. This type of locomotion appears to be advantageous only for bacteria longer than about 0.6 μm. Orientation to stimuli in bacteria smaller than this would be too inefficient, which explains why no motile bacteria shorter than 0.8 μm have been found [111].

The orientation process of bacteria is random because it swims in a random direction after it tumbles. The changes in tumbling frequency, however, results in a higher probability that the net movement of the bacteria will be toward an attractant rather than away from it, or vice versa for repellents (Figure 13.11). This is surprisingly effective, and the bacterial concentrations around nutrient patches may be orders of magnitude higher than ambient bacterial concentrations [61–63,85,112]. In this way, bacteria may form dense clusters at optimal conditions with respect to parameters such as oxygen tension or nutrient concentrations. The chemotactic behavior of bacteria may as much as double the turnover rate of organic carbon in aquatic ecosystems.

For protozoa, this results in an environment of alternating feasts and famines, with patches of high bacterial concentrations dispersed among large areas of low concentrations. This makes it important for protozoa to be capable of chemotactic orientation to such patches so as to utilize the available food resources. In addition,

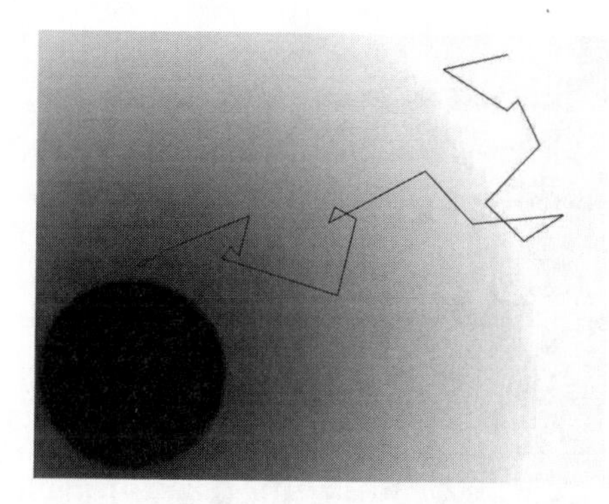

FIGURE 13.11 Orientation by random walk of a bacterium toward an attractant. As the bacterium swims up the concentration gradient, the frequency of tumbles decreases, resulting in longer "runs," whereas when it swims down the concentration gradient, the frequency increases. This results in an overall movement toward the attractant.

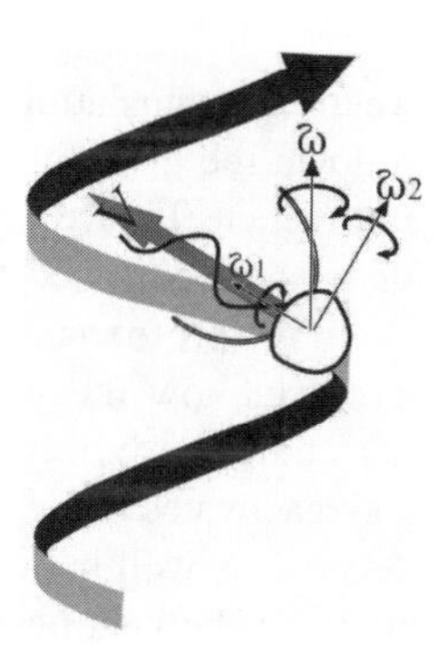

FIGURE 13.12 Swimming in a helix, in this case for the flagellate *Pteridomonas*. The swimming path of the organism and hence the helix is defined by the velocity vector (V) and the rotational vector (ω). is a sum of a component (ω_1) parallel to the velocity vector and a component (ω_2) perpendicular to the velocity vector. The axis of the helix will be parallel to ω.

protozoa may be phototactic, especially those that contain chloroplasts. Though some protozoa make use of the random-walk methods of orienting to stimuli [106], most protozoa are capable of a somewhat more precise orientation.

Most protozoa swim in a helical path (Figure 13.12). As pointed out by Jennings [113], this may be a way of swimming in a straight line, much in the manner of the spinning of a bullet, because the asymmetry of the organisms would otherwise tend to make them swim in circles. The helical swimming path also allows for an ingenious way of orienting to light or chemical stimuli. The helical movement of the protozoa can be resolved in a translational component and two rotational components, one parallel to and one perpendicular to the direction of motion (Figure 13.12). Theoretically this allows the organism to orient to stimuli if the relative velocities of these components are functions of the stimulus intensity [63,114–116].

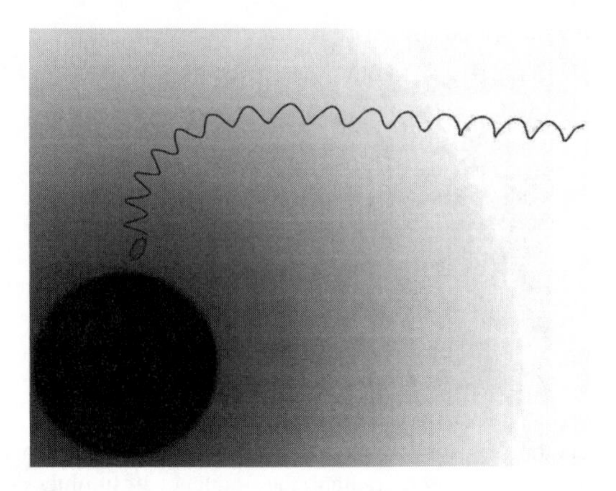

FIGURE 13.13 Orientation of a protozoa swimming in a helix toward an attractant. If the angle between $V\omega$ and is a function of the solute concentration, the axis of the helix will bend toward the source of the concentration gradient as long as the organism is not swimming directly toward or away from the source.

If, for instance, the angle between the translational and the rotational vectors increases as the organism moves down the gradient, the path of the organism will be bent toward the source of the attractant (Figure 13.13).

Numerous experimental studies have confirmed that this does indeed seem to be the main method by which most protozoa orient to stimuli (e.g., Refs. [106,117,118]). It is not clear however how the protozoa physically change the parameters of the helix. Numerous studies have been done on the reaction of whole organisms as well as on isolated and reactivated demembranated flagellar apparatuses in ambient ion concentrations. Hence, it is well known that changes in the intracellular Ca^{2+} level have marked effects on the beating pattern and frequency of cilia (e.g., Refs. [119,120]), flagella of flagellates (e.g., Refs. [121,122]), and spermatozoa from completely unrelated species (e.g., Refs. [123,124]). There is little doubt that this universal mechanism plays an important role in the chemo- and phototaxis of protozoa.

What is less clear is the step from the simple reversals and frequency changes observed with changing Ca^{2+} levels to the complex flagellar or ciliary reactions that are responsible for modifying the helical path and how these are achieved. This is only partially understood and only for a few species. Dinoflagellates make use of the transverse flagellum to control their rotational velocity [125]. *Paramecium* can alter the parameters of the helix by changing the direction and frequency of the three-dimensional movement of the cilia, the control of which seems to involve also H^+, Mg^{2+}, and ATP [118]. *Pteridomonas* seems to be capable of an accurate three-dimensional control of the axis of the planar waveform of the flagellum, resulting in a very precise control of swimming directions, but this is as yet incompletely understood [44]. For most species, the flagellar and ciliary mechanisms by which protozoa change the direction of the axis of the helix is not known. It is important for our understanding of the microbial ecosystems to gain a deeper understanding of how organisms orient physiologically to stimuli and of the factors that may affect this ability to orient.

Protozoa may react not only to light intensities and chemical gradients but also to gravity. In fact, the ecologically important vertical migrations of many organisms in aquatic environments seem driven as much by gravity as by a reaction to light intensities [126–128]. For most of the last century, gravitaxis was believed to be caused largely by a density gradient in the cell so that the lower, heavier end would passively align with the gravity vector [129–132]. It has been shown however that the ciliate *Laxodes* is capable of switching between positive and negative gravitaxis without any apparent structural changes in the cell [133], requiring more complicated mechanisms for perceiving gravity.

The asymmetrical shape of the organism may play an important role in the gravitational response of ciliates, and even slight changes in shape may have a profound effect on the orientation of the organism [107]. The prevailing theory today, however, is that at least some protozoa are capable of sensing gravity directly by the cytoplasmic contents of the cell exerting a pressure on the lower membrane, thus activating stretch-sensitive calcium-specific ion channels [105]. This hypothesis is supported, among other things, by the fact that changes in density, but not viscosity, of the fluid can reverse the gravitational response of *Euglena* [134] and that for this organism, inhibitors of stretch-sensitive ion channels inhibit gravitaxis [134,135]. The precise gravitational response of many organisms may in reality be a combination of this mechanism and shape and density distribution.

The extent to which the ability of an organism to orient to stimuli allows it to exploit patchy resources depends on a number of factors, including how long-lived and far away an average resource path is compared to the swimming speed and navigation abilities of the organism. Based on these parameters, Grünbaum [136] recently presented a simple numerical index, the Frost number, with which it is possible to predict roughly the availability of a patchy resource to a consumer using various forms of biased random walks. The Frost number is based on the mobility of individual organisms, yet provides a powerful tool in understanding ecological processes and first approximations of quantifications such as carbon turnover rates. As with many other aspects presented in this review, this shows how much our understanding of ecological processes can be improved by bringing our ecological investigations down to the level of the mechanics of the individual. Failing to do so will leave our understanding of ecological systems much poorer.

ACKNOWLEDGMENTS

I would like to thank Tom Fenchel for comments on the manuscript, which substantially improved the final version.

REFERENCES

1. Koch, A.L., The adaptive responses of *Escherichia coli* to a feast and famine existence, *Adv. Microbial Physiol.*, 6, 147, 1971.

2. Estep, K.W. et al., How important are oceanic algal nanoflagellates in bacterivory, *Limnol. Oceanogr.,* 31, 646, 1986.
3. Bird, D.F. and Kalff, J., Bacterial grazing by planktonic lake algae, *Science,* 231, 493, 1986.
4. Azam, F. et al., The ecological role of water-column microbes in the sea, *Mar. Ecol.-Prog. Ser.,* 10, 257, 1983.
5. Larsson, U. and Hagstrom, A., Phytoplankton exudate release as an energy-source for the growth of pelagic bacteria, *Mar. Biol.,* 52, 199, 1979.
6. Larsson, U. and Hagstrom, A., Fractionated phytoplankton primary production, exudate release and bacterial production in a Baltic eutrophication gradient, *Mar. Biol.,* 67, 57, 1982.
7. Thomas, J.P., Release of dissolved organic matter from natural populations of marine phytoplankton, *Mar. Biol.,* 11, 311, 1971.
8. Bell, W. and Mitchell, R., Chemotactic and growth responses of marine bacteria to algal extracellular products, *Biol. Bull.,* 143, 265, 1972.
9. Andersen, P. and Fenchel, T., Bacterivory by microheterotrophic flagellates in seawater samples, *Limnol. Oceanogr.,* 30, 198, 1985.
10. Fenchel, T., Ecology of heterotrophic microflagellates. 4. Quantitative occurrence and importance as bacterial consumers, *Mar. Ecol.-Prog. Ser.,* 9, 35, 1982.
11. Nielsen, T.G. and Richardson, K., Food-chain structure of the North-Sea plankton communities: seasonal variations of the role of the microbial loop, *Mar. Ecol.-Prog. Ser.,* 56, 75, 1989.
12. Sherr, B.F., Sherr, E.B., and Fallon, R.D., Use of monodispersed, fluorescently labeled bacteria to estimate *in-situ* protozoan bacterivory, *Appl. Environ. Microbiol.,* 53, 958, 1987.
13. Wright, R.T. and Coffin, R.B., Measuring microzooplankton grazing on planktonic marine-bacteria by its impact on bacterial production, *Microb. Ecol.,* 10, 137, 1984.
14. Goldman, J.C. and Caron, D.A., Experimental studies on an omnivorous microflagellate — implications for grazing and nutrient regeneration in the marine microbial food-chain, *Deep-Sea Res., A* 32, 899, 1985.
15. Drebes, G. and Schnepf, E., *Gyrodinium undulans* Hulburt, a marine dinoflagellate feeding on the bloom-forming diatom *Odontella aurita,* and on copepod and rotifer eggs, *Helgol. Meeresunters,* 52, 1, 1998.
16. Caron, D.A. et al., Nutrient cycling in a microflagellate food-chain. 2. Population-dynamics and carbon cycling, *Mar. Ecol.-Prog. Ser.,* 24, 243, 1985.
17. Goldman, J.C. et al., Nutrient cycling in a microflagellate food-chain. 1. Nitrogen dynamics, *Mar. Ecol.-Prog. Ser.,* 24, 231, 1985.
18. Goldman, J.C. and Dennett, M.R., Dynamics of prey selection by an omnivorous flagellate, *Mar. Ecol.-Prog. Ser.,* 59, 183, 1990.
19. Berg, H.C., Dynamic properties of bacterial flagellar motors, *Nature,* 249, 77, 1974.
20. Taylor, G., Analysis of the swimming of microscopic organisms, *Proc. R. Soc. Lond. Ser. A,* 209, 447, 1951.
21. Taylor, G., The action of waving cylindrical tails in propelling microscopic organisms, *Proc. R. Soc. Lond. Ser. A,* 211, 225, 1952.
22. Gray, J. and Hancock, G.J., The propulsion of sea-urchin spermatozoa, *J. Exp. Biol.,* 32, 802, 1955.
23. Hancock, G.J., The self-propulsion of microscopic organisms through liquids, *Proc. R. Soc. Lond. Ser. A,* 217, 96, 1953.
24. Brokaw, C.J., Non-sinusoidal bending waves of sperm flagella, *J. Exp. Biol.,* 43, 155, 1965.

25. Silverman, M. and Simon, M., Flagellar rotation and mechanism of bacterial motility, *Nature,* 249, 73, 1974.
26. Depamphilis, M.L. and Adler, J., Fine structure and isolation of hook-basal body complex of flagella from *Escherichia-coli* and *Bacillus-subtilis*, *J. Bacteriol.,* 105, 384, 1971.
27. Okrend, A.G. and Doetsch, R.N., Plasmolysis and bacterial motility — a method for study of membrane function, *Archiv Mikrobiol.,* 69, 69, 1969.
28. Vaituzis, Z. and Doetsch, R.N., Flagella of *Escherichia coli* spheroplasts, *J. Bacteriol.,* 91, 2103, 1966.
29. Machin, K.E., Control and synchronization of flagellar movement, *Proc. R. Soc. Lond. Ser. B-Biol. Sci.,* 158, 88, 1963.
30. Higdon, J.J.L., Hydrodynamic analysis of flagellar propulsion, *J. Fluid Mech.,* 90, 685, 1979.
31. Higdon, J.J.L., Hydrodynamics of flagellar propulsion — helical waves, *J. Fluid Mech.,* 94, 331, 1979.
32. Lowndes, A.G., The swimming of unicellular flagellate organisms, *Proc. Zool. Soc. London A.,* 114, 325, 1944.
33. Lowndes, A.G., Swimming of *Monas-stigmatica*, *Nature,* 155, 579, 1945.
34. Holwill, M.E.J., Motion of *Euglena viridis* — role of flagella, *J. Exp. Biol.,* 44, 579, 1966.
35. Chwang, A.T., Wu, T.Y., and Winet, H., Locomotion of *Spirilla*, *Biophys. J.,* 12, 1549, 1972.
36. Lighthill, J., Helical distributions of stokeslets, *J. Eng. Math.,* 30, 35, 1996.
37. Ramia, M., Numerical-model for the locomotion of *Spirilla*, *Biophys. J.,* 60, 1057, 1991.
38. Jahn, T.L., Landman, M.D., and Fonseca, J.R., Mechanism of locomotion of flagellates. 2. Function of mastigonemes of *Ochromonas*, *J. Protozool.,* 11, 291, 1964.
39. Brennen, C., Locomotion of flagellates with mastigonemes, *J. Mechanochem. Cell. Motil.,* 3, 207, 1976.
40. Taylor, G., Analysis of the swimming of long and narrow animals, *Proc. R. Soc. Lond. Ser. A,* 214, 158, 1952.
41. Holwill, M.E.J. and Sleigh, M.A., Propulsion by hispid flagella, *J. Exp. Biol.,* 47, 267, 1967.
42. Fauci, L.J., A computational model of the fluid dynamics of undulatory and flagellar swimming, *Am. Zool.,* 36, 599, 1996.
43. Gaines, G. and Taylor, F.J.R., Form and function of the dinoflagellate transverse flagellum, *J. Protozool.,* 32, 290, 1985.
44. Christensen-Dalsgaard, K.K., and Fenchel, T., Complex flagellar motions and swimming patterns of the flagellates *Paraphysomonas vestita* and *Pteridomonas danica*, *Protist,* 155, 79, 2004.
45. Gueron, S. and Liron, N., Ciliary motion modeling, and dynamic multicilia interactions, *Biophys. J.,* 63, 1045, 1992.
46. Gueron, S. and Levit-Gurevich, K., Computation of the internal forces in cilia: application to ciliary motion, the effects of viscosity, and cilia interactions, *Biophys. J.,* 74, 1658, 1998.
47. Blake, J., Hydrodynamic calculations on movements of cilia and flagella. 1. Paramecium, *J. Theor. Biol.,* 45, 183, 1974.
48. Wu, T.Y., Hydrodynamics of swimming at low Reynolds-numbers, *Fortschr Zool.,* 24, 149, 1977.

49. Kiorboe, T. and Visser, A.W., Predator and prey perception in copepods due to hydromechanical signals, *Mar. Ecol.-Prog. Ser.,* 179, 81, 1999.
50. Svensen, C. and Kiorboe, T., Remote prey detection in *Oithona similis*: hydromechanical versus chemical cues, *J. Plankton Res.,* 22, 1155, 2000.
51. Tiselius, P. and Jonsson, P.R., Foraging behavior of 6 calanoid copepods — observations and hydrodynamic analysis, *Mar. Ecol.-Prog. Ser.,* 66, 23, 1990.
52. Fenchel, T., Ecology of heterotrophic microflagellates. 2. Bioenergetics and growth, *Mar. Ecol.-Prog. Ser.,* 8, 225, 1982.
53. Fenchel, T. and Finlay, B.J., Respiration rates in heterotrophic, free-living protozoa, *Microb. Ecol.,* 9, 99, 1983.
54. Christensen-Dalsgaard, K.K. and Fenchel, T., Increased filtration efficiency of attached compared to free swimming flagellates, *Aquat. Microb. Ecol.,* 33, 77, 2003.
55. Gerritsen, J. and Strickler, J.R., Encounter probabilities and community structure in zooplankton — mathematical-model, *J. Fisheries Res. Board Canada* 34, 73, 1977.
56. Tanaka, T. and Rassoulzadegan, F., Full-depth profile (0–2000 m) of bacteria, heterotrophic nanoflagellates and ciliates in the NW Mediterranean Sea: vertical partitioning of microbial trophic structures, *Deep-Sea Res. Part II-Top. Stud. Oceanogr.,* 49, 2093, 2002.
57. Matz, C. and Jurgens, K., Interaction of nutrient limitation and protozoan grazing determines the phenotypic structure of a bacterial community, *Microb. Ecol.,* 45, 384, 2003.
58. Higdon, J.J.L., Generation of feeding currents by flagellar motions, *J. Fluid Mech.,* 94, 305, 1979.
59. Mathews, C.K. and Van Holde, K.E., *Biochemistry,* Benjamin/Cummings, Menlo Park, CA, 1996.
60. Eccleston-Parry, J.D. and Leadbeater, B.S.C., A comparison of the growth-kinetics of 6 marine heterotrophic nanoflagellates fed with one bacterial species, *Mar. Ecol.-Prog. Ser.,* 105, 167, 1994.
61. Bowen, J.D., Stolzenbach, K.D., and Chisholm, S.W., Simulating bacterial clustering around phytoplankton cells in a turbulent ocean, *Limnol. Oceanogr.,* 38, 36, 1993.
62. Blackburn, N., Fenchel, T., and Mitchell, J., Microscale nutrient patches in planktonic habitats shown by chemotactic bacteria, *Science,* 282, 2254, 1998.
63. Blackburn, N. and Fenchel, T., Modelling of microscale patch encounter by chemotactic protozoa, *Protist,* 150, 337, 1999.
64. Fenchel, T., Ecology of heterotrophic microflagellates. 3. Adaptations to heterogeneous environments, *Mar. Ecol.-Prog. Ser.,* 9, 25, 1982.
65. Rivier, A. et al., Growth of microzooplankton: a comparative study of bacterivotous zooflagellates and ciliates, *Mar. Microb. Food Webs* 1, 51, 1985.
66. Davidson, L.A., Ultrastructure, behavior, and algal flagellate affinities of the helioflagellate *Ciliophyrys-marina*, and the classification of the helioflagellates (Protista, Actinopoda, Heliozoea), *J. Protozool.,* 29, 19, 1982.
67. Kiorboe, T. and Titelman, J., Feeding, prey selection and prey encounter mechanisms in the heterotrophic dinoflagellate *Noctiluca scintillans*, *J. Plankton Res.,* 20, 1615, 1998.
68. Fenchel, T., Suspended marine bacteria as a food source, in *Flow of Material and Energy in Marine Ecosystems*, Fasham, M.J., Ed., Plenum Press, New York, 1984, p. 301.
69. Shimeta, J. and Jumars, P.A., Physical-mechanisms and rates of particle capture by suspension-feeders, *Oceanogr. Mar. Biol.,* 29, 191, 1991.

70. Monger, B.C. and Landry, M.R., Direct-interception feeding by marine zooflagellates — the importance of surface and hydrodynamic-forces, *Mar. Ecol.-Prog. Ser.,* 65, 123, 1990.
71. Monger, B.C., Landry, M.R., and Brown, S.L., Feeding selection of heterotrophic marine nanoflagellates based on the surface hydrophobicity of their picoplankton prey, *Limnol. Oceanogr.,* 44, 1917, 1999.
72. Shimeta, J., Diffusional encounter of submicrometer particles and small-cells by suspension feeders, *Limnol. Oceanogr.,* 38, 456, 1993.
73. Rothschild, B.J. and Osborn, T.R., Small-scale turbulence and plankton contact rates, *J. Plankton Res.,* 10, 465, 1988.
74. Kiørboe, T. and Saiz, E., Planktivorous feeding in calm and turbulent environments, with emphasis on copepods, *Mar. Ecol.-Prog. Ser.,* 122, 135, 1995.
75. Marrase, C. et al., Grazing in a turbulent environment — energy-dissipation, encounter rates, and efficacy of feeding currents in *Centropages-Hamatus*, *Proc. Natl. Acad. Sci. U.S.A.,* 87, 1653, 1990.
76. Peters, F. and Gross, T., Increased grazing rates of microplankton in response to small-scale turbulence, *Mar. Ecol.-Prog. Ser.,* 115, 299, 1994.
77. Shimeta, J., Jumars, P.A., and Lessard, E.J., Influences of turbulence on suspension-feeding by planktonic protozoa — experiments in laminar shear fields, *Limnol. Oceanogr.,* 40, 845, 1995.
78. Peters, F., Choi, J.W., and Gross, T., *Paraphysomonas imperforata* (Protista, Chrysomonadida) under different turbulence levels: feeding, physiology and energetics, *Mar. Ecol.-Prog. Ser.,* 134, 235, 1996.
79. Jorgensen, C.B., Fluid mechanical aspects of suspension feeding, *Mar. Ecol.-Prog. Ser.,* 11, 89, 1983.
80. Fenchel, T., Protozoan filter feeding, *Progr. Protistol.,* 1, 65, 1986.
81. Fenchel, T., Relation between particle-size selection and clearance in suspension-feeding ciliates, *Limnol. Oceanogr.,* 25, 733, 1980.
82. Fenchel, T., Suspension feeding in ciliated protozoa — functional-response and particle-size selection, *Microb. Ecol.,* 6, 1, 1980.
83. Smith, D.C. et al., Intense hydrolytic enzyme-activity on marine aggregates and implications for rapid particle dissolution, *Nature,* 359, 139, 1992.
84. Kiørboe, T. et al., Mechanisms and rates of bacterial colonization of sinking aggregates, *Appl. Environ. Microbiol.,* 68, 3996, 2002.
85. Fenchel, T., Eppur si muove: many water column bacteria are motile, *Aquat. Microb. Ecol.,* 24, 197, 2001.
86. Grossart, H.P. et al., Bacterial colonization of particles: growth and interactions, *Appl. Environ. Microbiol.,* 69, 3500, 2003.
87. Martinez, J. et al., Variability in ectohydrolytic enzyme activities of pelagic marine bacteria and its significance for substrate processing in the sea, *Aquat. Microb. Ecol.,* 10, 223, 1996.
88. Kiørboe, T. et al., Dynamics of microbial communities on marine snow aggregates: colonization, growth, detachment, and grazing mortality of attached bacteria, *Appl. Environ. Microbiol.,* 69, 3036, 2003.
89. Kiørboe, T. et al., Particle-associated flagellates: swimming patterns, colonization rates, and grazing on attached bacteria, *Aquat. Microb. Ecol.,* 35, 141, 2004.
90. Kiørboe, T., Marine snow microbial communities: scaling of abundances with aggregate size, *Aquat. Microb. Ecol.,* 33, 67, 2003.
91. Caron, D.A. et al., Heterotrophic bacteria and bacterivorous protozoa in oceanic macro-aggregates, *Science,* 218, 795, 1982.

92. Artolozaga, I. et al., Succession of bacterivorous protists on laboratory-made marine snow, *J. Plankton Res.,* 19, 1429, 1997.
93. Coppellotti, O. and Matarazzo, P., Ciliate colonization of artificial substrates in the Lagoon of Venice, *J. Mar. Biol. Assoc. U.K.,* 80, 419, 2000.
94. Lighthill, J., Flagellar hydrodynamics — Neumann, Jv Lecture, 1975, *SIAM Rev.,* 18, 161, 1976.
95. Jonsson, P.R., Johansson, M., and Pierce, R.W., Attachment to suspended particles may improve foraging and reduce predation risk for tintinnid ciliates, *Limnol. Oceanogr.,* 49, 1907, 2004.
96. Fukuda, H. and Koike, I., Feeding currents of particle-attached nanoflagellates — A novel mechanism for aggregation of submicron particles, *Mar. Ecol.-Prog. Ser.,* 202, 101, 2000.
97. Blake, J.R. and Otto, S.R., Filter feeding, chaotic filtration, and a blinking Stokeslet, *Theor. Comput. Fluid Dyn.,* 10, 23, 1998.
98. Orme, B.A.A., Otto, S.R., and Blake, J.R., Enhanced efficiency of feeding and mixing due to chaotic flow patterns around choanoflagellates, *IMA J. Math. Appl. Med. Biol.,* 18, 293, 2001.
99. Grunbaum, D., A model of feeding currents in encrusting bryozoans shows interference between zooids within a colony, *J. Theor. Biol.,* 174, 409, 1995.
100. Fenchel, T. and Glud, R.N., Veil architecture in a sulphide-oxidizing bacterium enhances countercurrent flux, *Nature,* 394, 367, 1998.
101. Grunbaum, D., Hydromechanical mechanisms of colony organization and cost of defense in an encrusting bryozoan, *Membranipora membranacea*, *Limnol. Oceanogr.,* 42, 741, 1997.
102. Kiørboe, T. and Thygesen, U.H., Fluid motion and solute distribution around sinking aggregates. II. Implications for remote detection by colonizing zooplankters, *Mar. Ecol.-Prog. Ser.,* 211, 15, 2001.
103. Brown, D.A. and Berg, H.C., Temporal stimulation of chemotaxis in *Escherichia coli*, *Proc. Natl. Acad. Sci. U.S.A.,* 71, 1388, 1974.
104. Fenchel, T., Microbial behavior in a heterogeneous world, *Science,* 296, 1068, 2002.
105. Hader, D.P. et al., Gravitaxis and graviperception in flagellates, in *Space Life Sciences: Gravity-Related Processes in Plants,* Pergamon, Oxford, 2003, p. 2181.
106. Fenchel, T. and Blackburn, N., Motile chemosensory behaviour of phagotrophic protists: mechanisms for and efficiency in congregating at food patches, *Protist,* 150, 325, 1999.
107. Roberts, A.M. and Deacon, F.M., Gravitaxis in motile micro-organisms: the role of fore-aft body asymmetry, *J. Fluid Mech.,* 452, 405, 2002.
108. Turner, L., Ryu, W.S., and Berg, H.C., Real-time imaging of fluorescent flagellar filaments, *J. Bacteriol.,* 182, 2793, 2000.
109. Larsen, S.H. et al., Change in direction of flagellar rotation is basis of chemotactic response in *Escherichia-coli*, *Nature,* 249, 74, 1974.
110. Adler, J., Chemoreceptors in bacteria, *Science,* 166, 1588, 1969.
111. Dusenbery, D.B., Minimum size limit for useful locomotion by free-swimming microbes, *Proc. Natl. Acad. Sci. U.S.A.,* 94, 10949, 1997.
112. Blackburn, N. and Fenchel, T., Influence of bacteria, diffusion and sheer on micro-scale nutrient patches, and implications for bacterial chemotaxis, *Mar. Ecol.-Prog. Ser.,* 189, 1, 1999.
113. Jennings, H.S., On the significance of the spiral swimming of organisms, *Am. Nat.,* 35, 369, 1901.

114. Crenshaw, H.C., Orientation by helical motion. 1. Kinematics of the helical motion of organisms with up to 6-degrees of freedom, *Bull. Math. Biol.,* 55, 197, 1993.
115. Crenshaw, H.C., Orientation by helical motion. 3. Microorganisms can orient to stimuli by changing the direction of their rotational velocity, *Bull. Math. Biol.,* 55, 231, 1993.
116. Crenshaw, H.C. and Edelsteinkeshet, L., Orientation by helical motion. 2. Changing the direction of the axis of motion, *Bull. Math. Biol.,* 55, 213, 1993.
117. Crenshaw, H.C., A new look at locomotion in microorganisms: rotating and translating, *Am. Zool.,* 36, 608, 1996.
118. Naitoh, Y. and Sugino, K., Ciliary movement and its control in paramecium, *J. Protozool.,* 31, 31, 1984.
119. Eckert, R., Bioelectric control of ciliary activity, *Science,* 176, 473, 1972.
120. Iwadate, Y. and Suzaki, T., Ciliary reorientation is evoked by a rise in calcium level over the entire cilium, *Cell Motil. Cytoskeleton,* 57, 197, 2004.
121. Hyams, J.S. and Borisy, G.G., Isolated flagellar apparatus of *Chlamydomonas* — characterization of forward swimming and alteration of waveform and reversal of motion by calcium-ions *in vitro*, *J. Cell Sci.,* 33, 235, 1978.
122. Holwill, M.E.J. and McGregor, J.L., Control of flagellar wave movement in *Crithidia-oncopelti*, *Nature,* 255, 157, 1975.
123. Cook, S.P. et al., Sperm chemotaxis — egg peptides control cytosolic calcium to regulate flagellar responses, *Dev. Biol.,* 165, 10, 1994.
124. Brokaw, C.J., Calcium and flagellar response during chemotaxis of bracken spermatozoids, *J. Cell. Physiol.,* 83, 151, 1974.
125. Fenchel, T., How dinoflagellates swim, *Protist,* 152, 329, 2001.
126. Hader, D.P., Polarotaxis, gravitaxis and vertical phototaxis in the green flagellate, *Euglena-gracilis*, *Arch. Microbiol.,* 147, 179, 1987.
127. Kessler, J.O., Hill, N.A., and Hader, D.P., Orientation of swimming flagellates by simultaneously acting external factors, *J. Phycol.,* 28, 816, 1992.
128. Hemmersbach, R., Volkmann, D., and Hader, D.P., Graviorientation in protists and plants, *J. Plant Physiol.,* 154, 1, 1999.
129. Brinkman, K., No geotaxis in euglena, *Z. Pflanzenphysiol.,* 59, 12, 1968.
130. Dembowski, J., Die Vertikalbewegungen von *Paramecium caudatum*, *Arch. Protistenkd.,* 74, 153, 1931.
131. Fukui, K. and Asai, H., Negative geotactic behavior of *Paramecium-caudatum* is completely described by the mechanism of buoyancy-oriented upward swimming, *Biophys. J.,* 47, 479, 1985.
132. Wagner, H., On the effect of gravity upon the movements and aggregation of *Euglena viridis* and other micro-organisms., *Proc. R. Soc. Lond. Ser. A,* 201, 333, 1911.
133. Fenchel, T. and Finlay, B.J., Geotaxis in the ciliated protozoan *Loxodes*, *J. Exp. Biol.,* 110, 17, 1984.
134. Lebert, M. and Hader, D.P., How *Euglena* tells up from down, *Nature,* 379, 590, 1996.
135. Lebert, M. and Hader, D.P., Negative gravitactic behavior of *Euglena gracilis* can not be described by the mechanism of buoyancy-oriented upward swimming, in *Life Sciences: Microgravity Research,* 1999, p. 851.
136. Grünbaum, D., Predicting availability to consumers of spatially and temporally variable resources, *Hydrobiologia,* 480, 175, 2002.

14 The Biomechanics of Ecological Speciation

Jeffrey Podos and Andrew P. Hendry

CONTENTS

14.1 INTRODUCTION

This book addresses the interplay of biomechanics and ecology. Ecology has long been recognized as an important factor in evolutionary diversification and speciation. Architects of the neo-Darwinian synthesis, particularly Mayr [1] and Dobzhansky [2], argued that spatial variation in ecological parameters should facilitate divergent trajectories of adaptive evolution among populations, at least among populations that are able to maintain some degree of reproductive isolation. This insight was overshadowed for several decades by attention to genetic mechanisms of divergence and stochastic models of speciation. Empirical and conceptual advances in recent years, however, have spurred a renewed emphasis on ecological causes of evolutionary diversification and speciation [3–5].

It thus seems timely to consider how biomechanics, through its interface with ecology, might affect the processes of evolutionary diversification and speciation. The possibilities here are admittedly broad. For the purposes of this chapter, we focus on a "by-product" model of speciation. This model features two stages. In the first, adaptive divergence of phenotypic traits drives, as an incidental consequence, divergence in mechanisms that mediate the expression and production of mating displays. Second, resulting divergent evolution of display behavior facilitates reproductive isolation, further adaptive divergence, and, ultimately, speciation. In this chapter we evaluate this model's conceptual foundations, review supporting empirical evidence, and outline some of its evolutionary implications. We begin with a more detailed explanation of the by-product model.

Consider an ecological resource that takes two discrete forms, $Resource_A$ and $Resource_B$, with the frequencies of these forms varying nonrandomly in space. Now consider an animal species possessing some morphological, behavioral, or physiological trait used for exploiting this resource. Assume that different trait values are best suited for the different resource forms, say $Trait_A$ for $Resource_A$ and $Trait_B$ for $Resource_B$. As long as these trait values are heritable, and as long as dispersal among sites with different resources is somewhat limited — thus restricting gene flow — natural selection should favor the evolution of $Trait_A$ in sites where $Resource_A$ predominates, and the evolution of $Trait_B$ in sites where $Resource_B$ predominates. Our example thus far follows the well-established logic of adaptive divergence in response to natural selection in distinct ecological environments [1–3,6–10].

Now consider the possibility that evolution of the trait in question also influences, as a by-product of selected changes in morphology, physiology, or behavior, the kind of mating displays these animals can express or produce. For example, individuals possessing $Trait_A$ might be constrained to produce a particular display variant, $Display_A$, whereas those possessing $Trait_B$ might necessarily produce another, $Display_B$. Possible biomechanical causes of correlated evolution among adaptive traits and mating displays are detailed later in this chapter. As male displays begin to diverge between sites, females would be expected to evolve, through sexual selection, divergent preferences that mirror the changes in display structure [11,12]. (We assume, for present purposes, that only males display and that females use displays to guide mate choice.) With sufficient time and sufficient limits on gene flow, $Resource_A$ environments should thus evolve populations wherein males possess $Trait_A$ and produce $Display_A$, and wherein females respond preferentially to $Display_A$. At the same time, $Resource_B$ environments should support populations that evolve the other suite of characteristics ($Trait_B$ and $Display_B$). If individuals from these two ecological environments then come into secondary contact, the probability of mating should be diminished, and speciation thus initiated.

The above scenario for divergence is conceptualized, for the sake of argument, as occurring in allopatry (separate and isolated locations) or parapatry (separate but not isolated locations). In the remainder of this chapter, "populations" or "environments" are thus envisioned as geographically distinct, and "migrants" as individuals that move between populations or environments. It is important to point out, however, that many of the same processes could in principle occur in sympatry (populations diverging in the same physical location). Under sympatric divergence, different groups of individuals may specialize on different resources within a common geographical location. Here "populations" would refer to sympatric groups using the distinct resource "environments," and "migrants" would be individuals that switch resources. It is not our intention to distinguish between these geographical scenarios because we are more interested in general mechanisms.

Our chapter continues with a brief overview of modes of speciation, focusing in particular on a distinction between ecologically dependent and ecologically independent isolating barriers. Attention to this distinction will be helpful later as we evaluate the possible impacts of adaptive divergence and concomitant evolution in the biomechanical bases of display behavior.

14.2 MODES OF SPECIATION

A traditional method for categorizing speciation events is on the basis of geography, i.e., as occurring in allopatry, sympatry, or parapatry [4,8,13]. Another way to categorize speciation events is by identifying "isolating barriers" that initiate and maintain separation between incipient species [2,4,14]. Isolating barriers are typically categorized as occurring before mating (premating), after mating but before zygote formation (postmating prezygotic), or after zygote formation (postzygotic). These categories are then further parsed into nested subcategories. Although identifying isolating barriers is a critical part of any research on speciation, we do not discuss these subcategories further because they have been reviewed elsewhere [4,14]. Instead, we focus our attention on one of the ultimate and long-standing questions of speciation: What are the initial causes of reproductive isolation (and thus of speciation) among diverging, incipient species?

Schluter [15] proposed that the initial causes of speciation can be divided into four general modes: hybridization and polyploidy, genetic drift, uniform natural selection, and divergent natural selection. Sexual selection is not considered a separate mode but rather as a potential contributor within each. Under uniform natural selection, in which different populations are exposed to similar ecological environments, divergence occurs as different advantageous mutations arise and spread to fixation in different populations. These mutations may be incompatible when brought together by interpopulation mating, thus causing reproductive isolation among different populations adapted to similar environments [4,16]. Conversely, under divergent natural selection, similar mutations may arise in multiple populations, but different mutations will be favored and therefore retained in different ecological environments. Adaptive divergence may then lead to initial reproductive isolation in a number of ways. For example, mutations favored in one environment might confer reduced fitness in alternative environments, perhaps also favoring individuals that mate assortatively. Schluter [3,15] refers to this latter mode of speciation — by-product effects of divergent natural selection — as "ecological speciation." This is the arena within which we consider the role of biomechanics.

Isolating barriers that arise through ecological speciation (or other speciation modes for that matter) may be manifest in two general ways. On the one hand, adaptation to different environments may lead to reproductive isolation that depends directly on features of those environments, such as the availability and distribution of food resources. Such isolating barriers are therefore considered "extrinsic," "conditional," "environment dependent," or "ecologically dependent" [4,17,18]. To illustrate, male displays are often optimized through natural selection for effective transmission in the particular environments that animals inhabit [19–21]. Songbirds living in forested environments, for example, tend to evolve mating songs with lower frequencies and lower rates of note repetition, as adaptations that minimize degradation by reverberation [21]. Optimization of transmission properties in local habitats may thus diminish the effectiveness of particular songs when sung in alternative environments. If the signaling environments inhabited by a species are sufficiently divergent, and the signals are differentially effective in these environments, then females may mate preferentially with males from local environments. A locality-dependent process of reproductive isolation would thus be initiated [22]. Isolation barriers in this example would be premating. Ecologically dependent *postmating* barriers are also feasible. Consider, for example, hybrids with phenotypes that are intermediate to parental phenotypes. Hybrids may suffer lower levels of fitness in either parental environment because of the difficulty in accessing resources on which parental types are specialized [e.g., 23]. However, as conceptualized in Figure 14.1A, hybrids with intermediate phenotypes might enjoy higher fitness, relative to either parental type, in "intermediate" environments, e.g., in which resource parameters fall in between those in parental environments [e.g., 24].

On the other hand, adaptation to divergent environments may lead to reproductive isolation that is manifest independently of environmental features. Such isolating barriers are considered "intrinsic," "unconditional," "environment independent," or "ecologically independent" [4,17,18]. Ecologically independent *premating* barriers may arise if traits under divergent selection are also used in mate choice and in ways that do not depend on the mating environment. In stickleback fishes, for example, divergent selection between benthic and limnetic morphs has fostered the evolution of differences in body size. Laboratory mating studies suggest that body size plays an important role in mate choice, in that females appear to choose males with body sizes similar to their own [25,26]. Ecological independence of body size as a mating cue is illustrated by the observation that body size cues are effective not only in the field but also under laboratory conditions, in which natural variation in environmental transmission properties is not present [e.g., 27,28]. As a generalized example of ecologically independent *postzygotic* barriers, hybrids of diverging lineages can have genetic incompatibilities that are expressed equivalently (or near equivalently) in any environment. Under such circumstances, hybrids would experience low fitness in nature even if intermediate environments are present (Figure 14.1B), and perhaps even under benign laboratory conditions — although such barriers may be stronger under more stressful conditions [4]. Many studies have demonstrated genetic incompatibilities in hybrids [4], although we are not aware of any conclusively attributing the resulting isolation to divergent natural selection.

Exploring the distinction between ecologically dependent and ecologically independent isolating barriers is useful because it speaks to the integrity of species in the face of environmental perturbation. Ecologically independent barriers may be more powerful and robust because they should persist even if the environment changes, at least during initial stages of divergence. In contrast, ecologically dependent barriers may collapse immediately after environments change and could therefore represent a more fragile and tenuous route to speciation. For example, in a long-term study of Darwin's finches on Daphne Major Island, environmental changes resulted in increased relative fitness for hybrids, which has led to the

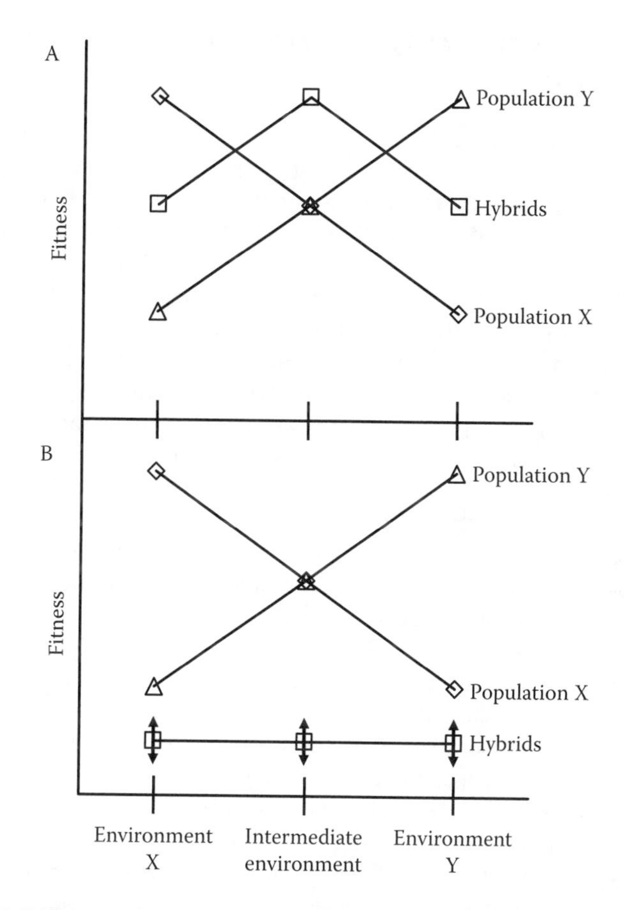

FIGURE 14.1 Differences between (A) ecologically dependent and (B) ecologically independent reproductive isolation. Assumed is a case whereby population X individuals are adapted to environment X, and population Y individuals are adapted to environment Y. When isolation is ecologically dependent, hybrids (with intermediate phenotypes) should have a higher fitness than parental types in intermediate environments. When isolation is ecologically independent, hybrids should have lower fitness than parental types in all environments, although the specific value for hybrid fitness relative to parental fitness could vary considerably (indicated by the arrows). Of course, both types of isolation may act at the same time, generating any number of intermediate scenarios for hybrid fitness.

breakdown of ecologically dependent isolating barriers and to the morphological convergence of formerly distinct species [29]. And yet, ecologically dependent barriers can evolve very quickly, simply because adaptive divergence can be very rapid in nature [reviews: 30,31]. As examples, insect herbivores adapting to introduced host plants have evolved ecologically dependent barriers after less than a few hundred generations [32,33; see also 34]. In addition, ecologically dependent barriers should be particularly widespread, and may therefore cause initial reductions in gene flow that allow subsequent ecologically independent barriers to evolve [35].

14.3 BIOMECHANICS AND ECOLOGICAL SPECIATION

As ecological speciation progresses, the adaptive divergence of phenotypic traits can presumably strengthen any number of isolating barriers, both pre- and postmating. For the purposes of this chapter, we focus our attention on the role of biomechanics in premating isolation, and specifically in relation to mating displays. Mating displays include ritualized movements such as visual or vocal signaling, or the presentation of static traits such as exaggerated morphological characters [36]. Many dynamic displays, particularly those under intense sexual selection, appear to be costly or to require high levels of biomechanical proficiency in their production. Indeed, a number of studies have shown that dynamic mating displays require large energy investments [37–40] or are underpinned by adaptations for rapid neuromuscular output [41–43]. High costs or high levels of required proficiency help to ensure that dynamic displays are "honest," because such displays tend to provide a reliable indication of a male's genetic or phenotypic quality [44–46].

The crux of the argument we develop here is that adaptive divergence in phenotypic traits (morphology, physiology, or behavior) may influence, as a secondary consequence, the nature or strength of biomechanical constraints on mating displays. Numerous examples are discussed. Insofar as displays are costly or challenging to produce, even minor divergence in biomechanical systems could influence an animal's ability to produce these displays. This divergence could potentially influence reproductive isolation. Four lines of evidence would ideally be gathered to demonstrate that adaptive divergence influences mating displays and therefore speciation. First, a trait related to biomechanical performance should be shown to diverge adaptively among populations and species. Second, the corresponding variation in performance should be shown to influence mating displays. Third, variation in these displays should be shown to influence mate choice. Fourth, the resulting mate choice should be shown to influence speciation. No study has yet systematically examined these criteria for a single taxon, although the conceptual relationships between divergence, signal variation, and speciation have been considered previously at length [e.g., 11,47,48]. We now review three broad classes of biological adaptations — in body size, locomotion, and feeding — that may affect, as a secondary consequence, the biomechanical bases and expression of mating displays.

14.3.1 Mating Displays and Body Size

Body size evolves in response to a wide array of environmental factors. Cold temperatures, for example, tend to favor larger body sizes, as illustrated by Bergmann's rule [larger body sizes at higher latitudes; 49–51]. In homeothermic animals, this trend might arise because of a positive relationship between body size and the ability to retain metabolic heat [52]. Large animals may also be favored in highly seasonal or unpredictable environments because they are comparatively resistant to starvation [53]. Moreover, body size tends to evolve in response to varying selection on life-history traits such as fecundity, reproductive rate, and dispersal. For example, selection for increased fecundity favors comparatively large body sizes, whereas selection for rapid offspring production often favors small body sizes [54,55].

Body size influences myriad aspects of organismal physiology and biomechanics [52,56]. Traits involved in communication are no exception. The maximum size of ornaments used in visual signaling is necessarily limited by body size [36,57]. Peacock tails or deer antlers, for example, are constrained to sizes and masses that can be effectively carried and displayed. Body size also shapes acoustic signals because of positive scaling between body size and the mass of acoustic source tissues [36,58]. Darwin's finches of the Galápagos Islands, for example, show a positive, nearly isometric relationship between body mass and syrinx (sound source) volume [59], and larger-bodied finches tend to sing at correspondingly lower vocal frequencies [60]. Similar relationships between vocal frequency and body size have been demonstrated in numerous taxa, especially anurans and birds [e.g., 61–66].

Body size influences how animals are able to execute visual and acoustic displays because of tradeoffs between body size and agility. Size–agility tradeoffs have been documented within some birds and butterflies. In these groups, the frequency, duration, and even the effectiveness of male aerial displays tend to be highest in species or individuals with the smallest body sizes [67–71]. This pattern is consistent with demonstrated negative impacts of body size on flight agility [67]. Negative impacts of body size on display production may help to explain selection for small body size ("reversed" sexual dimorphism) in species of birds and insects where males use aerial displays [68,69,72]. Body size also influences electric organ discharges (EODs) in electric fishes. Larger-bodied fishes can support larger populations of electrocytes, thus augmenting EOD intensity, and also express greater charge separation distances in their electrogenic organs, thus enhancing EOD range [36].

The first two criteria in support of the by-product speciation model are thus clearly met — body size has been shown to undergo adaptive divergence through natural selection, and body size variation can influence the expression of mating displays. Many lines of available evidence also support a role for body size in mate choice. In some taxa, females have been shown to express preferences for males with body sizes similar to their own [73,74]. In other taxa, females express general preferences for larger males, although intrasexual competition can limit female access to larger males and thus result in patterns of size-assortative mating [75]. Because body size is often highly correlated with aspects of behavior and courtship, which in turn provide proximate cues in mate choice, divergent selection on body size would seem to be relatively effective in promoting assortative mating [e.g., 76].

14.3.2 Mating Displays and Locomotion

Complex and highly specialized adaptations for locomotion are prevalent throughout the animal kingdom, and often entail substantial modification of broad suites of traits [77]. In terrestrial vertebrates, rapid sprint speeds are enabled by adaptations in limb length, aerobic capacity, and efficiency of pulmonary gas exchange [78]. In fishes that use their caudal fin for routine propulsion, sustained swimming is typically associated with fusiform bodies and high aspect ratio lunate tails, whereas burst swimming is typically associated with deep bodies and large fins, particularly in the caudal area [79]. In aerial vertebrates, powered (flapping) flight requires numerous adaptations including reduced body weight, aerodynamic body shape, broad lift surfaces, and efficient flight muscles [80]. In humans, selection for endurance running may have favored a broad suite of traits including springlike leg tendons, skeletal stabilization, plantar arches, forearm shortening, and expanded venous circulation for thermoregulation [81].

The ecological bases of locomotory adaptation are perhaps best studied through comparison of closely related species or populations. Many studies could be cited to this effect [82]; here we provide two representative examples. The first concerns *Anolis sagrei*, a lizard found throughout the Caribbean. In the late 1970s and early 1980s, this species was introduced to islands that contained no lizards. Ten to fourteen years later, the introduced populations were sampled and found to have undergone substantial divergence in hind- and forelimb length [83]. Moreover, these changes correlated positively with the mean diameter of available perches on the experimental islands, consistent with functional studies of limb length and locomotor efficiency [83,84]. The second example concerns *Gambusia affinis*, the western mosquitofish. Langerhans et al. [85] found that mosquitofish populations under high risk of predation have evolved comparatively large caudal regions, small heads, and elongate bodies, all of which are thought to improve escape ability. Interestingly, these "fast-start" adaptations may impair prolonged swimming ability, which could explain the retention of the opposing suite of traits in low-risk populations [see also 86].

Adaptive divergence in locomotory traits might, in turn, influence mating displays, given that displays often include, and sometimes even amplify, motor patterns used during normal locomotion. Courting displays in waterfowl, for example, include wing flapping, swimming, and changes in head posture similar to those that occur before flight [87,88]. Some other displays, such as courtship flights of hummingbirds and "strut" displays of grouse, are dominated by locomotion [89,90]. Indeed, a major preoccupation in early ethology was to explain ritualization, the process wherein common locomotory patterns become incorporated into stereotyped display sequences [91,92]. Beyond providing raw material for display patterns, selection for locomotory traits may also fine-tune animals' abilities to perform mating displays. The evolution of complex hummingbird flight displays, for instance, was presumably facilitated by selection for agile flight capabilities in other contexts, such as for food and territory defense. Another possible example concerns crickets and other orthopterans that produce acoustic signals through stridulation of the wings. The divergence of flight anatomy and biomechanics (e.g., wing size, flight muscle properties) presumably could influence the kinds of acoustic signals these animals produce and evolve.

Operationally it can be very difficult to study biomechanical impacts of locomotion on dynamic displays, simply because it is difficult to quantify the kinetics and dynamics of display movements in an animal that itself is moving through space [e.g., 89]. It is thus no surprise that most studies of display biomechanics have focused on animals that signal while stationary. An alternative approach is to study the biomechanical bases of multimodal signals, i.e., signals that involve multiple sensory channels. In a recent study of brown-headed cowbirds, for instance, Cooper and Goller [93] studied mating displays that feature simultaneous vocal output and wing movements. Analysis of dynamic changes in air sac pressure, wing movements, and vocal features provide evidence for a biomechanical interaction between wing movements and vocal displays. Specifically, wing position appears to constrain the timing of vocal output via biomechanical influences on the respiratory system [93]. Similar interactions between wing movements and breathing could presumably influence the evolution of vocal signals produced during flight [see also 94,95].

As in the previous section, the first two criteria for biomechanically driven ecological speciation are well supported. Morphological and physiological parameters certainly diverge adaptively through natural selection, and variation in locomotory performance certainly affects the expression of mating displays. There are few data, however, that directly support a link in any given system between locomotory adaptations, resulting divergence in mating displays, and mate choice. A promising model system on this front is the threespine stickleback, for which differences among sympatric morphs in body size and behavior are quite pronounced. While some attention has been given to causes and mating consequences of variation in body size in sticklebacks [73], less is known about intermorph differences in swimming performance, or about how such differences might affect mating displays and patterns. The intricacy and complexity of mating displays in this species, which has captivated behavioral biologists since Tinbergen [92], increases the likelihood that intermorph differences in display performance would be influenced by divergent selection on swimming performance.

14.3.3 Mating Displays and Feeding

Animals have evolved a wide range of morphological and behavioral adaptations for feeding [e.g., 96–99]. Fishes, for example, employ an impressive diversity of feeding modes including suction feeding, ram feeding, and prey capture through jaw protrusion [100–103]. Studies of variation within and among closely related species illustrate how ecological conditions may promote adaptive divergence in these traits and behaviors. Variation within a species in preferred prey (i.e., "resource polymorphisms"), for example, is sometimes mirrored by genetically based variation in feeding morphology, which in turn may provide the raw material for incipient speciation [3,104; but see 105]. The link between adaptation to alternative food resources and speciation is indeed evident in many classic adaptive radiations, including fishes in postglacial temperate regions [3], African cichlids [106,107], Galápagos finches [6,108], and Hawaiian honeycreepers [109].

In a majority of cases, feeding adaptations likely have little proximate impact on the biomechanics of display production. This is because the two functions often show little if any overlap in their mechanical and anatomical bases. This is certainly true for many familiar displays, such as plumage or color pattern in birds and fishes. In some taxa, however, feeding and mating adaptations make use of the same morphological structures. When they do, feeding and display functions may interact on both organismal and evolutionary scales. In an intriguing example, male giraffes use their long necks not only for foraging on high branches but also as weapons during intrasexual competition for females [110]. Feeding and display functions may sometimes oppose each other in biomechanical function. Male fiddler crabs, for example, use their claws during feeding and during displays to females; the small and agile claws are best for feeding and the large and conspicuous claws most useful for display. In response to this tradeoff, fiddler crabs have "assigned" each function to a different claw [111]. In other cases, however, feeding and display functions do not involve redundant structures, and morphological or biomechanical tradeoffs cannot be circumvented. One such case, on which we now focus, concerns overlap between feeding adaptations and mechanisms of vocal production in songbirds.

A primary feature in the radiation of songbirds is the exploitation of divergent feeding niches through divergence in the size, form, and function of beaks [108,109,112]. This divergence likely affects the evolution of vocal mating signals, i.e., songs, because of the recently identified contribution of beaks to vocal mechanics [113–115]. One prediction is that divergence in beak and vocal tract volume, and thus in vocal tract resonance properties, should affect the evolution of vocal frequencies. This is because larger-volume vocal tracts are best suited for low-frequency sounds, whereas small-volume vocal tracts are best suited for high frequency sounds [113,116]. In support of this prediction, fundamental frequency has been shown to vary negatively with beak length in neotropical woodcreepers [65]. Another prediction is that the evolution of increased force application, such as that required to crack larger and harder seeds, may detract from a bird's vocal performance capabilities [117]. Force–speed tradeoffs are a common feature of mechanical systems, and can be attributed to both biomechanical and muscular properties [118,119]. In the evolution of some kinds of "superfast" muscles, such as those used for sound production in the toadfish swimbladder, elevated rates of crossbridge detachment during contraction necessarily preclude strong force application [119]. The evolution of bite force in granivorous birds is expected to affect the rapidity with which they can adjust beak gape, with increases in bite force diminishing maximum rates of gape adjustment, and vice versa.

Of particular relevance to this latter prediction is the expression of song features that rely on changes in beak gape in their production. Gape changes are tightly correlated with changes in fundamental frequency [e.g., 120–124] and with the resonance function of the vocal tract filter [115,116]. Tradeoffs between beak gape speed and force, either at the level of jaw muscles or force transmission mechanics, could thus impede the evolution of high-performance songs, especially for strong biters. Recognition of this relationship suggests that two song features in particular, trill rate and frequency bandwidth, should be influenced by beak size evolution because the production of these features requires rapid beak gape cycling [125]. Some (but not all) available data support this prediction [126–130]. The nature of this relationship is illustrated in an ongoing study of a population of medium ground finches, *Geospiza fortis*, at El Garrapatero on Santa Cruz Island, Galápagos. This population shows a bimodal distribution of small and large beak sizes, with few intermediates [130] (A.P. Hendry et al., unpublished data). In this population, morphological variation is correlated closely with bite force capacities and with the frequency bandwidth of song, in directions predicted by biomechanical models of beak and vocal tract function (Figure 14.2) [130,131].

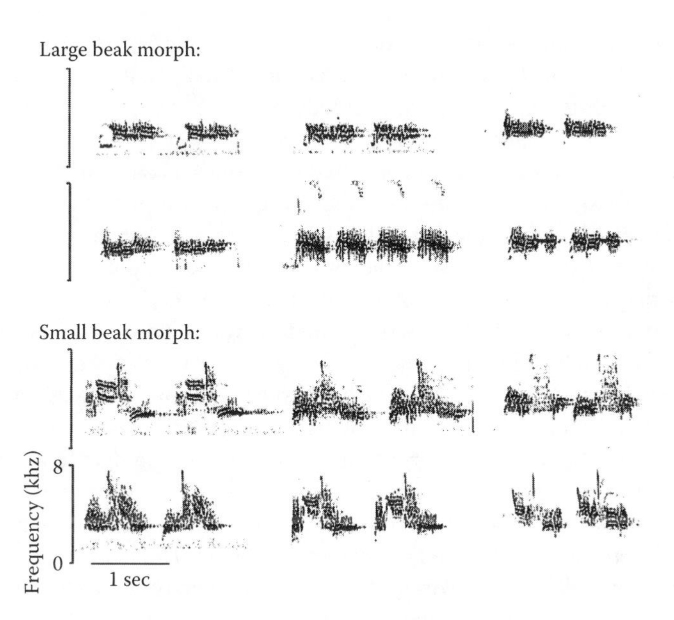

FIGURE 14.2 Sound spectrograms from 12 male medium ground finches (*Geospiza fortis*) from El Garrapatero, Santa Cruz Island, Galápagos. Songs of large and small beak morphs differ significantly in frequency bandwidth, with the smaller beaked birds producing wider frequency bandwidth (and thus higher performance) songs. This is consistent with a hypothesis of vocal tract constraints on song production and evolution in these birds [130]. (Figure modified from Huber, S.K. and Podos, J., *Biol. J. Linn. Soc.*, in press.)

For Darwin's finches, three of the four criteria for demonstrating the role of biomechanics in ecological speciation are well met, which to our eye gives this system excellent potential in testing the model of speciation presented in this chapter. First, the adaptive divergence of feeding morphology has been linked to variation in local ecological conditions, with a degree of precision rarely achieved in natural systems [112,132–134]. Second, beak divergence has been shown to influence some fundamental song features, such as trill rate and frequency bandwidth [121,128,130]. Third, female choice is highly dependent on song as a mating signal [135–137]. It remains to be seen, however, whether those song features mechanically linked to variation in beak form and function also have a significant impact on mate and species recognition. Playback studies offer one method for addressing this question empirically. If such a link is present, it follows that beak adaptation could affect the evolution of reproduction isolation in these birds.

14.4 ECOLOGICAL DEPENDENCE AND THE EVOLUTION OF ISOLATING BARRIERS

As discussed earlier, isolating barriers in ecological speciation may be ecologically dependent or ecologically independent. The former class of barriers may evolve more quickly, but should also be more susceptible to immediate decay in the face of environmental change [35]. We suggest that isolating barriers caused by divergence in biomechanical systems, at least as conceptualized in this chapter, will be characterized substantially by ecological independence. This is because the production and evolution of mating displays usually depends more on intrinsic biomechanical properties of the organism, as compared to extrinsic ecological properties of the environment (assuming the traits involved are genetically based). In other words, the intrinsic nature of biomechanical systems suggests that the second criterion discussed in the examples above — adaptive divergence in phenotypes influencing mating displays — will be expressed partly independently of diversity in environments. Hummingbirds selected for large bodies, for example, will necessarily be constrained to produce less agile flight displays, regardless of the environments in which they display. The same would hold true for many other dynamic mating displays, such as the electric signals of fishes, visual displays in waterfowl, and calling in crickets and anurans, for which the *potential* for producing different kinds of displays is at least partially independent of the particular environments inhabited. For this reason we expect that isolating barriers based on biomechanical performance will initially persist in the face of environmental change. The same barriers, however, might ultimately decay if environmental changes persist long enough to cause evolutionary convergence in formerly distinct species.

We do not mean to imply that displays are not influenced by variation in environmental factors — all phenotypic traits are subject to such impacts. For example, water temperature and conductivity influence the electric organ discharges of some fishes [138], and the physical structure of vegetation in terrestrial systems influences the propagation of acoustic signals in birds [21]. Some isolating barriers may be modified by location-dependent effects, even if they are not strictly ecologically dependent. Divergence of song in oscine songbirds depends in part on learning and thus on the features of available song models [139]. Different populations of white-crowned sparrows, for example, learn local song types even though they are capable of learning and producing song types from different locales [140]. Learned female preferences for local song types might thus facilitate speciation.

Recognition that isolating barriers may include both ecologically dependent and ecologically independent components leads us to a general suggestion for future investigations of ecological speciation, as a complement to our broader suggestion of exploring mechanical and functional links between adaptation, mating displays, and reproductive isolation. Instead of showing that any given isolating barrier is ecologically independent or dependent, it might be preferable to partition relative effects between the two components. For example, one can envision a scenario in which between-type matings are found to be half as likely as within-type matings *outside* of the ecological context (e.g., during encounters in the laboratory), and in which between-type matings are found to be one-quarter as likely as within-type matings *inside* the ecological context (e.g., during encounters in nature). We could then conclude that the ecologically dependent component of isolation is one-half as strong as the ecologically independent component (0.25/0.5), and thus causes one-third of the total isolation [0.25/(0.5 + 0.25)].

14.5 POSITIVE FEEDBACK LOOPS

Ecological speciation may be strengthened by positive feedback loops between adaptive divergence and barriers that reduce gene flow [14,35,141]. To elaborate, an initial reduction in gene flow between two ecological environments, such as that due to geographical separation, could allow some adaptive divergence to take place. This partial adaptive divergence could then further reduce gene flow (i.e., ecological speciation), which could allow further adaptive divergence, which could further reduce gene flow, and so on. We expect that phenotypic divergence and corresponding tradeoffs and constraints in biomechanical systems will contribute strongly to such feedback loops. Consider our example of song production and evolution in Darwin's finches. As populations begin to adapt to different resources, we have argued that corresponding changes in beak sizes lead to divergence in the types of songs different populations are able to sing. Insofar as this divergence in vocal "potential" leads to differences in realized song structure [139], individuals moving between environments could experience reduced mating success, particularly given that local females appear to adjust their preferences to match local song types [137, see also 142]. Reduced mating success of migrants between environments could further reduce gene flow between populations, thereby allowing further adaptive divergence and hence facilitating the positive feedback process.

14.6 DUAL FITNESS CONSEQUENCES FOR ECOLOGICAL SPECIATION

The contribution of ecological trait divergence to speciation depends on the "success" (i.e., fitness) of migrants between environments [14,35], as well as the success of any resulting hybrids (Figure 14.1). Success in both cases depends on the ability to exploit available resources, and is evident in the survival, growth, and fecundity of individuals before reproduction, i.e., natural selection, as well as on their ability to obtain mates, i.e., sexual selection. Both natural and sexual selection can be important in isolating populations from different ecological environments. With respect to natural selection, ecological and life-history performance (e.g., in resource use, growth, survival, and fecundity) is often low for migrants between environments relative to residents [14,32,33,143], or for hybrids in either parental environment relative to pure forms [23,33]. With respect to sexual selection, mating probabilities are often low for individuals from different environments relative to individuals from similar environments [26,28,144] or for hybrids relative to pure forms [145].

Isolating barriers manifest through either natural or sexual selection have been discussed at length [3,4], but the traits on which they act are usually, if implicitly, considered distinct entities with largely independent implications for speciation. And yet both types of barriers could be correlated if the trait(s) under natural and sexual selection share a common biomechanical basis. In such cases, adaptive divergence of a single trait can have dual fitness consequences. Beak size evolution in Darwin's finches, to illustrate, should have consequences for both feeding ability (natural selection) and mate acquisition (sexual selection). The possibility of such correlated isolating barriers is worth considering because they could increase the importance of a small subset of traits to ecological speciation.

14.7 PERFORMANCE AND MATING DISPLAY PRODUCTION

The relevance of biomechanical constraints to ecological speciation depends on the relationship between performance abilities (i.e., the *potential* to produce certain displays) and actual signaling behavior (i.e., displays *as realized in nature*). Performance is increasingly recognized as a critical link between morphology and behavior [77,146,147] but is rarely considered in the studies on production, development, and evolution of animal displays [e.g., 148,149]. We have argued here that as populations begin to diverge in adaptive traits, corresponding changes in biomechanical systems may influence the potential for animals to perform different kinds of displays. Such changes, however, will not necessarily cause immediate changes in the structure of mating displays. Correlated changes in performance abilities and displays are most likely to occur under two conditions [see 139]. First, signals in their initial state should be produced at or near some performance limitation. For example, particularly rapid and intricate aerial displays in a hummingbird may challenge baseline flight performance capacities. Such displays are likely to be favored by sexual selection because they can help females accurately assess variation in quality among males [37]. Second, adaptive divergence (e.g., in morphological traits) would need to occur in directions that run counter to performance limits, e.g., by imposing phenotypic tradeoffs or by barring further evolutionary modifications. In our hummingbird example, adaptive evolution toward reduced body sizes and increased flight performance (e.g., under selection for stronger resource defense) should not detract from display performance abilities, whereas selection for large body size and thus against flight agility would immediately detract from display performance. Of course, the former case could still cause reproductive isolation if the population evolving higher performance capacities then performed correspondingly more intricate mating displays.

14.8 CONCLUSION

Many recent studies provide evidence for assortative mating resulting from adaptive divergence, and thus support a strong role for ecology in speciation. It has also become clear that the evolutionary divergence of signals plays an important role in assortative mating, insofar as signal divergence facilitates reproductive isolation. We have argued here that, in some cases, phenotypic divergence can drive, as an incidental by-product, the biomechanics of display behavior and the structure of associated signals, and thus augment the likelihood of ecological speciation.

REFERENCES

1. Mayr, E., *Systematics and the Origin of Species*, Columbia University Press, New York, 1942.
2. Dobzhansky, T., *Genetics and the Origin of Species*, Columbia University Press, New York, 1951.
3. Schluter, D., *The Ecology of Adaptive Radiation*, Oxford University Press, Oxford, 2000.
4. Coyne, J.A. and Orr, H.A., *Speciation*, Sinauer Associates, Inc, Sunderland, 2004.
5. Orr, M.R. and Smith, T.B., Ecology and speciation, *Trends Ecol. Evol.,* 13, 502, 1998.
6. Lack, D., *Darwin's Finches*, Cambridge University Press, Cambridge, 1947.
7. Darwin, C., *On the Origin of Species by Means of Natural Selection, or the Preservation of Favoured Races in the Struggle for Life*, John Murray, London, 1859.
8. Mayr, E., *Animal Species and Evolution*, Harvard University Press, Cambridge, 1963.
9. Simpson, G.G., *The Major Features of Evolution*, Columbia University Press, New York, 1953.
10. Endler, J.A., *Geographic Variation, Speciation, and Clines*, Princeton University Press, Princeton, NJ, 1977.
11. Ptacek, M.B., The role of mating preferences in shaping interspecific divergence in mating signals in vertebrates, *Behav. Proc.,* 51, 111, 2000.
12. Butlin, R.K. and Ritchie, M.G., Behavior and speciation, in *Behavior and Evolution*, Slater, P.J.B. and Halliday, T.R., Eds., Cambridge University Press, Cambridge, 1994, p. 43.
13. Bush, G.L., Sympatric speciation in animals: new wine in old bottles, *Trends Ecol. Evol.,* 9, 285, 1994.
14. Nosil, P., Vines, T.H. and Funk, D.J., Reproductive isolation caused by natural selection against immigrants from divergent habitats, *Evolution*, 59, 705, 2005.
15. Schluter, D., Ecology and the origin of species, *Trends Ecol. Evol.,* 16, 372, 2001.
16. Turelli, M., Barton, N.H., and Coyne, J.A., Theory and speciation, *Trends Ecol. Evol.,* 16, 330, 2001.
17. Rundle, H.D. and Whitlock, M.C., A genetic interpretation of ecologically dependent isolation, *Evolution,* 55, 198, 2001.
18. Rice, W.R. and Hostert, E.E., Laboratory experiments on speciation: what have we learned in 40 years, *Evolution,* 47, 1637, 1993.
19. Boughman, J.W., Divergent sexual selection enhances reproductive isolation in sticklebacks, *Nature,* 411, 944, 283, 2001.
20. Endler, J.A. and Basolo, A.L., Sensory ecology, receiver biases and sexual selection, *Trends Ecol. Evol.,* 13, 415, 1998.
21. Wiley, R.H., Associations of song properties with habitats for territorial oscine birds of eastern North America, *Am. Nat.,* 138, 973, 1991.
22. Slabbekoorn, H. and Smith, T.B., Habitat-dependent song divergence in the little greenbul: an analysis of environmental selection pressures on acoustic signals, *Evolution,* 56, 1849, 2002.
23. Rundle, H.D., A test of ecologically dependent postmating isolation between sympatric sticklebacks, *Evolution,* 56, 322, 2002.
24. Wang, H., McArthur, E.D., Sanderson, S.C., Graham, J.H. and Freeman, D.C., Narrow hybrid zone between two subspecies of big sagebrush (Artemisia tridentata: Asteraceae). 4. Reciprocal transplant experiments, *Evolution,* 51, 95, 1997.

25. Schluter, D. and Nagel, L.M., Parallel speciation by natural selection, *Am. Nat.,* 146, 292, 1995.
26. Rundle, H.D., Nagel, L., Boughman, J.W., and Schluter, D., Natural selection and parallel speciation in sympatric sticklebacks, *Science,* 287, 306, 2000.
27. Nagel, L. and Schluter, D., Body size, natural selection, and speciation in sticklebacks, *Evolution,* 52, 209, 1998.
28. McKinnon, J.S., Mori, S., Blackman, B.K., David, L., Kingsley, D.M., Jamieson, L., Chou, J., and Schluter, D., Evidence for ecology's role in speciation, *Nature,* 429, 294, 2004.
29. Grant, P.R. and Grant, B.R., Unpredictable evolution in a 30-year study of Darwin's finches, *Science,* 296, 707, 2002.
30. Hendry, A.P. and Kinnison, M.T., The pace of modern life: measuring rates of contemporary microevolution, *Evolution,* 53, 1637, 1999.
31. Stockwell, C.A., Hendry, A.P., and Kinnison, M.T., Contemporary evolution meets conservation biology, *Trends Ecol. Evol.,* 18, 94, 2003.
32. Sheldon, S.P. and Jones, K.N., Restricted gene flow according to host plant in an herbivore feeding on native and exotic watermilfoils (Myriophyllum: Haloragaceae), *Int. J. Plant Sci.,* 162, 793, 2001.
33. Via, S., Bouck, A.C., and Skillman, S., Reproductive isolation between divergent races of pea aphids on two hosts. II. Selection against migrants and hybrids in the parental environments, *Evolution,* 54, 1626, 2000.
34. Hendry, A.P., Wenburg, J.K., Bentzen, P., Volk, E.C., and Quinn, T.P., Rapid evolution of reproductive isolation in the wild: evidence from introduced salmon, *Science,* 290, 516, 2000.
35. Hendry, A.P., Selection against migrants contributes to the rapid evolution of ecologically-dependent reproductive isolation, *Evol. Ecol. Res*., 6, 1219, 2004.
36. Bradbury, J.W. and Vehrencamp, S.L., *Principles of Animal Communication*, Sinauer Associates, Sunderland, MA, 1998.
37. Kodric-Brown, A. and Brown, J.H., Truth in advertising: the kinds of traits favored by sexual selection, *Am. Nat.,* 124, 309, 1984.
38. Reinhold, K., Greenfield, M.D., Jang, Y.W., and Broce, A., Energetic cost of sexual attractiveness: ultrasonic advertisement in wax moths, *Anim. Behav.,* 55, 905, 1998.
39. Ryan, M.J., Energy, calling, and selection, *Am. Zool.,* 28, 885, 1988.
40. Taigen, T.L. and Wells, K.D., Energetics of vocalization by an anuran amphibian (*Hyla versicolor*), *J. Comp. Physiol. B.,* 155, 163, 1985.
41. Rome, L.C., Syme, D.A., Hollingworth, S., Lindstedt, S.L., and Baylor, S.M., The whistle and the rattle: the design of sound producing muscles, *Proc. Natl. Acad. Sci. U.S.A.,* 93, 8095, 1996.
42. Bostwick, K.S. and Prum, R.O., High-speed video analysis of wing-snapping in two manakin clades (Pipridae: Aves), *J. Exp. Biol.,* 206, 3693, 2003.
43. Elemans, C.P.H., Spierts, I.L.Y., Muller, U.K., van Leeuwen, J.L. and Goller, F., Superfast muscles control dove's trill, *Nature,* 431, 146, 2004.
44. Zahavi, A., Mate selection: A selection for a handicap, *J. Theor. Biol.,* 53, 205, 1975.
45. Grafen, A., Biological signals as handicaps, *J. Theor. Biol.,* 144, 517, 1990.
46. Walther, B.A. and Clayton, D.H., Elaborate ornaments are costly to maintain: evidence for high maintenance handicaps, *Behav. Ecol.,* 16, 89, 2005.
47. West-Eberhard, M.J., Sexual selection, social competition, and speciation, *Quart. Rev. Biol.,* 58, 155, 1983.
48. Ryan, M.J. and Rand, A.S., Species recognition and sexual selection as a unitary problem in animal communication, *Evolution,* 47, 647, 1993.

49. Graves, G.R., Bergmann's rule near the equator: latitudinal clines in body size of an Andean passerine bird, *Proc. Natl. Acad. Sci. U.S.A.,* 88, 2322, 1991.
50. de Queiroz, A. and Ashton, K.G., The phylogeny of a species-level tendency: species heritability and possible deep origins of Bergmann's rule in tetrapods, *Evolution,* 58, 1674, 2004.
51. Ashton, K.G. and Feldman, C.R., Bergmann's rule in nonavian reptiles: turtles follow it, lizards and snakes reverse it, *Evolution,* 57, 1151, 2003.
52. Schmidt-Nielsen, K.S, *Scaling: Why Is Animal Size So Important?* Cambridge University Press, Cambridge, 1984.
53. Arnett, A.E. and Gotelli, N.J., Bergmann's rule in larval ant lions: testing the starvation resistance hypothesis, *Ecol. Entomol.,* 28, 645, 2003.
54. Roff, D.A. *The Evolution of Life Histories*, Chapman & Hall, New York, 1992.
55. Blanckenhorn, W.U., The evolution of body size: what keeps organisms small? *Quart. Rev. Biol.,* 75, 385, 2000.
56. Calder, W.A., *Size, Function, and Life History*, Harvard University Press, Cambridge, 1984.
57. Ord, T.J. and Blumstein, D.T., Size constraints and the evolution of display complexity: why do large lizards have simple displays? *Biol. J. Linn. Soc.,* 76, 145, 2002.
58. Fletcher, N.H., A simple frequency-scaling rule for animal communication, *J. Acoust. Soc. Am.,* 115, 2334, 2004.
59. Cutler, B., *Anatomical Studies of the Syrinx of Darwin's Finches*, San Francisco State University, San Francisco, 1970.
60. Bowman, R.I., The evolution of song in Darwin's finches, in *Patterns of Evolution in Galápagos Organisms,* Bowman, R.I., Berson, M., and Leviton, A.E., Eds., American Association for the Advancement of Science, San Francisco, 1983, p. 237.
61. Ryan, M.J. and Brenowitz, E.A., The role of body size, phylogeny, and ambient noise in the evolution of bird song, *Am. Nat.,* 126, 87, 1985.
62. Wallschläger, D., Correlation of song frequency and body weight in passerine birds, *Experientia,* 36, 69, 1980.
63. Martin, W.F., Evolution of vocalization in the toad genus *Bufo*, in *Evolution in the Genus Bufo,* Blair, W.F., Ed., University of Texas Press, Austin, 1972, p. 279.
64. Ryan, M.J., *The Tungara Frog*, University of Chicago Press, Chicago, 1985.
65. Palacios, M.G. and Tubaro, P.L., Does beak size affect acoustic frequencies in woodcreepers? *Condor,* 102, 553, 2000.
66. Bertelli, S. and Tubaro, P.L., Body mass and habitat correlates of song structure in a primitive group of birds, *Biol. J. Linn. Soc.,* 77, 423, 2002.
67. Andersson, M. and Norberg, R.A., Evolution of reversed sexual size dimorphism and role partitioning among predatory birds, with a size scaling of flight performance, *Biol. J. Linn. Soc.,* 15, 105, 1981.
68. Blomqvist, D., Johansson, O.C., Unger, U., Larsson, M., and Flodin, L.A., Male aerial display and reversed sexual size dimorphism in the dunlin, *Anim. Behav.,* 54, 1291, 1997.
69. Szekely, T., Freckleton, R.P., and Reynolds, J.D., Sexual selection explains Rensch's rule of size dimorphism in shorebirds, *Proc. Natl. Acad. Sci. U.S.A.,* 101, 12224, 2004.
70. Hernández, M.I.M. and Benson, W.W., Small-male advantage in the territorial tropical butterfly *Heliconius sara* (Nymphalidae): a paradoxical strategy? *Anim. Behav.,* 56, 533, 1998.
71. Kemp, D.J. and Wiklund, C., Fighting without weaponry: a review of male-male contest competition in butterflies, *Behav. Ecol. Sociobiol.,* 49, 429, 2001.

72. McLachlan, A.J. and Allen, D.F., Male mating success in Diptera: advantages of small size, *Oikos* 48, 11, 1987.
73. Boughman, J.W., Rundle, H.D., and Schluter, D., Parallel evolution of sexual isolation in sticklebacks, *Evolution* 59, 361, 2005.
74. Rahman, N., Dunham, D.W., and Govind, C.K., Size-assortative pairing in the big-clawed snapping shrimp, *Alpheus heterochelis*, *Behaviour,* 139, 1443, 2002.
75. Harari, A.R., Handler, A.M., and Landolt, P.J., Size-assortative mating, male choice and female choice in the curculionid beetle *Diaprepes abbreviatus*, *Anim. Behav.,* 58, 1191, 1999.
76. Crespi, B.J., Causes of assortative mating in arthropods, *Anim. Behav.,* 38, 980, 1989.
77. Irschick, D.J. and Garland, T. Jr., Integrating function and ecology in studies of adaptation: investigations of locomotor capacity as a model system, *Ann. Rev. Ecol. Syst.,* 32, 367, 2001.
78. Jones, J.H. and Lindstedt, S.L., Limits to maximal performance, *Ann. Rev. Physiol.,* 55, 547, 1993.
79. Webb, P.W., Body form, locomotion and foraging in aquatic vertebrates, *Am. Zool.,* 24, 107, 1984.
80. Norberg, U.M., *Vertebrate Flight: Mechanics, Physiology, Morphology, Ecology and Evolution*, Springer-Verlag, Berlin, 1990.
81. Bramble, D.M. and Lieberman, D.E., Endurance running and the evolution of *Homo*, *Nature,* 432, 345, 2004.
82. Irschick, D.J., Measuring performance in nature: implications for studies of fitness within populations, *Integ. Comp. Biol.,* 43, 396, 2003.
83. Losos, J.B., Warheit, K.I., and Schoener, T.W., Adaptive differentiation following experimental island colonization in *Anolis* lizards, *Nature,* 387, 70, 1997.
84. Losos, J.B. and Irschick, D.J., The effect of perch diameter on escape behaviour of *Anolis* lizards: Laboratory predictions and field tests, *Anim. Behav.,* 51, 593, 1996.
85. Langerhans, R.B., Layman, C.A., Shokrollahi, A.M., and DeWitt, T.J., Predator-driven phenotypic diversification in *Gambusia affinis*, *Evolution,* 58, 2305, 2004.
86. Langerhans, R.B. and DeWitt, T.J., Shared and unique features of evolutionary diversification, *Am. Nat.,* 164, 335, 2004.
87. Lorenz, K.Z., The comparative method in studying innate behaviour patterns, *Symp. Soc. Exp. Biol.,* 4, 221, 1950.
88. Heinroth, O. and Heinroth, K., *Aus dem Leben der Vogel*, Springer-Verlag, Berlin, 1955.
89. Dantzker, M.S., Deane, G.B., and Bradbury, J.W., Directional acoustic radiation in the strut display of male sage grouse *Centrocercus urophasianus*, *J. Exp. Biol.,* 202, 2893, 1999.
90. Gibson, R.M., Female choice in sage grouse: the roles of attraction and active comparison, *Behav. Ecol. Sociobiol.,* 39, 55, 1996.
91. Daanje, A., On locomotory movements in birds and the intention movements derived from them, *Behaviour,* 3, 48, 1950.
92. Tinbergen, N., *The Study of Instinct*, Clarendon Press, Oxford, 1951.
93. Cooper, B.G. and Goller, F., Multimodal signals: enhancement and constraint of song motor patterns by visual display, *Science,* 303, 544, 2004.
94. Boggs, D.F., Seveyka, J.J., Kilgore, D.L.J., and Dial, K.P., Coordination of respiratory cycles with wingbeat cycles in the black-billed magpie (*Pica pica*), *J. Exp. Biol.,* 200, 1413, 1997.
95. Huber, S.K., Coordination of vocal production and flight in the cockatiel *Nymphaticus hollandicus*, *Integ. Comp. Biol.,* 42, 1246, 2002.

96. Liem, K.F., Evolutionary strategies and morphological innovations: cichlid pharyngeal jaws, *Syst. Zool.,* 22, 425, 1973.
97. Yanega, G.M. and Rubega, M.A., Hummingbird jaw bends to aid insect capture, *Nature,* 428, 615, 2004.
98. Schwenk, K., *Feeding: Form, Function, and Evolution in Tetrapod Vertebrates*, Academic Press, San Diego, 2000.
99. Dumont, E.R., Bats and fruit: an ecomorphological approach, in *Bat Ecology,* Kunz, T.H. and Fenton, M.B., Eds., University of Chicago Press, Chicago, 2003, p. 398.
100. Lauder, G.V. and Liem, K.F., Prey capture by *Luciocephalus pulcher*: implications for models of jaw protrusion in teleost fishes, *Env. Biol. Fishes,* 6, 257, 1981.
101. Liem, K.F., Adaptive significance of intra- and interspecific differences in the feeding repertoires of cichlid fishes, *Am. Zool.,* 20, 295, 1980.
102. Westneat, M.W. and Wainwright, P.C., Feeding mechanism of *Epibulus insidiator* (Labridae, Teleostei): evolution of a novel functional system, *J. Morphol.,* 202, 129, 1989.
103. Norton, S.F. and Brainerd, E.L., Convergence in the feeding mechanics of ecomorphologically similar species in the Centrarchidae and Cichlidae, *J. Exp. Biol.,* 176, 11, 1993.
104. Smith, T.B. and Skúlason, S., Evolutionary significance of resource polymorphisms in fishes, amphibians, and birds, *Ann. Rev. Ecol. Syst.,* 27, 111, 1996.
105. Mittelbach, G.C., Osenberg, C.W., and Wainwright, P.C., Variation in feeding morphology between pumpkinseed populations: phenotypic plasticity or evolution? *Evol. Ecol. Res.,* 1, 111, 1999.
106. Liem, K.F., Modulatory multiplicity in the feeding mechanism in cichlid fishes, as exemplified by the invertebrate pickers of Lake Tanganyika, *J. Zool., Lond.,* 189, 93, 1979.
107. Streelman, J.T., Alfaro, M., Westneat, M.W., Bellwood, D.R., and Karl, S.A., Evolutionary history of the parrotfishes: biogeography, ecomorphology, and comparative diversity, *Evolution,* 56, 961, 2002.
108. Grant, P.R., *Ecology and Evolution of Darwin's Finches*, Princeton University Press, Princeton, NJ, 1999.
109. Amadon, D., The Hawaiian honeycreepers, *Bull. Am. Mus. Nat. Hist.,* 95, 151, 1950.
110. Simmons, R. and Scheepers, L., Winning by a neck: sexual selection in the evolution of giraffe, *Am. Nat.,* 148, 771, 1996.
111. Crane, J., *Fiddler Crabs of the World. Ocypodidae: Genus Uca.* Princeton University Press, Princeton, NJ, 1975.
112. Bowman, R.I., Morphological differentiation and adaptation in the Galápagos finches, *Univ. Calif. Publ. Zool.,* 58, 1, 1961.
113. Nowicki, S., Vocal-tract resonances in oscine bird sound production: evidence from birdsongs in a helium atmosphere, *Nature,* 325, 53, 1987.
114. Beckers, G.L., Suthers, R.A., and ten Cate, C., Pure-tone birdsong by resonance filtering of harmonic overtones, *Proc. Nat. Acad. Sci. U.S.A.,* 100, 7372, 2003.
115. Nowicki, S. and Marler, P., How do birds sing? *Music Perception,* 5, 391, 1988.
116. Fletcher, N.H. and Tarnopolsky, A., Acoustics of the avian vocal tract, *J. Acoust. Soc. Am.,* 105, 35, 1999.
117. Nowicki, S., Westneat, M.W., and Hoese, W.J., Birdsong: motor function and the evolution of communication, *Sem. Neurosci.,* 4, 385, 1992.
118. Herrel, A., O'Reilly, J.C., and Richmond, A.M., Evolution of bite performance in turtles, *J. Evol. Biol.,* 15, 1083, 2002.

119. Rome, L.C., Cook, C., Syme, D.A., Connaughton, M.A., Ashley-Ross, M., Klimov, A., Tikunov, B., and Goldman, Y.E., Trading force for speed: why superfast cross-bridge kinetics leads to superlow forces, *Proc. Natl. Acad. Sci. U.S.A.*, 96, 5826, 1999.
120. Westneat, M.W., Long, J.H., Hoese, W., and Nowicki, S., Kinematics of birdsong: functional correlation of cranial movements and acoustic features in sparrows, *J. Exp. Biol.*, 182, 147, 1993.
121. Podos, J., Southall, J.A., and Rossi-Santos, M.R., Vocal mechanics in Darwin's finches: correlation of beak gape and song frequency, *J. Exp. Biol.*, 207, 607, 2004.
122. Williams, H., Choreography of song, dance and beak movements in the zebra finch (*Taeniopygia guttata*), *J. Exp. Biol.*, 204, 3497, 2001.
123. Goller, F., Mallinckrodt, M.J., and Torti, S.D., Beak gape dynamics during song in the zebra finch, *J. Neurobiol.*, 59, 289, 2004.
124. Hoese, W.J., Podos, J., Boetticher, N.C., and Nowicki, S., Vocal tract function in birdsong production: experimental manipulation of beak movements, *J. Exp. Biol.*, 203, 1845, 2000.
125. Podos, J., A performance constraint on the evolution of trilled vocalizations in a songbird family (Passeriformes: Emberizidae), *Evolution*, 51, 537, 1997.
126. Grant, B.R. and Grant, P.R., Simulating secondary contact in allopatric speciation: an empirical test of premating isolation, *Biol. J. Linn. Soc.*, 76, 545, 2002.
127. Slabbekoorn, H. and Smith, T.B., Does bill size polymorphism affect courtship song characteristics in the African finch *Pyrenestes ostrinus*? *Biol. J. Linn. Soc.*, 71, 737, 2000.
128. Podos, J., Correlated evolution of morphology and vocal signal structure in Darwin's finches, *Nature*, 409, 185, 2001.
129. Seddon, N., Ecological adaptation and species recognition drives vocal evolution in neotropical suboscine birds, *Evolution*, 59, 200, 2005.
130. Huber, S.K. and Podos, J., Beak morphology and song features covary in a population of Darwin's finches, *Biol. J. Linn. Soc.*, in press.
131. Herrel, A., Podos, J., Huber, S.K., and Hendry, A.P., Bite performance and morphology in a population of Darwin's finches: implications for the evolution of beak shape, *Funct. Ecol.*, 19, 43, 2005.
132. Price, T.D., Grant, P.R., Gibbs, H.L., and Boag, P.T., Recurrent patterns of natural selection in a population of Darwin's finches, *Nature*, 309, 787, 1984.
133. Grant, P.R. and Grant, B.R., Predicting microevolutionary responses to directional selection on heritable variation, *Evolution*, 49, 241, 1995.
134. Boag, P.T. and Grant, P.R., Intense natural selection in a population of Darwin's finches (Geospizinae) in the Galápagos, *Science*, 214, 82, 1981.
135. Grant, B.R. and Grant, P.R., Hybridization and speciation in Darwin's finches: the role of sexual imprinting on a culturally transmitted trait, in *Endless Forms: Species and Speciation*, Howard, D.J. and Berlocher, S.H., Eds., Oxford University Press, Oxford, p. 404, 1998.
136. Grant, P.R. and Grant, B.R., Mating patterns of Darwin's finch hybrids determined by song and morphology, *Biol. J. Linn. Soc.*, 60, 317, 1997.
137. Grant, P.R. and Grant, B.R., Hybridization, sexual imprinting, and mate choice, *Am. Nat.*, 149, 1, 1997.
138. Silva, A., Quintana, L, Ardanaz, J.L., and Macadar, O., Environmental and hormonal effects upon EOD waveform in gymnotiform pulse fish, *J. Physiol.*, 96, 473, 2002.
139. Podos, J., Huber, S.K., and Taft, B., Bird song: the interface of evolution and mechanism, *Ann. Rev. Ecol. Evol. Syst.*, 35, 55, 2004.

140. Marler, P. and Tamura, M., Culturally transmitted patterns of vocal behavior in sparrows, *Science,* 146, 1483, 1964.
141. Crespi, B.J., Vicious circles: positive feedback in major evolutionary and ecological transitions, *Trends Ecol. Evol.,* 19, 627, 2004.
142. Podos, J., Discrimination of intra-island song dialects by Darwin's finches, *Proc. Royal. Soc. London B,* manuscript in review.
143. Nosil, P., Reproductive isolation caused by visual predation on migrants between different environments, *Proc. R. Soc. Lond. B,* 271, 1521, 2004.
144. Nosil, P., Crespi, B.J., and Sandoval, C.P., Host-plant adaptation drives the parallel evolution of reproductive isolation, *Nature,* 417, 441, 2002.
145. Vamosi, S.M. and Schluter, D., Sexual selection against hybrids between sympatric stickleback species: evidence from a field experiment, *Evolution,* 53, 874, 1999.
146. Arnold, S.J., Morphology, performance and fitness, *Am. Zool.,* 23, 347, 1983.
147. Garland, T.J. and Losos, J.B., Ecological morphology of locomotor performance in squamate reptiles, in *Ecological Morphology: Integrative Organismal Biology,* Wainwright, P.C. and Reilly, S.M., Eds., University of Chicago Press, Chicago, 1994, p. 240.
148. Podos, J. and Nowicki, S., Performance limits on birdsong, in *Nature's Music: The Science of Bird Song,* Marler, P. and Slabbekoorn, H., Eds., Elsevier Press, New York, 2004, p. 318.
149. Van Hooydonck, B., Herrel, A., Van Damme, R., and Irschick, D.J., Does dewlap size predict male bite performance in Jamaican *Anolis* lizards? *Funct. Ecol.,* 19, 38, 2005.

Index

D

E

F

G

H

I

J

K

L

M

N

Q

R

S

T

V

W

X

Y